Open Pit Mine Planning and Design

John T. Crawford, III
William A. Hustrulid
Editors

Editorial Committee

C. D. Broadbent	**M. K. McCarter**
J. T. Crawford, III	**B. L. Seegmiller**
W. A. Hustrulid	**M. Snedeker**

S. P. Winkelmann

Published by

of the

American Institute of Mining, Metallurgical, and Petroleum Engineers, Inc.
New York, New York • 1979

Printed in the United States of America
by Port City Press, Inc., Baltimore, Maryland

Library of Congress Catalog Card Number 79-52269
ISBN 0-89520-253-0

Preface

This book is a product of the "Open Pit Minc Planning and Design" workshop which was held in Denver, CO, in conjunction with the Annual Meeting (1978) of the American Institute of Mining, Metallurgical, and Petroleum Engineers. It contains the papers presented during the workshop as well as the discussion by the participants.

The philosophy of the workshop and the book was basically simple. It has been the feeling of many of us, both working in the mining industry and training new mining engineers for the industry, that we as engineers are very good at providing the minerals that our nation and the world need, but we are not as good about writing down the engineering principles and the rules of thumb, which form the basis for our planning and design.

Some of the reasons that have been given in the past for not writing down carefully what we do and why we do it are the following: (1) every property is different, and what is true in one place is not necessarily true somewhere else; (2) it is all done by computer now; and (3) the material is company confidential. It is for these and possibly some additional reasons that we have a shortage of good current technical information available to our industry today.

Much of our published literature contains either the broad brush type of overview or the detailed description of some very specialized operations. Although the papers are very interesting and contribute to our overall understanding, they rarely discuss the engineering principles involved.

Some time ago an official of an explosives company was asked about how they train their sales people to design rounds and help their customers with blasting problems. He replied, "Well, we take the new engineer and put him with the representative on the route, and he goes around with this man for a period of about two years trying to learn all the things this man has learned during his career in the explosives industry. Then after this period he is ready to go out and tell the world and the mining companies about how they should do blasting."

Although many things can only be learned through experience, there is much that doesn't require or justify the time and expense of a one-on-one apprenticeship either with the experienced man or with life. Many costly mistakes and pitfalls could have been avoided if something in the way of directions could have been read beforehand.

We must have the man write down his experiences so that we can all gain from his knowledge.

It is however the rare individual who is able to write down carefully and thoroughly in a paper what he knows about a subject, what he doesn't know too well, i.e., the judgment factors, etc., and then indicate how he can proceed, in spite of uncertainties, to the bottom line, which is the solution—a design, a plan, a program.

We think that you will find that the authors of the chapters contained in this book have succeeded rather well in putting such material together for us in a form which will be useful for many, many years to come. Their contributions are well-balanced presentations of theory and practical applications done in an understandable manner. This contribution to our industry by these authors, their sponsoring companies, and the workshop participants is very much appreciated.

THE ORGANIZING COMMITTEE

Participants

MARVIN P. BARNES
Marvin Barnes & Associates
Salt Lake City, Utah

CARL D. BROADBENT
Manager of Mining Engineering
Kennecott Minerals Co.
Salt Lake City, Utah

RICHARD D. CALL
Vice President
Pincock, Allen & Holt, Inc.
Tucson, Arizona

THOMAS R. COUZENS
Manager, Mine Planning
Pincock, Allen & Holt, Inc.
Tucson, Arizona

JOHN T. CRAWFORD, III
Manager, Operations Control—Mining
Kennecott Minerals Co.
Salt Lake City, Utah

ALAN DAHLSTRAND
Director of Planning
Butte Operations
The Anaconda Co.
Butte, Montana

RODERICK K. DAVEY
Technical Superintendent
Utah Copper Div.
Kennecott Minerals Co.
Bingham Canyon, Utah

GERALD C. DOHM, JR.
Manager of Mine Development
Minerals Exploration Co.
Union Oil of California
Tucson, Arizona

R. H. HEINEN
Senior Mining Engineer
Dravo Corp.
Denver, Colorado

DON HENDRICKS
Chief Operations Engineer
Berkeley Pit
The Anaconda Co.
Butte, Montana

WILLIAM E. HUGHES
Senior Mine Systems Engineer
Ford, Bacon, and Davis
Salt Lake City, Utah

WILLIAM A. HUSTRULID
Professor of Mining Engineering
Colorado School of Mines
Golden, Colorado

MICHAEL B. KAHLE
Manager of Development
Phelps Coal Co.
Dallas, Texas

BENJAMIN C. KOSKINIEMI
Manager of Mining and Geology
AMAX Arizona, Inc.
Tucson, Arizona

Stanley J. Lefond
Industrial Minerals Consultant
Evergreen, Colorado

Michael K. McCarter
Professor
University of Utah
Salt Lake City, Utah

Stanley D. Michaelson
Consulting Mining Engineer
Salt Lake City, Utah

Donald C. Myntti
Manager
Maintenance Engineering
AMAX Coal Co.
Evansville, Indiana

Alan C. Noble
Manager
Scientific and Engineering Systems
United Nuclear Corp.
Albuquerque, New Mexico

R. M. (Mike) Robb
Senior Mining Engineer
United Nuclear Corp.
Albuquerque, New Mexico

Dermot M. Ross-Brown
Senior Mining Engineer
Science Applications, Inc.
Fort Collins, Colorado

Ferris E. Sainsbury
President
Boyles Brothers Drilling Co.
Salt Lake City, Utah

Fred J. Scheaffer
Senior Mining Engineer
Morrison-Knudsen Co., Inc.
Boise, Idaho

Ben L. Seegmiller
Seegmiller Associates
Salt Lake City, Utah

Marianne Snedeker
Manager of Publications
Society of Mining Engineers of AIME
Littleton, Colorado

Bruce T. Stanley
Mine Evaluation Engineer
AMAX Exploration, Inc.
Denver, Colorado

Franklin J. Stermole
Professor of Mineral Economics
Colorado School of Mines
Golden, Colorado

Stephen P. Winkelmann
Rail Operations Drilling and Blasting General Foreman
Utah Copper Div.
Kennecott Minerals Co.
Bingham Canyon, Utah

Robert F. Winkle
Vice President—Mining
Pincock, Allen & Holt, Inc.
Tucson, Arizona

Contents

Contents

The Committee, editors

1 Open Pit Mining—Past, Present, and Future

Stanley D. Michaelson
Consulting Mining Engineer

Stanley D. Michaelson retired in 1976 as chief engineer, Metal Mining Div. and director of the Engineering Center for Kennecott Copper Corp. An active consultant on mining and metallurgical assignments, Mr. Michaelson is also currently adjunct professor of mining engineering at the University of Utah. He received his B.S. degree in mining engineering from Lehigh University in 1934 and an honorary E.M. degree from the Montana School of Mines in 1961.

Mr. Michaelson received AIME's Robert H. Richards Award in 1962 and was named a Distinguished Member of the Society of Mining Engineers of AIME in 1975. He has served as a Director and as a Vice-President on the AIME Board of Directors and is a Past President of the Society of Mining Engineers. He became an Honorary Member of AIME in 1978.

INTRODUCTION

One primary objective of this chapter is to present a historical review of early to present 20th century open pit mining practices, focusing particularly on the introduction of higher capacity and more reliable mining and transport equipment and more economic pit designs. Unfortunately, however, the combination of these advances has not been able to overcome the industry's declining trends in average ore grades, waste/ore ratios, and unit productivities and costs (in terms of salable end products). Therefore, this chapter also presents some hopefully provocative concepts of the generic types of innovation that open pit mining needs to develop, debug, and adopt. Without sound breakthrough techniques, the industry simply cannot reach and retain the needed long-term viability that is dependent on assured adequate financial returns for this capital intensive business, despite mining's inherent cyclical marketing problems, its increasing environmental constraints, and its political and technical risks.

HISTORICAL PERSPECTIVE

It is a well-known fact that open pit mining is not a 20th century innovation. The earliest relatively large-scale mining for outcropping native copper sometime between 5000 and 15,000 BC was the precursor of today's open pits. That period's primitive equivalent of today's drilling and blasting was the cyclical application of fire and water; today's shovel loading and truck, rail, or conveyor haulage was then done by manual labor with stone, wooden, and bronze tools for excavating, and animals and human beings for haulage. All of these were done at a probable insignificant labor and material cost when compared to the absolute costs for our relatively sophisticated current open pit mining systems.

Progress in pit mining technology through the centuries was exceedingly slow like the grinding action of the wheels of the legendary gods and was limited to minor changes in manually operated simple mechanical wheeled tools until the steam engine began to be applied in the early 1900's to shovel loading and rail haulage equipment. Shown in Figs. 1 and 2 are haulage methods in use about 1892 in Bingham Canyon. The tram in Fig. 2 served the Telegraph, Galena, and Commercial mines.

The real start of large-scale open pit mining was in the 1904-1907 period when D. C. Jackling pioneered the now world-renowned Bingham Canyon copper pit in Utah. Fig. 3 shows how the Bingham mine appeared a few years ago.

Fig. 4 shows the Bingham Canyon mine site before the initial pit development in the 1904-1906 period. Underground bulk sampling done in adits provided ore to the pilot mill. The data from churn drilling, the underground work, and the pilot operation indicated to Jackling that his concept of large-scale mechanized open pit copper mining would be feasible for Bingham's lower grade ore than any mined elsewhere at the time. Financing was arranged, equipment procured, and pit work began. Figs. 5 and 6 show the initial pit in the 1906-1907 period. The original pit equipment used, including some full-revolving steam shovels purchased from the Panama Canal project, is shown in Fig. 7.

The original churn drills were gradually replaced with larger and increasingly heavier rotary rigs. Blasting practices evolved from squib fired hand-loaded and tamped holes through sequential electrical and cordite fired charges in base sprung holes (Fig. 8) to mechanized loading and shooting of slurry explosives (Figs. 9-11) in holes drilled with sophisticated self-propelled and powered rotary large-diameter blasthole rigs.

Fig. 1. Haulage in Bingham Canyon about 1892.

Fig. 2. Ore tram going down the canyon about 1892. Tram was built about 1872 by Duncan McVichie to serve Telegraph, Galena, and Commercial mines.

Loading progressed in stages from the steam shovels (Fig. 7) to electrically powered shovels. The bucket capacities have steadily increased from 3.8 to 4.6 m^3 (5 to 6 cu yd), Fig. 14, to 5.3 to 6.9 m^3 (7 to 9 cu yd), to 9.1 to 11.4 m^3 (12 to 15 cu yd), and currently to the 19.1-m^3 (25-cu yd) shovels on overburden (Figs. 12 and 13). As unit shovel capacity increased, dozers were added for shovel cleanup and bench servicing.

For ore haulage, electric locomotives (Fig. 14) replaced the steam-driven units and these in turn are being replaced in many areas by diesel-electric locomotives to avoid the high operating costs involved with an overhead trolley and pantograph power supply.

Waste haulage went through a similar evolution, first to increasingly larger side-dump railcars (Fig. 15), to relatively small rear-dump mechanical drive trucks, then to increasingly larger diesel-generator/electric-wheel or traction-motor driven trucks. New units are in the 227-t (250-st) live load capacity range (Fig. 16).

Waste leaching and cementation of the dissolved copper from the mineralized overburden have followed a generally parallel pattern of capacity increase in facilities and design improvements for solution distribution, solution collection, avoidance of surface and ground water contamination, and for the precipitation and drying process. Fig. 17 shows Bingham's early labor-intensive concrete scrap-filled launder cement plant and its wet hard-to-handle product. In contrast, the new automated cone precipitation and filtration plant produces a drier high-grade cement copper that is readily suitable for smelting (see Fig. 18). Some other operations now use solvent extraction and electrowinning to produce directly salable cathodes from leach solutions.

The major remaining problems in waste leaching that need resolution are improving waste dump construction methods and leach solution distribution to leach all the mineralized material in the dump by avoiding solution channeling, perched solution tables, blowouts, and dump void plugging by basic iron salt precipitation. Answers to these problems are essential to improve

Fig. 3. Kennecott's Bingham pit as it appeared in 1972.

Fig. 4. Upper Bingham Canyon and Copperfield in 1904 before the initial pit development.

current low recoveries of copper from waste. There is also considerable potential in the successful development of waste leaching techniques that would recover the precious elements, radioactive elements, and other valuable metals present in the waste rock and overburden at many open pits.

There have been no basic changes in the fundamentals of early manual pit design methods. However, improvements in the reliability of the underlying geological, mineralogical, and rock mechanics information needed for pit design and the adoption of computerized modeling, simulation, calculating, plotting, and mapping, coupled with improved and more detailed capital and operating cost information, can now provide operators with frequent and cheaply produced short- and long-range alternative pit designs for making mine planning and scheduling decisions. A major part of this book will concentrate on many of the better and currently available techniques of pit designing for optimum mining operations under given combinations of ore grade, recovery of values from waste, capital and operating costs, product prices, and other management mandates or constraints.

The growth of the open pit mining industry over the past 75 years with its attendant equipment, pit design, and management methods improvements poses some logical questions for the future. Have these been able to overcome the industry's gradually increasing stripping ratios and decreasing average ore grades? Have they really improved productivity enough in terms of tons mined per unit of labor or salable product costs to maintain satisfactory returns on the risk capital that must be attracted to keep older mines competitive and to open new ones? Unfortunately, the answers to

Fig. 5. Initial pit development in the 1906-1907 period.

Fig. 6. Boston Consolidated at Q level, July 1907. Note steam shovel in left foreground. Harry Shipler in C. C. Parson's collection, Utah Historical Society.

date are largely negative and if the added capital and operating effects of mining's environmental, land use, safety, and other regulatory costs are added, it becomes obvious that the industry's ability to compete in the capital market is currently very poor and likely to get worse.

The figures given in Table 1 substantiate my claim that present trends in open pit mining are shoving the industry downhill, and that radical breakthroughs in equipment and methods are a vital requirement for long-term viability.

Never since 1910 have productivity per man-hour figures increased to meet labor and material cost increases even without the effects of decreasing ore grades and increasing stripping ratios, and this latter effect equates to a negative figure of about 1.5% per year. Furthermore, recent productivity in ore and waste tonnage per man-hour has been falling further behind the increases in direct operating costs between 3.6 and 8.0% each year since 1971. If we consider the additional material that must be mined to produce the same poundage or tonnage of salable end products from

Fig. 7. Shovel loading at Utah Copper about 1907. Some revolving steam shovels were purchased from the Panama Canal project.

Fig. 8. Springing toeholes in waste.

Fig. 10. Slurry explosives are mechanically loaded and shot.

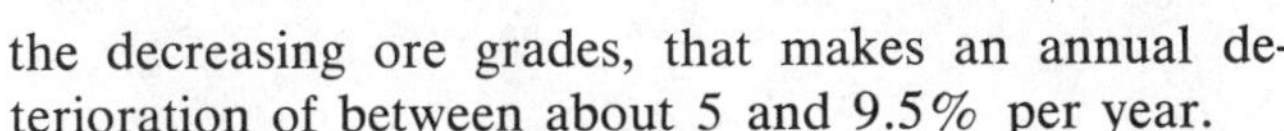

the decreasing ore grades, that makes an annual deterioration of between about 5 and 9.5% per year.

Similar adverse trends can be shown for investment return figures and almost any other figure that shows the yearly status of the open pit mining industry. These trends must be reversed if North American mining is to produce its future products at reasonable prices and remain a viable business which can attract its essential capital needs.

Obviously, with these recent trends in open pit mining practices proving inadequate to guarantee affirmative answers to the two questions given previously, we must face up to radical surgery in our mine planning, equipment development and selection, and to our operating and maintenance practices. In other words, we need a whole new look at how to recover metals from our open pits.

The conference on "Productivity in Mining" at the University of Missouri-Rolla, in May 1974, reviewed the trends described previously in considerable detail. The situation has not changed since then but instead has been aggravated by higher inflation rates and decreasing productivity trends worldwide. The same conference also presented and published a number of concepts directed toward reversing the adverse effects of poor productivity and increasing unit cost forecasts. A number of these were later published in *Engineering and Mining Journal* (October 1974) under the title "Wanted: New Systems for Surface Mining." An excerpt from this paper follows to try to provoke the generic kinds of thinking that pit engineering and open pit mining must develop and adopt for their future.

Fig. 9. Slurry explosives used in modern blasting methods.

Fig. 11. Blasting pattern at Bingham.

Fig. 12. Modern shovel and haulage equipment currently used by Kennecott. Photograph by Don Green.

THE FUTURE OF PIT DESIGN

This unacceptable productivity situation leads to the question: "How and where do we improve our productivity?" The necessary directions are not easy to program and will be even more difficult to implement. Nevertheless, certain promising concepts have evolved or are appearing on the horizon. Some innovative concepts are in limited use, and many more will be conceived and developed to reality.

Average current cost figures for metalliferous surface mining show that about 10% of the total direct cost of material mined is attributable to breaking rock (drilling

Fig. 14. Electric locomotives which are gradually being replaced by diesel-electric units.

Fig. 13. Shovel loading an ore train.

Fig. 15. Side-dump railcars.

Fig. 16. Kennecott's new haulage units are in the 227-t (250-st) live load capacity range. Photograph by Don Green.

and blasting), 16% is for shovel loading, and close to 45% is accounted for by ore and overburden haulage. The average industrywide stripping ratio is estimated at about 2:1.

Obviously then, the one variable which has the greatest potential for improving productivity and reducing ore cost is the stripping ratio, which now requires an average of three tons of material to be mined to send one ton of ore to the mill. The next greatest target for productivity improvement is a haulage system that can run reliably with fewer operators and less maintenance than trucks. Other areas of potential improvement lie in loading, and drilling and blasting operations.

The difficult problems in improving productivity for all these unit operations would literally be dissolved by the successful development of in situ leaching or solution mining techniques for recovering all presently mined minerals. However, this is an impractical dream for mineral deposits whose geological structures do not permit solution distribution, dissolution of desired values, or subsequent recovery of the solubilized mineral or metal. Iron, manganiferous, and many other

Fig. 17. Bingham's early labor-intensive concrete scrap-filled launder cement plant.

Fig. 18. High grade cement copper readily suitable for smelting.

ores are not now amenable to such treatment, nor are they likely to become amenable in the foreseeable future because of their marketing specifications. Others, such as certain copper, gold, uranium, and a few nonmetallic ores, are targets for solution mining. This chapter, however, is limited to a discussion of mining concepts requiring physical removal and transport of ore and overburden.

The Potential in Slope Control

Applications of recently developed pit slope engineering techniques have clearly demonstrated that some, and probably most, open pit slopes can be steepened safely to decrease stripping ratios and make more ore available to economic mining. As an example, Fig. 19 shows part of the Kimbley pit in Nevada, which was safely steepened by mining from a 45° pit wall to an overall slope of about 60° during a joint Kennecott-USBM slope design research program.

Fig. 19. Kimbley pit in Nevada had 60° slope after caving.

The same design concepts that apply to steepening slopes can be utilized for predictable controlled caving of slopes as an alternative to drilling and blasting under certain conditions. As an added incentive, slope caving produces a much finer muck pile than conventional methods, perhaps because the caving mechanism creates an essentially autogenous crushing action. Fig. 20

Table 1. Annual Production and Cost Increases (per Ton* Material) Compounded Annually

Period	Bingham mine rail-truck pit Tonnage,%	Productivity,%	Best US truck pit, Productivity,%	Labor & material cost, %
1910–1975	4.17	4.92		4.93
1971–1976	nil	3.52		10.82
1971–1977			6.24	9.82
	Decreases in ore grades, range for major copper pits			
1910–1977	1.35% to 1.86% per annum			

* Metric equivalent: 1 st × 0.907 184 7 = t.

Fig. 20. Broken rubble in caved slope at Kimbley.

shows the caving that was induced at the Kimbley pit to verify slope design and caving techniques. Leaching of the finely broken rubble has since yielded low-cost cement copper.

The Kimbley pit programs show how research and development efforts and novel techniques can be developed into more highly productive mining plans and operating methods for open pit properties. Other examples, while not offering immediate answers to increase open pit mining productivity, illustrate that programs like those at the Kimbley pit are the generic types of R&D effort that can yield the kinds of technical breakthroughs needed for improved productivity.

Fig. 21 shows a nonconventional open pit layout with a potential for simplifying the application of improved haulage methods. Such a slot or wedge-type mine design for deposits that are suitably shaped and sited can provide opportunities for low-cost, high-production truck haulage on uniform and relatively straight grade-haul profiles or for continuous conveyor haulage. Cer-

Fig. 21. Slot, wedge, or rectangular pit designs are suitable for belts.

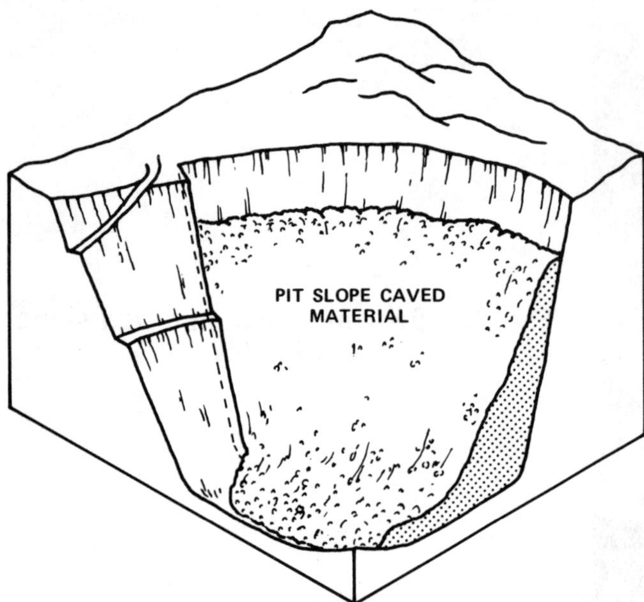

Fig. 22. Variants of slope caving: caved pit wall.

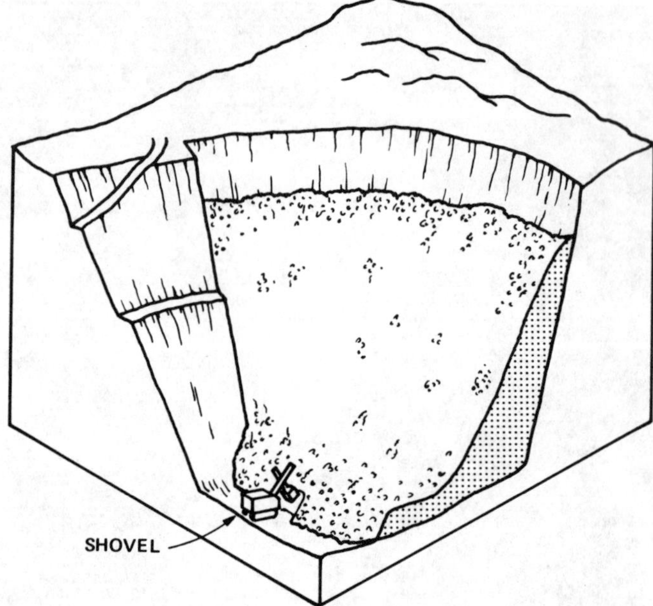

Fig. 24. Variants of slope caving: ore mining of caved pit wall.

tain other deposits can be developed with a rectangular or square mining plan to provide similar improvements in haulage.

Promoting Pit Wall Caving

Either of these presently unusual mining plans and many conventional circular mining plans could go a step or two further and use advanced pit slope technology for controlled pit wall caving. Such a combined scheme would produce, by autogenous crushing from caving material, fine enough material in some deposits for either in-place leaching or for belt conveyor haulage. Other deposits would need already proven mobile oversize scalping devices or portable crushers ahead of the belt systems.

Figs. 22 through 26 illustrate variants of such a plan. The sequence shows conversion of a conventional pit to pit wall caving. This might be followed by in-place leaching, conventional shovel loading, or lower cost dragline loading. In the latter case, much of the material is delivered to the haulage system at elevations much higher than the dumping height of a shovel,

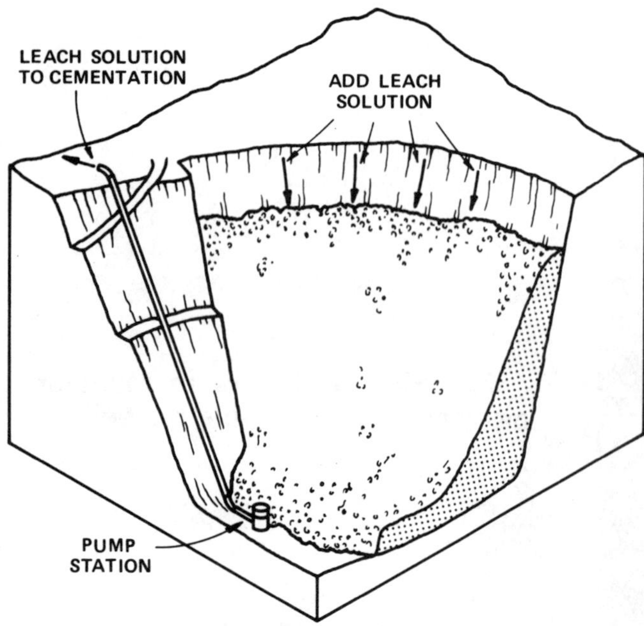

Fig. 23. Variants of slope caving: leaching caved pit wall rubble.

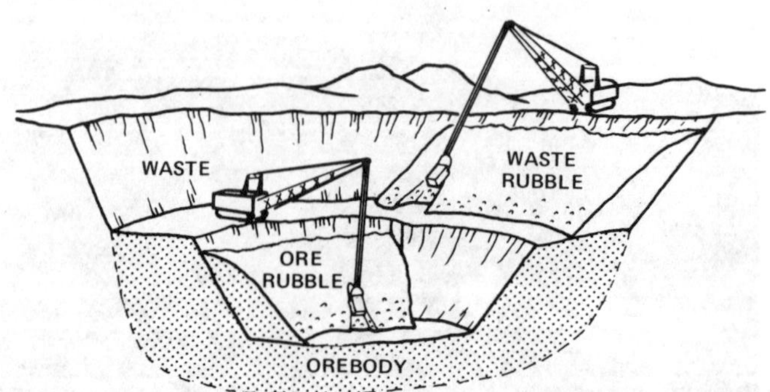

Fig. 25. Variants of slope caving: caved pit wall dragline mining scheme.

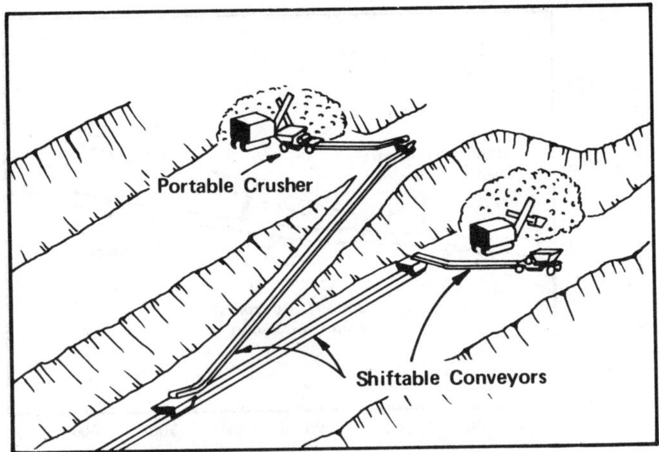

Fig. 26. Use of shiftable conveyors between shovel and main conveyor, a plan for elimination of trucks.

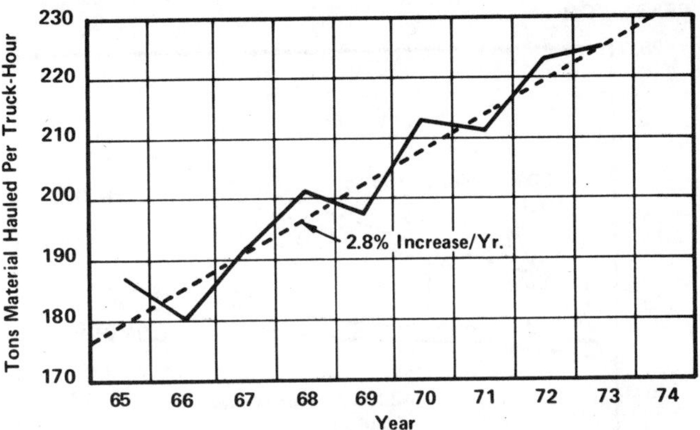

Fig. 27. Truck productivity. Metric equivalent: 1 st×0.907 184 7=t.

thereby minimizing the need for a portion of an adverse grade-haul profile. As previously stated, truck haulage accounts for about 45% of total direct mining costs. Table 2 shows annual costs and operating data over the 1965-1973 period for several large truck fleets. From these data and by extrapolation of Fig. 27, it is apparent that there is limited potential for major productivity improvement of conventional truck haulage systems at existing mines.

For long, adverse hauls, self-propelled scalpers or crushers and belt conveyors could yield a breakthrough

Table 2. Combined Operating Data for Several Major Haulage Truck Fleets*

	1965	Jan.-July 1967	1969	1971	1973	% change, 1965-1973
Operating costs, $ per hr						
Labor	3.671	3.897	4.502	5.017	5.751	+ 57
Other supplies	0.319	0.389	0.548	0.690	0.402	+ 26
Lube	0.365	0.416	0.421	0.450	0.451	+ 24
Tires and tubes	3.169	4.030	4.567	5.452	4.926	+ 55
Fuel	1.012	1.158	1.399	1.575	2.027	+100
Maintenance and repair	7.434	13.060	13.261	17.028	16.332	+122
Subtotal	15.970	22.950	24.698	30.212	29.889	+ 87
Depreciation	—	4.065	2.902	3.239	3.136	+ 33
Total cost per hr	15.970	27.015	27.600	33.451	33.025	+107
Total tons hauled (000)†	160,040	80,775	192,156	179,122	207,175	+ 29
Tons hauled per operating hr	187	191	197	211	225	+ 20
% increase in tons hauled	100	51	120	112	129	+ 29
% increase in operating hr	100	50	114	100	110	+ 10
% availability, 24-hr basis	64.3	67.1	64.5	61.0	63.5	− 1
% utilization 24-hr basis	53.9	58.4	59.9	52.2	54.3	− 8
% utilization/availability	83.8	87.0	92.9	85.6	85.5	+ 2
Number of haulage trucks	182	194	233	241	235	+ 29
% increase in fleet units	100	107	128	132	129	+ 29
Average weighted capacity (tons per unit)	64.4	71.3	78.6	78.7	81.7	+ 27
% increase in fleet capacity	100	118	156	162	164	+ 64
Average age of fleet, yr	2.88	3.74	4.48	6.33	7.83	+172
Average haul distance, ft†	6260	7208	7524	7466	8100	+ 29
% increase in cost per ton hauled	100	170	164	186	175	+ 75
% increase in total mining cost per ton material	100	96	118	118	125	+ 25

* Yearly weighted average of all truck models in fleet.
† Metric equivalents: 1 st × 0.907 184 7 = t; 1 ft × 0.3048 = m.

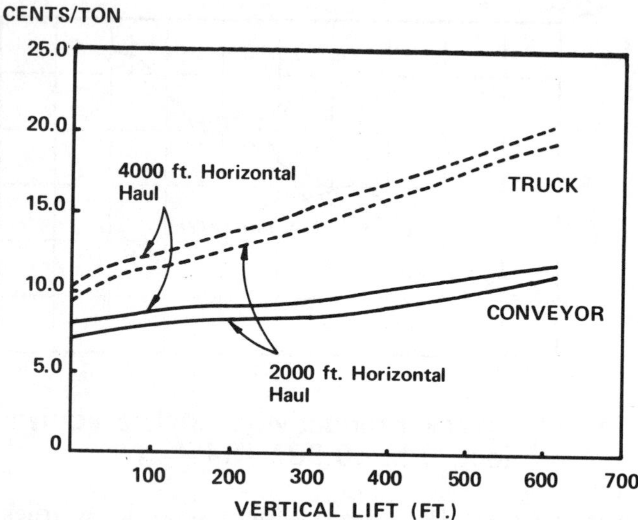

Fig. 28. Ore haulage and crushing (comparison of truck and conveyor costs). Metric equivalents: 1 ft×0.304 8=m; 1 st×0.907 184 7=t.

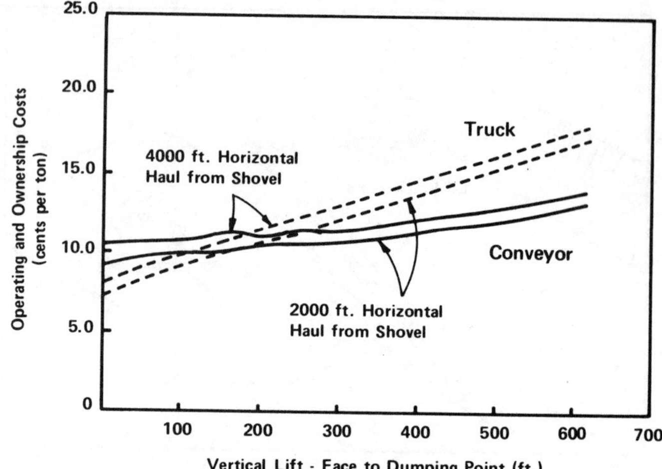

Fig. 29. Waste haulage and crushing (comparison of truck and conveyor costs). Metric equivalents: 1 ft×0.304 8=m; 1 st×0.907 184 7=t.

in productivity improvement over trucks. Such systems are already in fairly extensive use on soft materials, and others are being developed or are in partial use on harder rock. Fig. 26 shows such a scheme where trucks are totally eliminated. A variant might use shuttle trucks between the loader and a shiftable crusher-conveyor. Relative costs for typical truck and conveyor haulage systems as applied in ore and overburden are presented in Figs. 28 and 29, respectively. The potential productivity improvement and overall cost reduction are impressive.

Where mineralized overburden is amenable to leaching, a crusher-conveyor system provides further opportunity for using a belt stacker to build highly permeable leach dumps. Such loosely knit dumps will accelerate leaching and yield higher recoveries than possible from compacted and relatively impermeable truck or rail-built leach dumps. One specific example of a longer-range mining development program was described in the *Christian Science Monitor* of Apr. 12, 1974. The article reported that Soviet scientists had been commissioned, as a part of a major project, to design and develop robots to remotely control equipment "to help extract mineral wealth from the wastes of Northern Siberia." Undoubtedly, the major objective of this project is to obtain high productivity under difficult or otherwise uneconomical mining conditions.

CONCLUSION

It seems obvious that combinations of unconventional, or more correctly, nontraditional mining techniques can provide starting points for initiating and directing action programs to achieve the productivity jumps so evidently needed by the surface mining industry.

Walter B. Wriston, in a speech before the Economic Club of Detroit, very aptly quoted historian Barbara Tuchman: "The doomsayers work by extrapolation; they take a trend and extend it, forgetting that the doom factor, sooner or later, generates a coping mechanism. . . ." US surface mining productivity trends are disappointing, but by recognizing this condition, we can use our creativity to develop and apply innovative technology. In so doing, the surface mining industries can recapture favorable productivity trends of the past, notably those which were initiated when D. C. Jackling introduced large-scale open pit copper mining about 70 years ago, and in this generation when truck haulage began to replace in-pit mine railroads. The time is ripe for another cyclical change in mining.

Contents

Introduction to Section 2, Subsections 2A, 2B

Optimal exploitation of a mineral property requires a well-planned and executed exploration and development program. Technically, the development stage has been reached when sufficient information is available to show that the extent, character, and grade of a deposit justify commercial exploitation. In a practical sense, exploration and/or development usually continue throughout the life of the property and often provide basic motivation for revision of mining plans and reassessment of production equipment and processing plant requirements.

The initial and continuing acquisition of information pertaining to a mineral deposit is most often accomplished through a drilling program. In order for the engineer to provide his employer with maximum information, in a timely manner, at minimum cost, it is imperative that he be aware of the advantages and disadvantages of the various drilling methods. In addition, it is well for the engineer to be informed of the potential pitfalls which can be created in drilling contracts, particularly where an expert is not consulted as to best methods of accomplishing the goals of a specific program. These areas as well as other aspects of drilling programs are presented in the first chapter of this section entitled, "The Roll of the Drilling Contractor in Open Pit Mine Planning and Design" by Ferris Sainsbury.

An appropriate drilling plan requires careful consideration in selecting a hole pattern. The hole pattern is dictated by the probable geometrical shape of the deposit being sought and the techniques to be used in data processing. A common failing of many drilling programs is inadequate attention given to data acquisition. Often holes are drilled with the sole purpose of determining grade or presence of water, etc., and no attempt is made to gather other available information. Such a shortsighted approach is often wasteful and usually results in high costs and future duplication of effort. The chapter entitled, "Development Drilling" by Richard D. Call reviews methods of increasing the effectiveness of drilling programs and offers some helpful suggestions on how to avoid loss of potentially important data.

Once data has been acquired for a deposit, it must be presented in a logical and easily accessible manner. The most common form for data storage is the block model. This model is generally a three-dimensional computer representation of the mineralized volume and surrounding area. Details concerning the size and shape of blocks, limitations imposed by available data processors, and considerations in formatting are covered by Bruce T. Stanley in his chapter entitled, "Principles of Ore-Body Modeling."

The last two chapters of this section, "Mineralization Interpolation Techniques" by William E. Hughes and Roderick K. Davey and "Drill-Hole Interpolation—Estimating Mineral Inventory" by Marvin P. Barnes address the subject of conventional and geostatistical extension functions. An extension function is a technique or mathematical relationship by which sample values can be extended to surrounding volumes. These concepts are commonly employed in assigning values to elements within a block model for which actual data are not available. The representation of a mineral deposit in the form of grade, geometry, mineralogical, geological, and metallurgical properties, properly stored in a block or other suitable model, provides the basic tool for open pit mine planning and design.

2A Development Drilling

M. K. McCarter, editor

Contents

2 The Role of the Drilling Contractor

Ferris E. Sainsbury
Boyles Brothers Drilling Co.

Ferris E. Sainsbury is the president and general manager of Boyles Brothers Drilling Co. He joined Boyles Bros. in 1951; he rose through the ranks from field superintendent; manager of Boyles' Chile operations; manager of Boyles' South American operations comprising Peru, Chile, Argentina, and Bolivia; general manager of the drilling division in Salt Lake City; vice-president and general manager; to his current position in the company.

He became a member of AIME in 1957. Currently, he serves on the Executive Committee of the Utah Section and is a member of the Board of Directors of the Utah Mining Association.

As most mining companies utilize contractors rather than in-house drilling departments, it is difficult for most mining engineers to become familiar with drilling techniques. This chapter will explain the various drilling options available to the mining industry through the services of a drilling contractor. It will also suggest improvements in program planning and specification writing by the mining company representatives to enable them to derive maximum benefit from the contractor.

Basically we can divide most drilling programs into two categories: (1) exploration of a possible ore body and (2) delineation of a known ore body with emphasis on boundaries and continuity of grade.

Emphasis will be on pit development. However, information available from the first hole drilled on a new property must eventually be considered in the final analysis. So, every drilling program should be planned and conducted to give maximum information from initial discovery to actual production. As either program must be accomplished utilizing the same drilling methods, we will briefly discuss the four basic methods available at this time. They are: (1) churn or cable tool drilling, (2) conventional rotary or down-the-hole hammer drilling, (3) reverse circulation drilling, and (4) core drilling.

Any one or a combination of all these methods may be utilized in any drilling program. Local conditions will determine those best suited for a specific program. This chapter will not attempt to give in-depth descriptions, but a brief resume of the potential of each method will be given so that better decisions can be made as to which methods provide the desired information at the least cost.

Churn Drilling

Churn drilling is not widely used in the western USA but is frequently employed in the mid USA lead and zinc areas. Churn drilling is particularly well adapted in enabling a hole to be completed in areas abundant in solution caverns or abandoned underground workings. Churn drilling is especially effective in getting a hole down through loose uncemented river boulders and gravel, as casing can be driven to within a few inches of the bottom as the hole advances. Progress generally is rather slow compared to other methods of drilling. Samples are recovered as a chip in slurry form from an open hole. They do not give good information relative to minor fracturing or faulting, and they are exposed to contamination from the open portion of the hole. However, one advantage gained through churn drill-hole samples is that no circulating coolant is required, so any soluble minerals are retained in a relatively small amount of water, enabling the sampler to utilize a press or a filter so when evaporation occurs, one can recover most of the soluble mineral. Drills are large and quite heavy, requiring roads to drill sites. Very little water is required; risk of underground contamination is minimal; and surface contamination is usually less than other methods of drilling.

Rotary or Down-The-Hole Hammer Drilling

Rotary or down-the-hole hammer drilling is widely used in exploration when the need for an accurate, uncontaminated sample is secondary to the need for drilling holes rapidly and at least cost. Most uranium and much coal drilling is done in this manner, as down-the-hole instruments can be used to verify the presence of these and some other minerals. Either method is most effective in medium soft, nonsticky formations containing little or no water. Under these circumstances, using high pressure air, advances of 30 to 60 m/hr (100 to 200 fph) are not uncommon. A shallow water table with excessive inflow of water limits the economic depth a hole can be drilled using air. When this occurs, a changeover to water or mud must be made. This means the end to down-the-hole hammer drilling, as an efficient hydraulic down-the-hole hammer has not been developed. Rotary tricone drilling is possible to great depths using mud; however, the rate of penetration in hard rock may be so slow that it is more economical to change over to coring. Contamination of the sample is probable and is contingent upon the nature of the formation being drilled. In homogeneous, solid formations, fairly accurate samples may be recovered. In badly broken, alternately hard and soft layered formations, much contamination should be anticipated. In some formations, it may be impossible to recover a sample at all as the air or water escapes into the formation carrying the sample with it.

Costs for down-the-hole hammer or mud rotary drilling vary greatly depending on the formations, location, and depth of the ore body. Some uranium exploratory drilling in western Colorado or eastern Utah costs only 15 to 20% of prices charged for drilling igneous or volcanic formations. There are formations where higher prices are bid for rotary drilling than for core drilling.

Reverse Circulation Drilling

Reverse circulation down-the-hole hammer or tricone drilling utilizing Duo Tube* or similar equipment is a fairly recent method of drilling that enables a com-

* Tradename, Drillco Industrial Div., Smith International, Inc.

pany to recover virtually uncontaminated samples of both loosely cemented and solid formations. This method is particularly adept in getting holes down through caving or sluffing formations, through zones where lost circulation would normally occur, or in holes having a great inflow of water. Drilling is done using either air or water as the coolant and the medium for transportation of the sample to the surface.

This reverse circulation method employs the use of dual tube rods which consist of two concentric tubes, one within the other. The coolant is introduced into the space between the tubes and forced under pressure to an adaptor near the hammer or bit where it is exhausted. It travels around the outside and underneath the bit, entering the inner tube through which it returns to the surface. Samples are captured and, if the coolant is air or air with very little water, these samples give a fairly accurate sampling of the ore body as only a very short section of the hole is exposed to action of the air on the formation. Contamination is minimal. If the coolant is water, a problem exists in that the sample may not be accurate unless all the water is captured, filtered, and evaporated to recover those elements which have gone into solution during transport of the sample to the surface. In some cases, the handling of the samples has been a time-consuming and costly part of the drilling.

An advantage of this method is that the air or water is contained within the two tubes. This virtually eliminates hole-wall deterioration so common in single tube rotary or down-the-hole hammer work. A second advantage is the elimination of most lost circulation problems. Unfortunately, there have been quite a few strings of dual tube rods stuck and abandoned when drilling has been done in sticky clay or running sand.

As in all drilling methods, costs are contingent upon the formations being drilled, hole depth, and location, but as a rule, per meter (foot) cost for reverse circulation drilling is more than conventional down-the-hole hammer or tricone drilling, but less than core drilling.

Reverse Circulation Drilling Employing Core Bits

A few years ago, a secondary reverse circulation drilling method was developed. This method utilized coring bits and was successful in some formations. Core, like the rotary chips, is carried up the inner string of rods by the coolant. At the surface, it is exhausted with the coolant and is then captured.

There are two major negative aspects to this method of drilling. The core is exposed to excessive washing action by the coolant while traveling to the surface. This allows both mineral and waste to go into solution, making an accurate analysis of ore grade difficult. In sticky or caving formations, many strings of rod are stuck and abandoned. These disadvantages discourage the extensive use of this drilling method. A few contractors employ this technique on special projects, but their costs and degree of success are not available.

Core Drilling

Although this is one of the slower and more expensive methods of drilling, it is the one most used whenever a virtually undisturbed, uncontaminated sample is required. From the core, the geologist can obtain maximum information on mineral content and grade. He can observe solution action, fracturing, faulting, minute intrusions, and alteration. With core taken from an oriented core barrel, formation dip and strike can be ascertained. By using a split inner tube core barrel, samples of badly faulted or fractured formations can be recovered and displayed in a manner not possible in any other way. In some formations where the core must be preserved unexposed to air, a triple-tube core barrel can be employed.

The development of the small diameter wire-line system and the adaptation of bentonite drilling muds in the mid 1950's opened a new era in diamond core drilling. Core recovery in faulted, fractured, or loosely cemented formations rose from the 50 to 60% average to the current average of 85 to 95%. Grinding of core is eliminated and the quality of the core greatly improved. Core recovered now more closely resembles the in-place condition, and contamination from caving hole is virtually eliminated. Greater depths and hole completion with larger sized core is now possible. The need for cementing or reaming has been greatly reduced. Lastly, the per meter (foot) cost of diamond drilling is less as the driller can get more meters (feet) per shift with less work on his part.

Core drilling today falls into two categories: (1) sampling of the entire hole and (2) sampling of selected portions of the hole. These categories involve different techniques, and frequently, different equipment, so a brief description will be presented for each.

Sampling the Entire Hole

If the entire hole is to be sampled, most contractors will install a short collar pipe through the overburden into bedrock. Initial coring will be done with a size large enough to permit one or more reductions in hole size. This will insure hole completion to the depth specified. It is very important that the contractor be advised as soon as possible about the intended depth of the hole. This will enable him to use different techniques to assure that the hole will be completed to the desired depth with maximum core recovery and least cost per meter (foot). In most formations where holes in excess of 300 or 400 m (1000 or 1500 ft) are scheduled, a 102-mm (4-in.) collar pipe would be installed. Coring would then commence with NC or similar size equipment. This will make a 95-mm (3.75-in.) hole and recover a 61-mm (2.40-in.) core. A reduction to NX hole [82 mm (3.25 in.)] will be made when in-hole conditions dictate. These in-hole conditions could be excessive loss of drilling mud or water, open voids caused by solution action or old mine workings, and extreme faulting or fracturing. Conversely, a reduction in hole size should be considered if extremely hard, solid rock indicates that no abnormal adverse conditions will be encountered lower in the hole.

Most contractors have at least four sizes of rods and casing available to enable them to make reductions as needed and still assure hole completion to the specified depth. Utilization of these various sizes will save money in 90% of the holes drilled. When caving or squeezing ground is encountered, the contractor can expend a considerable amount of time and money in cementing, reaming, or recovering lost circulation. Although these costs are usually charged back to the customer, the revenue generated is only about 70% of the amount the contractor normally obtains from drilling operations. So costs go up while profits go down. If one is reasonably sure that a reduction in size will not adversely affect core recovery or the contractor's ability to complete the hole, every consideration should be given to making the reduction instead of expending money cementing or reaming.

Occasionally, the requirement for large core or the uncertainty of conditions that might be encountered deeper in the hole makes a reduction in hole size unadvisable. When these conditions exist, frequent conferences should be held to constantly evaluate current conditions, and techniques should be adapted to cope with them. Core drilling, under the best of conditions, is an expensive method of obtaining information. Close cooperation between the contractor and the customer is imperative to assure reasonable costs and profits.

Sampling of Selected Portions of the Hole

As stated earlier, coring is slow and expensive, and unless a core sample is required from the entire length of the hole, consideration should be given to using a combination of one or more of the aforementioned noncoring drilling methods and coring only that portion of the hole where core is required. Generally, this will greatly reduce the per meter (foot) cost and will speed up the drilling program. All the noncore methods listed give a hole sufficiently large enough to accommodate the standard four core sizes. If intermittent coring is required, any of the rotary drills can utilize a core barrel on standard rods and can core a section of the hole without casing it. This section would then be reamed and noncore drilling of the previous size resumed.

A commonly used combination involves the use of rotary down-the-hole hammer or tricone bit drilling through previously explored or barren formations with core drilling done in the remainder of the hole. Chip sampling is done in the rotary portion to assure the rock is indeed barren. The hole may or may not be cased before coring is done, depending on how much footage is to be cored. Much coal drilling does not require the hole be cased as only 30 to 60 m (100 to 200 ft) of coring is done in each hole. Generally, wire-line rods and core barrel are used to assure maximum core recovery. Air or bentonite mud may be used for coring. In base metal drilling, where a continuous core is required from a greater percentage of the hole, the rotary portion of the hole is usually cased, the rotary drill moved off, and a diamond drill moved onto the hole. Wire-line equipment and bentonite mud are usually used for the coring. This combination results in quicker hole completion, adequate sampling of barren formations, and minimum total cost.

Let's review an example. Suppose the target is a primary mineral deposit underlying a secondary zone that may have a possibility for leaching. An effective exploration program could utilize reverse circulation rotary drilling through this upper zone followed by core drilling of the primary deposit. The result is that both zones are adequately sampled to ascertain if leaching of the upper zone will be practical and to establish ore grades in the primary deposit. Costs are usually less than 100% core drilling, and the hole is completed in less time.

Specification Shortcomings

Drilling, regardless of the method employed, is a sizable expenditure in the evaluation of any ore body. All too often those costs are higher than necessary because the geologist has not specified exactly what his needs are or what he expects to accomplish from the drilling program. He may specify that a particular method be employed when another would be less costly. He may insist on maintaining a large hole size when a reduction would save him thousands of dollars and still give him the sample he requires. In a drilling program, indecision can cost thousands of dollars. The following example will illustrate the problem.

We expect to core drill holes to 610 m (2000 ft) in depth but "maybe" we'll go as deep as 1220 m (4000 ft). So instead of mobilizing a 2B drill costing $40,-000, the contractor must mobilize a 4B drill costing $52,500. Ancillary equipment, rods, and casing sent to the job will be doubled or tripled. Mobilization and demobilization costs are increased. The contractor, not being assured a hole will bottom at 610 m (2000 ft), maintains a larger hole size and fights in-hole problems longer than necessary. All of these factors greatly increase the cost of every meter (foot) drilled in every 610-m (2000-ft) hole. Per meter (foot) costs on the "maybe we'll drill 1220 m (4000 ft) but didn't" can be 15% to 20% more than those of the program where only 610-m (2000-ft) holes were considered.

If one's knowledge of drilling is limited, the resources of a drilling contractor or a competent consultant should be utilized in the preparation of the drilling program specifications. Poorly written specifications or the use of specific equipment, methods, or techniques can add thousands of dollars to the cost of the drilling program. Recently, two adjoining coal properties were drilled by the same contractor on hourly paid contracts. Both were for utility companies with zero drilling experience. One utility company consulted the contractor who had drilled thousands of meters (feet) in the area. Together they developed a comprehensive rotary and core drilling program. The second utility company retained a consulting firm with experience in petroleum exploration, but not in mineral exploration, to direct the project. Imposition of needless oil well drilling techniques by the consultant in what was basically a routine rotary core program resulted in per meter (foot) costs 405% higher than those on the adjoining property. Drilling contractors prefer to have money spent on additional drilling rather than expended needlessly.

Special Drilling Applications

Frequently, in a property evaluation study, the need for some unconventional drilling methods may be required. Specialized services have been developed to enable samples to be recovered under adverse circumstances.

On one project, a drilling contractor developed a method of adapting a split-tube drive sampler to an air-driven down-the-hole hammer. This enabled him to recover virtually undisturbed samples in silts and loosely compacted sands and gravels at various depths in rotary holes being drilled on a prospective mine location. Even though the blow count of a free dropping hammer cannot be recorded, the seconds or minutes required to advance the split tube 304 mm (1 ft) can be recorded. Using this information, along with the samples, a good estimate can be made as to whether or not blasting of the overburden is required or if it can be ripped and stripped.

On another property, major faulting fairly deep in a hole was encountered. Weeks were spent trying to recover some representative samples of the fault gouge. All that was recovered were buckets of slimy sludge that gave little, if any, information. In desperation, a deflection wedge was set above the fault, the deflected hole was drilled and cased to the faulted area to seal off the water, and a small down-the-hole hammer and split-tube drive sampler were then used to recover some beautiful samples all the way through the fault. Coring was then resumed below the fault.

Of major importance in many pit planning programs is the need to produce bulk samples of the ore for pilot plant testing or metallurgical sampling. Frequently, tons of samples must be secured from various depths that may be 100, 200, or even 500 or more meters (500, 1000, or even 2000 or more feet) below the surface. To sink a shaft and mine out a few tons for testing is terribly expensive, time-consuming, and in many cases, the abandoned shaft is a nuisance forever if open pit mining is done. An alternative being used with increasing frequency is to secure bulk samples from drill holes. A number of programs have been undertaken where the bulk sample was secured either as chips from a rotary hole or as core, using a large diameter core barrel.

The depth from which the sample is to be recovered dictates the method to be employed. In shallow depths, it is relatively easy to collect tons of chip samples from either the conventional rotary, down-the-hole hammer drilling, or reverse circulation methods. On one project, the bulk sample zone was only 50 to 100 m (200 to 300 ft) below the surface. Large diameter [305-mm (12-in.)] holes spaced 3 m (10 ft) apart were drilled. Compressed air was used which made the sampling very easy. A minimum of one hole was completed each 10-hr shift so the cost to the customer per ton of sample was reasonable.

In 1977, a project in Pennsylvania required recovery of bulk coal samples from three seams at depths between 230 and 300 m (750 and 1000 ft). To accomplish this task, a 102-mm (4-in.) vertical hole was drilled to total depth. Casing was installed with a window wedge in it. NC wire-line coring was then done through the coal seam. The NC wire-line rods were pulled back just inside the casing, the wedge was turned a few degrees, and the seam recored. Using this procedure, six passes were made through the seam without pulling the wire-line rods. The casing with the window wedge was then pulled up to the next seam and the process repeated. By using this process, the seams were recored with less than 3 m (10 ft) of waste rock per wedging. Costs were much less than would have been possible using mining procedures, and the customer was able to get a percentage of his bulk sample from many different areas in the coal deposit.

Vein deposit type ore bodies are frequently difficult to evaluate unless a series of holes are drilled through them at various depths. Today, it is common for a drilling contractor to make three or four intersections of the ore body by using oriented deflection wedges or a Dyna-Drill.† Either method can be used in vertical or inclined holes. The Dyna-Drill makes a more uniform deflection but cannot recover core so, before the vein is intersected, a return to conventional drilling is required after the deflection has been made. Oriented wedging is time-consuming; the deflection is not the rounded curve of the Dyna-Drill, but core can be obtained for the entire length of the deflected hole.

Drilling Contractor Involvement

In petroleum exploration, most states require that hole logs be submitted on all holes drilled, and, in most cases, this information eventually becomes available to the public. In mineral exploration, such logs are not required, and much valuable information may be lost. However, much information is recorded in an informal way by both the drilling contractor and the geologist in charge of the project. Some contractors retain drill reports on almost every project they have been on. If a property is to be drilled again and the same contractor is employed, these reports can offer valuable advice on how the drilling should be done to avoid in-hole problems and reduce costs. Invitations are frequently received to bid on work for Company C on a property which was drilled a few years ago for Companies A and B. Sometimes it becomes apparent from the drilling specifications that the drilling contractor knows much more about the property than does Company C. The contractor cringes when someone specifies the use of down-the-hole hammer drilling to 450 m (1500 ft) when inflow of 757 L/min or 1135 L/min (200 or 300 gpm) of water at 100 or 200 m (500 or 600 ft) in depth was encountered in previously drilled holes on that property; or when someone specifies a reduction to BX size at 450 m (1500 ft) when that same contractor knows there is major faulting 50 or 60 m (160 or 200 ft) deeper in the formation. Contractors will respect and preserve the confidentiality of the program. The contractor should be taken into confidence as soon as possible on every program; his experience and know-how are of great advantage. This procedure will save time and money for the company, and it will enable the contractor to do a better job at minimum cost to the company with a probable increase in profits for him.

† Tradename, Dyna-Drill Co. Div., Smith International, Inc.

3 Development Drilling

Richard D. Call
Pincock, Allen & Holt, Inc.

Richard D. Call is currently vice president and head of geological engineering and rock mechanics with Pincock, Allen & Holt. He received a B.A. in physics from Williams College, an M.A. in geology from Columbia University, and a Ph.D. in geological engineering from the University of Arizona. He has extensive experience in soils engineering, mineral exploration, slope design, slope monitoring, and other areas of geotechnical research. Recently, Dr. Call has served as a member of the review board for the Canadian government's CANMET Pit Slope project and is currently engaged in slope design and rock mechanics for a wide variety of mining operations on a worldwide basis.

Introduction

Data obtained from development drilling provide the basic input for open pit mine planning and design. Development drilling is defined as delineation of the size, mineral content, and disposition of an ore body by drilling boreholes (Thrush, 1968).

The objectives of development drilling are: (1) determining the geometry of the mineralization, (2) determining the grade and tonnage, (3) obtaining samples for metallurgical testing, and (4) obtaining geotechnical data for mine design.

To accomplish these objectives, an appropriate drilling plan must include the number of holes, the spacing and orientation of the holes, and a suitable data collection program. The data collection program covers sample collection, hole logging, and data presentation. This chapter will present some guidelines for development drilling and data collection techniques. The operational aspects of drilling will be covered in another chapter.

Geologic Interpretation and Statistics

The importance of geologic interpretation in development drilling cannot be overemphasized. With the widespread use of mathematical models for ore reserve calculation, it is often assumed that the computer center needs only the assays to come up with a statistical analysis of the data and an accurate estimate of the reserves. Although there are statistical methods for making inferences and for assessing the precision of the estimates, use of these methods is often restricted to certain methods of sampling and types of data.

Cochran, Mosteller, and Tukey (1954) distinguish between the target population and the sampled population in statistical analysis. The target population consists of all items about which inferences are to be made or from which conclusions are to be drawn. In the case of development drilling, the target population would be all of the minable units, e.g., each truckload of material within the pit limits. The sampled population, on the other hand, is the population which is actually sampled. Because of access limitations and restrictions imposed by any realistic drilling pattern, not all of the target population is included in the sampled population. The difference between sampled and target populations is important because "the step from sampled population to target population is based on subject matter knowledge and skill, general information, and intuition, but not on statistical methodology" (Cochran, Mosteller, and Tukey, 1954).

Thus, the validity of an analytical model must ultimately be determined by geologic interpretation, not by statistical tests. Statistics can be used as a guide for evaluation but cannot be substituted for sound judgment based on geologic information and reasoning.

A set of cross sections and a set of level maps are essential for geologic interpretation. Information such as surface geology and drill-hole intercepts should be plotted, without interpretation, on reproducible sheets. From these, copies can be made for use in interpretation. This process will maintain the distinction between observed fact and interpreted geology. Also, as new holes are drilled and as interpretations need to be revised, a new print of the factual sheet can be made. This avoids the messy procedure of erasing and redrafting the well-worn original.

Sections and level maps provide the best means to communicate information to mine planners on rock type and structure. Consequently, geologic sections and level maps should be drawn at the same scale as the mine planning maps. Couzens (1978) suggests 1 in.= 100 ft (1 mm=1.2 m) or 1 in.=200 ft (1 mm= 2.4 m) and 1:1000, 1:1250, or 1:2000 in metric ratios. Detailed geologic mapping and interpretation may be necessary on a large scale such as 1:500; however, maps on this scale can be reduced to a standard scale for use in planning.

To use sections and level maps in slope design, they must be extended beyond the edge of the ore body to include the rock in the pit wall. As a rule of thumb, one pit depth beyond the edge of the pit is sufficient.

Drilling Patterns

There are three basic types of drilling patterns: a systematic grid (Fig. 1a), a statistically random pattern (Fig. 1b), and an arbitrary hole location based on geologic evidence (Fig. 1c). There are relative advantages and disadvantages associated with each of these drilling patterns (Koch and Link, 1970; and Bailly, 1968).

Random sampling is a classical statistical technique for obtaining an unbiased sample of a population. With true uniform random sampling, every member of the population has an equal chance of being sampled. To apply random sampling to the drilling pattern, the x and y coordinates of the drill-hole locations are chosen from a table of random numbers so that every location has an equal chance of being drilled.

An obvious drawback in random sampling is inadequate areal coverage, as shown in Fig. 1b. There is also a theoretical objection. For random sampling and subsequent statistical analysis to be valid, all the samples must be drawn from a single population, and each must be statistically independent. This is rarely the case in drilling an ore body as there are usually several types

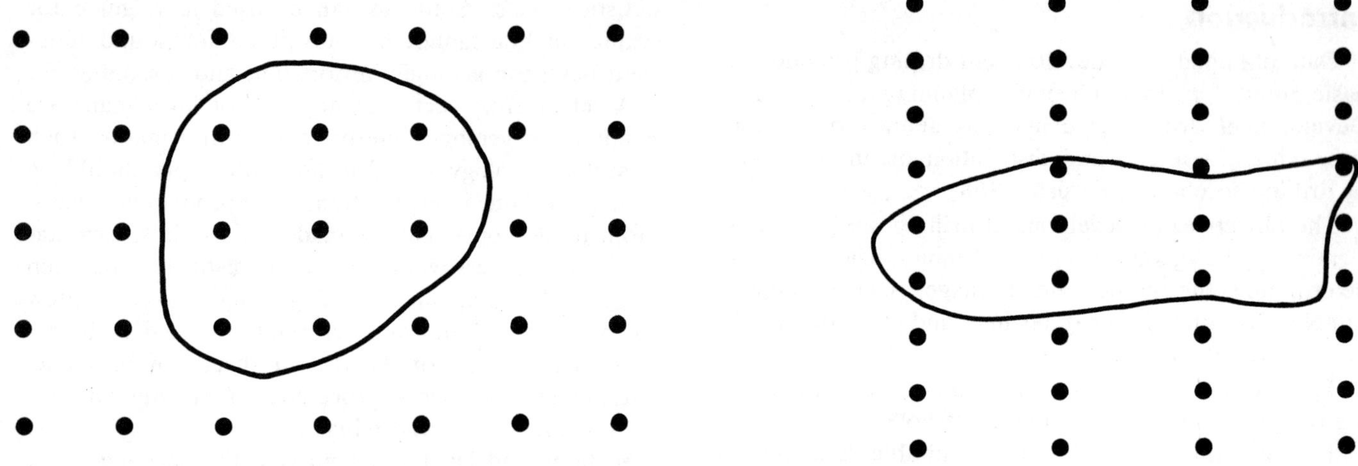

1a. Systematic drilling patterns.

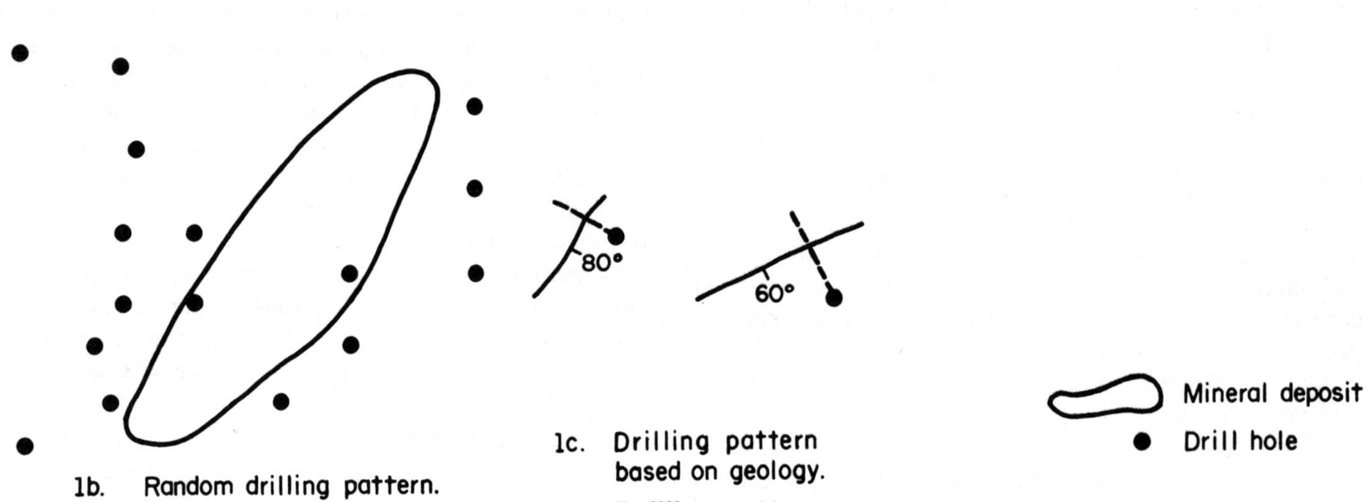

1b. Random drilling pattern.

1c. Drilling pattern based on geology.

Fig. 1. Drilling patterns.

of mineralization. Also, because of grade trends, the assays of two closely spaced holes are not independent of each other. For these reasons, random drill patterns are not suitable for development drilling.

Arbitrary drill-hole locations based on geologic reasoning are more appropriate at the exploration stage than during development drilling. Locating drill holes on the basis of alteration trends or favorable structural and lithological environments greatly enhances the probability of finding ore with a minimum amount of drilling. This is not the case with random or regular grid drilling. In development drilling, however, a representative sampling of the ore body is desired, and geologically controlled drilling tends to produce biased results. Submitting the assay results from a drill hole located in waste or low-grade areas does not make a geologist a candidate for promotion; thus, there is a temptation to cluster holes in the high-grade areas. Locating holes for geologic reasons cannot be ruled out of development drilling, however. Defining the edge of the structurally controlled ore body is best done by geologic drilling.

Grid drilling is generally the preferred pattern for obtaining a representative sample with good areal distribution. An advantage of grid drilling in geologic interpretation is that cross sections can be constructed with a minimum of projection. The chief disadvantage of grid drilling is that a regular spatial variation in the ore body could coincide with the drill-hole spacing. This would result in a major bias in the grade estimation. If the drill holes happen to coincide with high-grade zones, the grade of the deposit would be overestimated. Thus, it is prudent to break the pattern occasionally with a hole at an intermediate spacing and at a different angle, if possible.

Drill-Hole Spacing

The optimum spacing of drill holes is a trade-off between the confidence in tonnage and grade results, and the costs of drilling. As the drill spacing decreases, the costs increase geometrically. This increase is shown in Fig. 2 for drilling vertical holes into a flat-lying ore body. For this example, the cost per 1000 m^3 increases rapidly for spacings less than 150 m when ore to waste ratios are high and for spacings less than 50 m when drilling is in ore only.

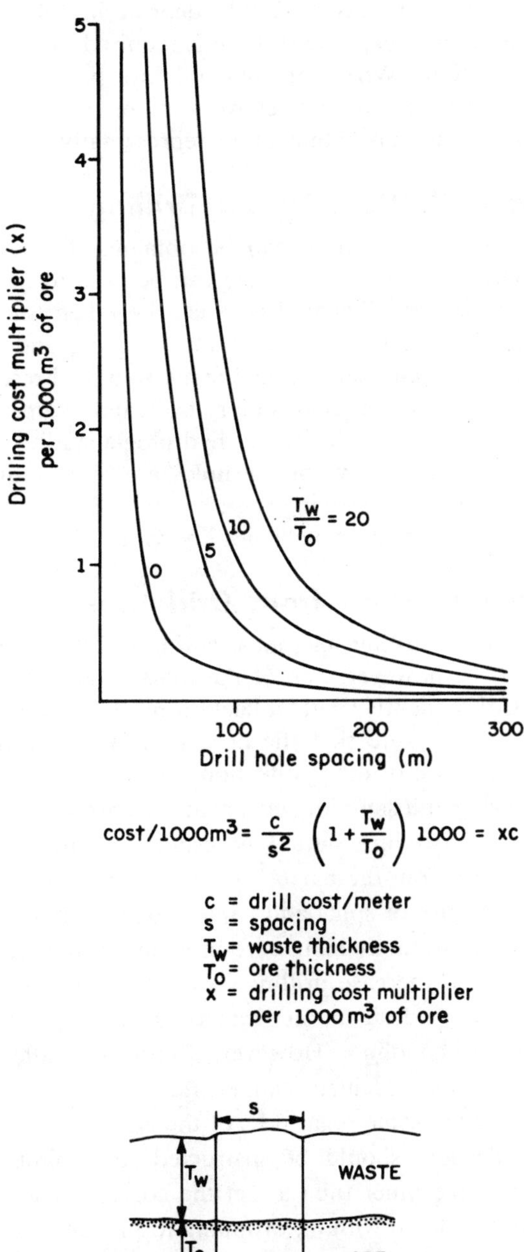

Fig. 2. Drilling cost related to drill-hole spacing.

There are two aspects to confidence in the ore estimates. One is the probability of missing a high- or low-grade zone, and the other is a statistical confidence of the mean problem.

When an ore zone consisting of small pods or high-grade zones is the target, there is a possibility of missing the ore zone entirely. The probability of missing a small circular target when drilling on a square grid is shown in Fig. 3. Again, detailed geologic interpretation is probably the most useful tool in locating deposits of this type.

The spacing of holes for grade definition within the boundaries of a large deposit can often be determined by the use of geostatistics. Initial drill holes should be sufficient for the calculation of a variogram in several horizontal directions and in the vertical direction.

The variogram is a graph that displays the relationship between the squared difference in sample grades and the distance between the sample points. Generally, the greater the distance between samples, the greater the expected square difference in their grades, i.e., the farther apart two samples are, the more statistically independent they are likely to be.

The variogram is a directional plot in that the distances between samples are measured in a single direction. By calculating variograms in several directions, the directional variability of the deposit can be measured.

Fig. 4 is an example variogram from a uranium deposit (Knudsen and Kim, 1977). The curve fit to the

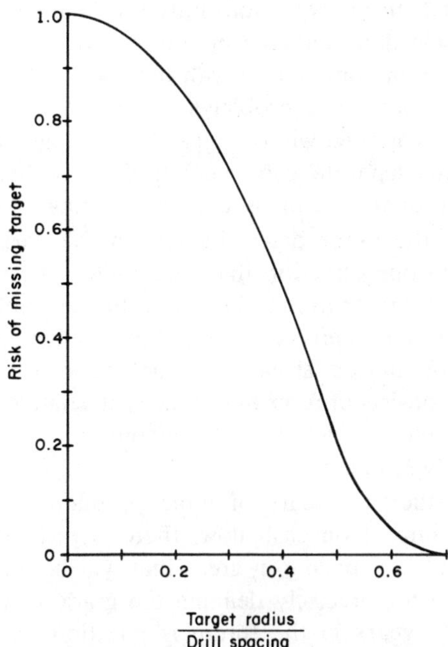

Fig. 3. Probability of missing circular target with a square drilling grid.

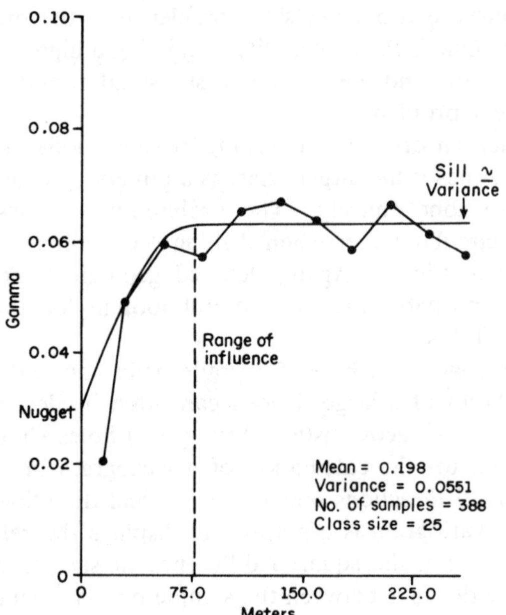

Fig. 4. Example variogram from uranium (Knudsen and Kim, 1977).

data indicates the mean squared difference for samples separated by a given distance. The distance where the curve breaks over is the range or radius of influence of a sample. Samples spaced farther apart in the given direction than this range are considered statistically independent. The range is a good estimate of the required drill spacing for adequate grade definition.

Geostatistics is not a substitute for detailed recording of geologic data and interpretations. It is, however, a tool to use in conjunction with more conventional geologic methods. The problems become obvious if geostatistics is applied without regard to geology. Lumping of several separate geologic populations together for statistical analysis can produce misleading results.

When the range determined from the variogram is greater in one direction than in another, the holes can be spaced farther in that direction. For example, in roll front uranium deposits where the mineralization is controlled by buried stream channel deposits, the best pattern consists of a series of fences at relatively closely spaced holes across the channel but with greater distance between fences.

Since the first years of mine production have the greatest impact on cash flow, there is justification for using a closer spacing in areas that will be mined first. The need for precisely defining the grade of ore to be mined 20 years in the future is questionable because the estimated commodity price, mining costs, and other financial factors, such as taxes, approach pure speculation.

Types of Drilling

A number of drilling methods are available, such as diamond core, rotary, and churn, etc. Peters (1978) discusses types of drilling and summarizes their characteristics (Table 1). The drilling method to be used is a function of the type of information required, the costs involved, and the condition of the rock to be drilled. For example, in uranium deposits where the primary need is a hole for downhole geophysical logging, a low cost noncoring method is preferable. On the other hand, for a strataform sulfide deposit in folded and metamorphic rock, core drilling is required for geologic interpretation. Where ore minerals are predominantly along fractures and core recovery is low, reverse circulation drilling can obtain more representative samples.

Data Collection During Drilling

Pertinent information can be obtained during drilling. Drilling pressure and rate can be used to evaluate rock properties. Oriented core can be taken for slope design. Information on water loss or gain, and water level in holes, can help define the hydrology. Preserving the hole for subsequent water level and temperature measurements can greatly aid hydrologic studies.

Orientation surveys of the holes are often required. Drilling on 100-m spacing is questionable if the position of the bottom of the hole is not known within 200 m.

Data Collection from Drill Core

Of all the drilling methods, coring supplies the best data. It is, however, one of the most expensive methods. Consequently, data collection must be planned in advance to ensure that the core is in a form suitable for each phase of data collection. As most of the serious logging and sampling occurs at the core shed, a core handling procedure should be established for transfer of the core from the barrel to the core box to the core shed. The use of split- and triple-tube core barrels has advantages in minimizing core breakup; however, these systems also increase drilling cost. Data such as RQD (rock quality designation) and fracture frequency are affected by handling. However, if core is handled in a consistent and specified manner, the conclusions drawn should be the same regardless of the type of core barrel used. Drillers should be instructed to minimize the hitting of the inner tube to get the core out and to set core boxes down without dropping to prevent excessive breakage. All core boxes should have firmly fixed lids, especially during transport from the drill site to the core shed. In wet localities, waterproof core boxes should be specified.

Table 1. Drilling Methods and Normal Characteristics*

	Diamond core	Rotary	Reverse circulation	Downhole rotary	Downhole hammer	Percussion	Churn
Geologic information	Good	Poor	Fair	Poor	Poor	Poor	Poor
Sample volume	Small	Large	Large	Large	Large	Small	Large
Minimum hole diameter	30 mm	50 mm	120 mm	50 mm	100 mm	40 mm	130 mm
Depth limit	3000 m	3000 m	1000 m	3000 m	300 m	100 m	1500 m
Speed	Low	High	High	High	High	High	Low
Wall contamination	Variable	Variable	Low	Variable	Variable	Variable	Variable
Penetration—broken or irregular ground	Poor	Fair	Fair	Fair	Good	Good	Good
Site, surface, and underground	S + U	S	S	S + U	S + U	S + U	S
Collar inclination, range from vertical and down	180°	30°	0°†	30°	180°	180°	0°
Deflection capability	Moderate	Moderate	None	High	None	None	None
Deviation from course	High	High	Little	Little	Little	High	Little
Drilling medium, air or liquid	L	A + L	L	A + L	A	A + L	L
Cost per unit depth	High	Low	Moderate	Low	Low	Low	High
Mobilization cost	Low	Variable	Variable	Variable	Variable	Low	Variable
Site preparation cost	Low	Variable	Variable	Variable	Variable	Low	High

* Peters, 1978.

† Reverse circulation has recently been used at inclinations up to 40°.

Core Recovery and RQD

Core recovery is the length of core footage recovered divided by the length of core footage drilled expressed as a percent. Core recovery is needed to evaluate ore reserves and should be measured prior to core splitting.

The rock quality designation, RQD, which is a modified core recovery, should be measured in addition to core recovery. This consists of measuring the total length of those core pieces greater than 101.6 mm (4 in.) long for NX core in a run and expressing this length as a percent. For core diameters other than NX, the length of core measured should be twice the diameter of the core in order to minimize the effect of core diameter on RQD. Where a fracture breaks the core diagonally, the length should be measured along the centerline of the core. If the core is split longitudinally by a fracture, it should not be included in the +101.6-mm (+4-in.) lengths. For convenience, RQD can be measured over the assay interval or drill-run interval.

In conjunction with the RQD previously mentioned, three additional measurements should be taken: (1) the +25.4-mm (+1-in.) material, (2) the +0.3048-m (+1-ft) material, and (3) the longest piece. Measurements of core pieces that are +25.4 mm (+1 in.) and +0.3048 m (+1 ft) in length, in addition to the regular RQD [101.6 mm (4 in.)], can be used to estimate the distribution of fragment sizes. Work is in progress to correlate this estimated fragmentation with fragmentation in caving and in situ leaching operations. The longest piece measurement helps to define the extreme limit of the core fragment size distribution. The problem with the longest piece measurement is that the maximum length measured can only equal the length of the core box, unless the break can be defined as nonstructurally controlled. These additional measurements will help to define the range and distribution of core lengths.

Fig. 5 presents a format that can be used to measure core recovery and RQD data from drill core. It may be possible to incorporate this format into either drill log or assay data sheets. Because large amounts of data are produced, it is best that core recovery and RQD data be stored on computer magnetic tape along with drillhole number, interval, and rock type. It may be possible to include this information in with assay data storage tapes.

Core Photography

Prior to splitting, the core should be photographed. This gives a permanent record of the breakage. It is also a handy way to look at the core when reviewing the logs, particularly when the core is stored elsewhere. By building a frame to hold the camera and then taking

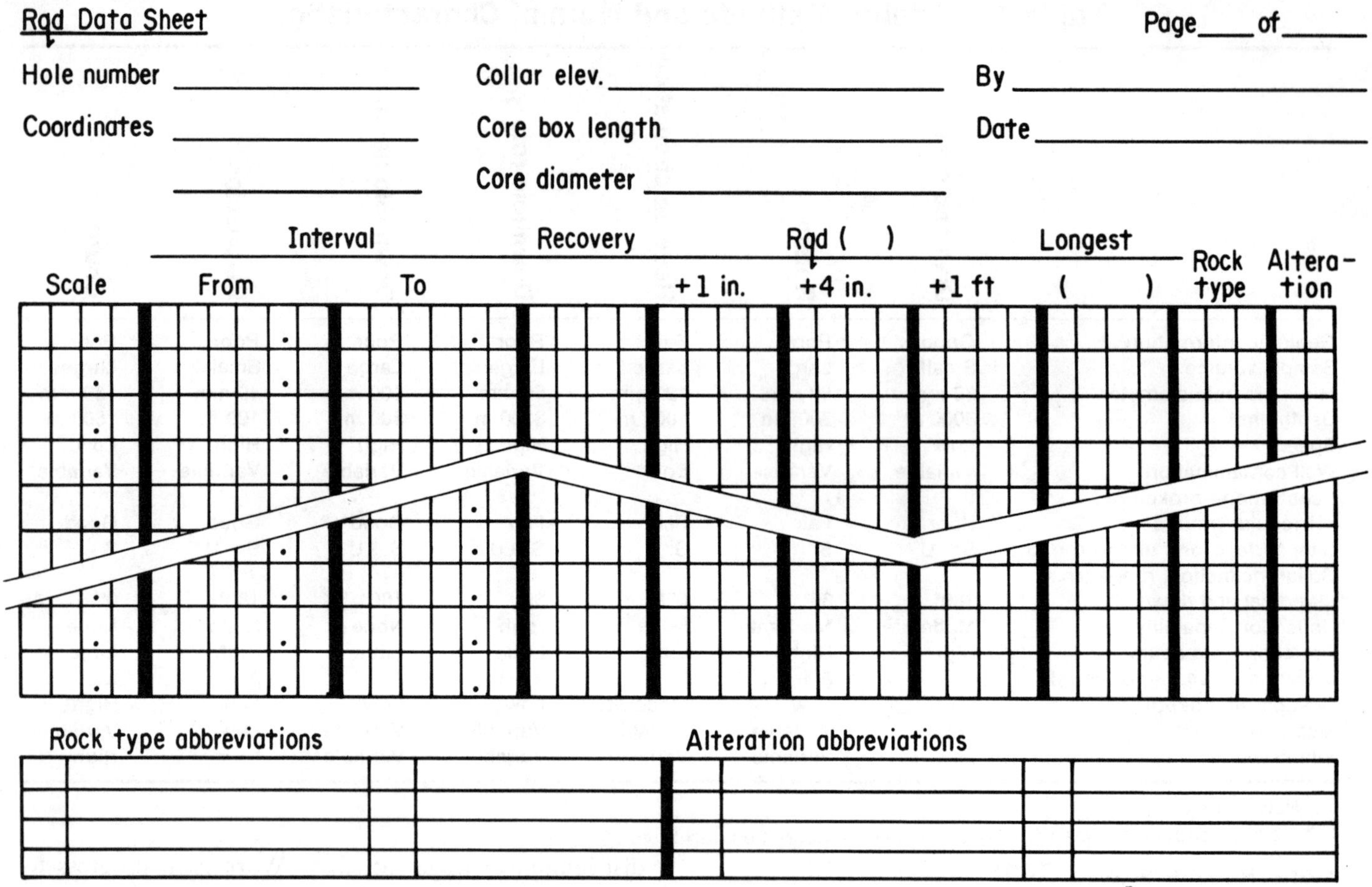

Rqd Data Sheet Page____ of_____

Hole number __________ Collar elev. __________ By __________

Coordinates __________ Core box length __________ Date __________

__________ Core diameter __________

Scale	Interval From	Interval To	Recovery	Rqd ( ) +1 in.	Rqd ( ) +4 in.	Rqd ( ) +1 ft	Longest ( )	Rock type	Alteration

Rock type abbreviations

Alteration abbreviations

Fig. 5. Example data collection form for core recovery and RQD. Metric equivalents: 1 in.×25.4=mm; 1 ft×0.304 8=m.

35-mm color slides of the boxed core, the time and cost involved are minimal. All photographs should include drill-hole number, core interval, and a scale. If the blocks identifying depth are dirty or otherwise illegible, they should be made readable.

Geologic Logging

The basic information required during geologic logging consists of lithology, alteration, mineralization, and structure. Logging forms need to be developed for each property. For good record keeping, each log should include the following information: project name, hole number, coordinates, collar elevation, hole bearing and inclination, hole diameter, date of logging, and name of person doing the logging. This information is usually recorded at the top of the log. The format for the body of the logging form depends on the type of deposit and personal preferences. Many logs contain columns for a sketch of the core, general remarks, and information on depth interval, recovery, assays, rock types, and structure.

During the logging process, it is important that the person doing the logging has enough room to lay out at least 30 m of core. The logging area should also be well-lighted.

As the core is logged, samples for petrography and mineralogy should be collected. These samples should include typical as well as atypical occurrences of the rock units.

Fracture Characterization

Several holes should be drilled to obtain oriented drill core for use in measuring joint dips, strikes, and spacings. These holes will aid in projecting the results from mapped exposures and in detecting structural changes with depth. Holes should be oriented to intersect the greatest number of structures. The number of holes depends on equipment, personnel availability, and complexity of geology. Fig. 6 shows a general format for logging oriented core.

For those holes that are not oriented, certain information should be noted: (1) fracture angle to core axis (dip), (2) filling type between fractures, and (3) shear strength along fracture surfaces. The dip mea-

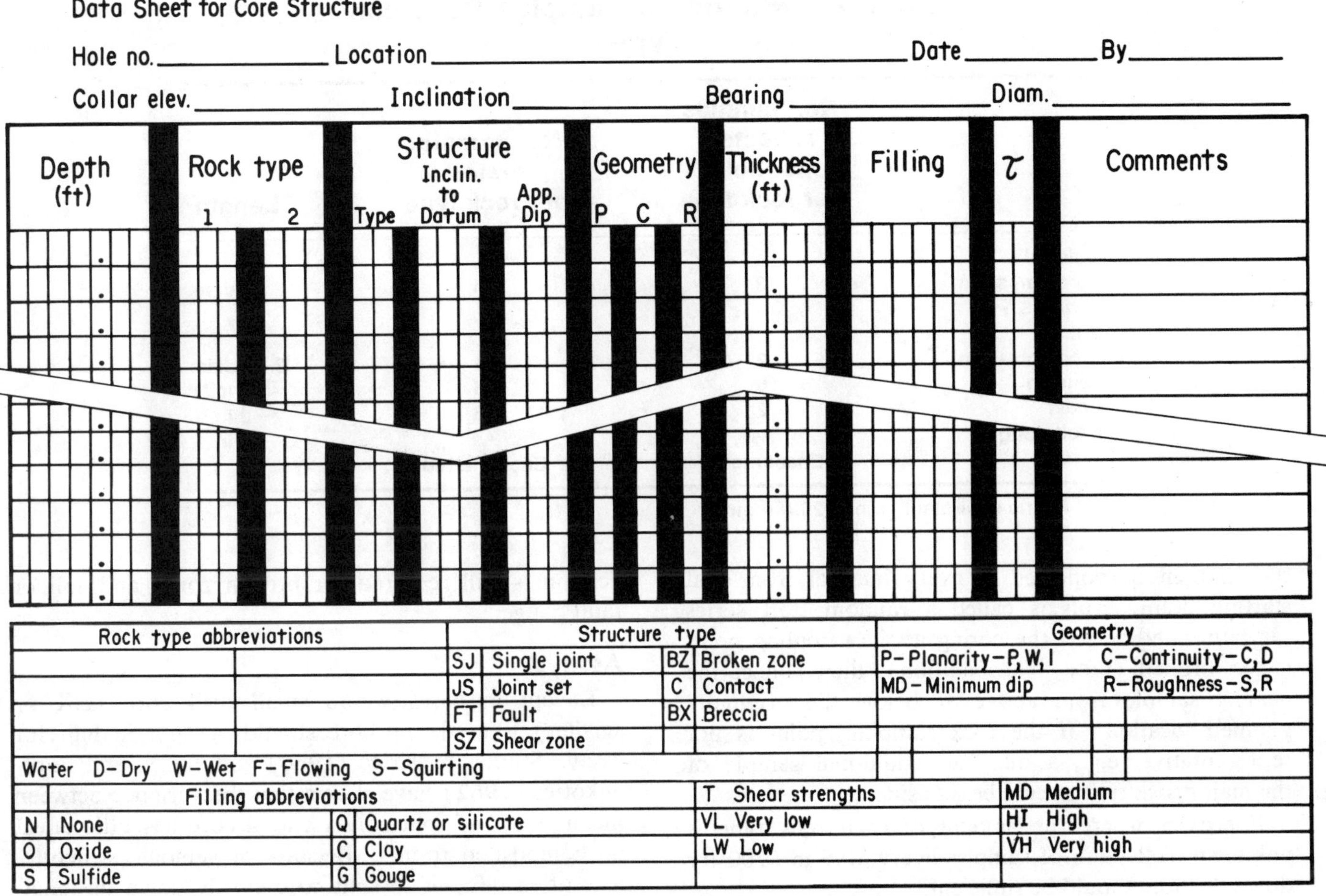

Data Sheet for Core Structure

Hole no.________ Location________ Date________ By________

Collar elev.________ Inclination________ Bearing________ Diam.________

Depth (ft)	Rock type		Structure			Geometry			Thickness (ft)	Filling	τ	Comments
	1	2	Type	Inclin. to Datum	App. Dip	P	C	R				

Rock type abbreviations		Structure type				Geometry	
		SJ	Single joint	BZ	Broken zone	P-Planarity-P, W, I	C-Continuity-C, D
		JS	Joint set	C	Contact	MD-Minimum dip	R-Roughness-S, R
		FT	Fault	BX	Breccia		
		SZ	Shear zone				

Water D-Dry W-Wet F-Flowing S-Squirting

Filling abbreviations				T Shear strengths	
N	None	Q	Quartz or silicate	VL Very low	MD Medium
O	Oxide	C	Clay	LW Low	HI High
S	Sulfide	G	Gouge		VH Very high

Fig. 6. Example data collection form for oriented core.

surements can be recorded in terms of the number of fractures per dip interval. Possible dip intervals are:

0-30
30-45
45-60
60-75
75-90

The percent of joints of a given shear strength within a drill interval should be measured. The following categories of shear strength are recommended: (1) very low—smooth joint and/or gouge filled, broken during coring; (2) low—has asperities, contains minimal fill, broken during coring; (3) medium—fracture easily broken by twisting with hand; (4) strong—takes great effort to break by twisting, or cannot be broken in this way, but has strength less than rock substance strength; and (5) very strong—fracture strength equal to or greater than rock substance strength. Although these measurements have no direct input to a stability analysis, they can be used to define zones of fracture characteristics throughout the deposit.

Samples for Rock Testing from Drill Core

For compressive and elastic properties (Young's modulus and Poisson's ratio), samples of core 2.5 times the core diameter are required. For small-scale direct shear and tensile strength, core samples 50.8 mm (2 in.) in length are required. When faults with appreciable gouge are encountered, a sample of gouge should be collected for shear testing. A piece of gouge 50.8 mm (2 in.) long would be best, but if this would affect the assay, a remolded test could be run on the gouge remaining after core splitting. Otherwise, all samples must be collected prior to splitting.

For each engineering rock type (different alteration phases would be considered different rock types), the number of samples listed in Table 2 is desirable. If the total footage of a rock type to be drilled is known (or a good estimate is available), then the number of core samples required can be divided by the depth to determine how frequently a sample should be collected. The first specimen should be collected at a random distance in the first interval. The following samples

Table 2. Number of Samples Per Rock Type

	No. Samples collected per hole per rock type	No. Samples tested per rock type	Length
Uniaxial compression w/ E & γ	3	24	2.5 x diam
Triaxial compression	3	24	2.5 x diam
Tension	6	48	~ 2 in.*
Shear	2	16	~ 2 in.
Fault gouge	As Encountered	As Encountered	~ 2 in.

* Metric equivalent: 1 in. x 25.4 = mm.

should then be collected at every interval from that starting point. This is called a random start series sampling method. If the coring at the sampling point is too broken (note this condition), then collect the nearest sample from above or below the originally planned location. If the rock sampling point is not representative, e.g., a dike, an additional sample of the major rock type should be set aside.

For cases where the amount of each rock type is unknown, collection of samples every 30.5 m (100 ft) down the hole should be sufficient.

By sampling each hole in this fashion, a collection of samples will be built up from which samples can be taken for the testing program. At each sample location a specimen whose length is 50.8 mm (2 in.) greater than 2.5 times the diameter should be collected.

Rock Hardness (Uniaxial Compressive Strength)

To evaluate relative variations within a deposit and to enable comparisons and discussions with other engineers, an estimate of the rock hardness or uniaxial compressive strength is required. Table 3 proposes one method for classifying soils (fault gouge) and rock hardness. This table is the result of work by Deere (1968), Terzaghi and Peck (1967), Jennings and Robertson (1969), and Piteau (1970). The system proposed in this table has the advantage of requiring only normal field equipment.

Another method for estimating the compressive strength is the point load test. The result of a point load test run on a core section is multiplied by the appropriate correction factor (Bieniawski, 1975) to estimate the uniaxial compressive strength. Because this system will only accommodate intact rock, a classification is still required for broken zones and soil or fault gouge.

Assays

Except in cases where small drill core (EX or smaller) is used, the core should be split in half for assay. Studies by Krige (1966) and Hazen and Berkenkotter (1962) have shown that the variance between mean assays of split core vs. total core has little effect on composited results. Because of geologic structure, it is often difficult to split core evenly down its longitudinal axis. In these cases, it may be necessary to saw cut the core. The splitting process leaves core for relogging at some future time when new or additional geologic information is needed. The saved split core can also be used for additional assaying. One technique that has proved useful is to saw cut the core three-quarters through and split the remainder. This provides the geologist with both a smooth and rough surface for logging.

The length of core to be assayed varies according to geologic breaks, high-grade zones, weights sample bags can carry, or drill-run intervals. Assay intervals usually run between 1.5 and 3 m. The key point to remember is that assay intervals can always be composited but cannot be broken down. The minerals assayed depend on the type of deposit (Waterman and Hazen, 1968).

In preparing the split core for assays, it is usually crushed, then split into quarters, of which one is ground and used for assaying. The remaining crushed rock should be saved for a metallurgist to perform Bond Work Index, flotation, and other metallurgical tests. This crushed material should be around 6.4 mm (¼ in.). With the crushed rejects, the metallurgist can

Table 3. Relationship Between Hardness or Consistency and Unconfined Compressive Strength ‡

Hardness	Consistency	Field identification	Approximate range of unconfined compressive strength, psi §
		Soils and fault gouge	
S1*	Very soft soil	Easily penetrated several millimeters (inches)§ by fist	<3.5
S2	Soft soil	Easily penetrated several millimeters (inches) by thumb	3.5–7
S3	Firm soil	Can be penetrated several millimeters (inches) by thumb with moderate effort	7–14
S4	Stiff soil	Readily indented by thumb but penetrated only with great effort	14–28
S5	Very stiff soil	Readily indented by thumbnail	28–56
S6	Hard soil	Indented with difficulty by thumbnail	>56
		Rock	
R0	Extremely soft rock	Indented by thumbnail	28–100
R1†	Very soft rock	Crumbles under firm blows with point of geologic pick; can be peeled by a pocket knife	100–1000
R2	Soft rock	Can be peeled by a pocket knife with difficulty; shallow indentations made by firm blow of geological pick	1000–4000
R3	Average rock	Cannot be scraped or peeled with a pocket knife; specimen can be fractured with single firm blow of hammer end of geological pick	4000–8000
R4	Hard rock	Specimen required more than one blow with hammer end of pick to fracture it	8000–16 000
R5	Very hard rock	Specimen required many blows of hammer end of geological pick to fracture it	16 000–32 000
R6	Extremely hard rock	Specimen can only be chipped with geological pick	> 32 000

* S1 to S6 (Terzaghi and Peck, 1967).
† R1 to R5 (Deere, 1968; and Jennings and Robertson, 1969).
‡ Modified by Piteau, 1970.
§ Metric equivalents: 1 in. x 25.4 = mm; 1 psi x 6.894 757 = kPa.

run a flotation test for all known ore conditions that will define the variability of the metallurgical nature of the ore. Bulk samples may still be used for pilot plant testing. Only one or two ore types will usually be represented in the bulk samples. Therefore, the variability of metallurgical characteristics of the ore will not be determined.

Core Storage

All core and assay rejects should be saved at least until mine planning is well along. The storage facility should protect the core, especially core in cardboard boxes, from the elements. Saving core is much less expensive than redrilling. When, if ever, the core is to be discarded, a skeleton of core should be retained. Skeletonizing usually consists of saving a 50.8- to 101.6-mm (2- to 4-in.) piece for every 0.6 m (2 ft) of core.

Data Collection for Noncoring Drill Methods

Noncoring drill methods generally have the advantage of lower costs than coring drill methods. Consequently, when core is not required, a noncoring drill method is preferable. Noncoring drill methods are commonly used for: (1) geophysical logging, (2) obtaining samples for assay and metallurgical testing, (3) defining ore contacts in extensive sedimentary deposits, (4) drilling through thick sections of overburden, and (5) hydrological testing. For holes used in geophysical logging or hydrological testing, there are minimum diameters and limitations on casing. Specific requirements should be supplied by the geophysicist or hydrologist who will conduct the testing.

Chips from a noncoring drill hole can be used for identifying general rock types. One of the better methods is to collect samples every 1.5 to 3 m and to glue a portion of the chips onto a board. This will ease making differentiations between rock units. After about 50 to 60 m, the intermixing of chips may make logging more difficult.

Final Comments

The preceding chapter briefly discussed determination of drill pattern and spacing, types of drilling methods, and required data collection. At first it may appear that a great deal of expensive data collection must be conducted; however, the cost of core drilling alone is the most expensive part of the program, $25 to $70 per meter. Conversely, all the data collection would usually be under $10 per meter. Because of the costs involved in driving an exploration shaft or adit, as much information should be determined from the drill core as possible.

The geologist responsible for development drilling should collect data not only for his purposes, but also for those of the rock mechanics engineer, the miner, and the metallurgist. Consequently, prior to or during the initial period of the drilling program, the geologist should meet with these people to determine what information they will need. Without this interaction, a lot of money could be spent to obtain an inadequate estimate of just tonnage and grade.

References

Bailly, P. A., 1968, "Exploration Methods and Requirements," *Surface Mining,* E. P. Pfleider, ed., AIME, New York.

Bieniawski, Z. T., 1975, "The Point Load Test in Geotechnical Practice," *Engineering Geology,* Vol. 9, Mar., pp. 1-11.

Cochran, W. G., Mosteller, F., and Tukey, J. W., 1954, "Principles of Sampling," A Report on Sexual Behavior in the Human Male, American Statistical Association, p. 18.

Couzens, T. R., 1978, "Aspects of Production Planning: Operating Layout and Phase Plans," *Open Pit Mine Planning and Design,* J. T. Crawford and W. A. Hustrulid, eds., AIME, New York, 1979.

Deere, D. U., 1968, "Geological Considerations," *Rock Mechanics in Engineering Practice,* K. G. Stagg and O. C. Zienkiewicz, eds., John Wiley & Sons, London, pp. 1-20.

Hazen, S. W., Jr., and Berkenkotter, R. D., 1962, "An Experimental Mine-Sampling Project Designed for Statistical Analysis," Report of Investigation 6019, US Bureau of Mines, 111 pp.

Jennings, J. E., and Robertson, A. M., 1969, "The Stability of Slopes Cut into Natural Rock," *Proceedings,* Vol. 2, 7th International Conference on Soil Mechanics and Foundation Engineering, Sociedad Mexicana de Mecanica de Suelos, Mexico, pp. 585-590.

Knudsen, H. P., and Kim, Y. C., 1977, *A Short Course on Geostatistical Ore Reserve Estimation,* University of Arizona, Tucson, AZ, p. 158.

Koch, G. S., Jr., and Link, R. F., 1970, *Statistical Analysis of Geological Data,* John Wiley & Sons, New York, 375 pp.

Krige, D. G., 1966, "Two-Dimensional Weighted Moving Average Trend Surfaces for Ore Valuation," *Proceedings,* Symposium on Mathematical Statistics and Computer Applications in Ore Valuation, South African Inst. of Mining and Metallurgy, pp. 13-79.

Peters, W. C., 1978, *Exploration and Mining Geology,* John Wiley & Sons, New York, p. 434.

Piteau, D. R., 1970, "Engineering Geology Contribution to the Study of Stability of Slopes in Rock with Particular Reference to DeBeers Mine," Ph.D. Thesis, University of Witwatersrand, Johannesburg, pp. 114-115.

Terzaghi, K., and Peck, R., 1967, *Soil Mechanics in Engineering Practice,* John Wiley & Sons, New York, 729 pp.

Thrush, P. W., 1968, *A Dictionary of Mining, Mineral, and Related Terms,* P. W. Thrush and the Staff of the USBM, eds., US Dept. of the Interior, Washington, D.C., p. 318.

Waterman, G., and Hazen, S., 1968, "Development Drilling and Bulk Sampling," *Surface Mining,* E. P. Pfleider, ed., AIME, New York.

2B Mineral Block Models

M. K. McCarter, editor

Contents

4 Mineral Model Construction: Principles of Ore-Body Modeling

Bruce T. Stanley
AMAX Inc.

Bruce T. Stanley is a 1970 graduate of Colorado School of Mines with an E.M. degree in mining. He was employed by Asarco as a mining engineer at their Mission unit, working with pit design and production reporting. Since 1971 he has been employed at AMAX's Henderson mine with duties involving mine design, planning, and production control. He is currently responsible for production and draw control, computerized mine monitoring and reporting systems, and ore reserves, upon which subjects he has published several papers concerning geostatistical techniques.

Introduction

A key point in the design and operation of a modern mining operation is the construction of what is called an ore-body model or block model. This model is a representation of reality constructed from predicted information. The blocks involved are merely subsets of the overall model which allows manipulation of the contained information on a local scale.

In general, block models enable mine planners to effectively select the most promising means of extracting the ore both physically and economically. The uses of a block model can be quite diversified, but one must realize that a single model that satisfies all curiosities and forms of expertise is difficult to construct.

Physical Limitations

Almost any and all information needed can be modeled by manual methods due to the great amount of flexibility involved. Consider, however, that the basic concept is to divide the entire ore body into smaller blocks which contain information pertinent to that block. Next, consider all of the various types of information possible for each block: (1) mineralogical, (2) geological, (3) statistical, (4) production, (5) financial, and (6) metallurgical.

It readily becomes apparent that for small operations hand-generated models for homogeneous ore bodies may be adequate, but for most large-scale operations with great amounts of capital expenditure, more sophisticated methods of data storage and information generating are required. Computerizing data manipulation can mean the difference between success and failure, between automatic procedures and personal interpretation, and between desired information and that which is humanly practical.

Data Modeled

When considering a block model, we must consider all the different things which can be modeled. From estimated grade to financial rate of return, many items may be chosen, depending upon particular interests. Each type of data may require different formats as well as storage capacities; therefore, careful planning must be exercised when defining the parameters. The following section emphasizes the peculiarities of block modeling with respect to geology and ore reserves.

Block Concept

The whole idea of ore-body modeling is centered around the fracturing of the ore body into units small enough to give interesting pictures of reality. One must realize that what is interesting to one person may be totally useless to another. Hence, the problem arises as to what size and shape the chosen blocks should take. These two problems are really based on a third concept of timeliness. Models will begin being generated as soon as any data at all is available, so it is important to be able to combine both mathematical backup and common sense when making decisions concerning size, shape, and timeliness.

Let us say that a block can be defined as the smallest basic volume of material to which it is practical to assign grade, tonnage, and geologic values. Parameters used in determining this basic size are: (1) grade variability, (2) geologic continuity, (3) machine-time capabilities, (4) slope stability, and (5) data storage limits.

In large porphyry-type deposits, grade distribution into a veinlet structure can mask the true continuity of the grade if very small blocks are chosen. The macrostructure of variation in grade may not be recognizable until a minimum sample spacing is achieved, and the attempt to model blocks that are any smaller is a useless task. One may determine a minimum block size in this respect by the use of geostatistics and related variograms. For example, if blocks are desired to be 6 m on a side, a 3-m sample interval may be too small to use for interpolation, and the resulting grade information may be grossly incorrect. If those 3-m samples were recombined to make composites of perhaps 9 m, results might show that a logical minimum-size block would have to be 15 m on a side. This is a very important concept and should be studied further in a course on geostatistics.

Some deposits are not only fractured on a small scale, but major fault zones and planes of weakness may exist. Where this is the case, block size could be directly affected if one wishes to apply the best possible fit to the available data. Geologically speaking, the existence of, say, a large chimney of highly silicious barren material may alter our opinion and change the basic block size to a 12-m cube. In this manner, a better mining sequence may be arranged due to the flexibility of smaller blocks (Fig. 1).

At the early stages of mine planning, the sizes and types of production equipment probably have not been selected, but a knowledgeable person should be able to estimate a most probable shovel, truck, and drill size. From this information, some basic production capabilities can be predicted with reference to machine-time-space requirements. Consider the following points: (1) optimum drill depth and penetration rate; (2) shovel capacity, loading time, required queuing, traffic area, and vertical digging limits, and (3) truck capacity and cycle time.

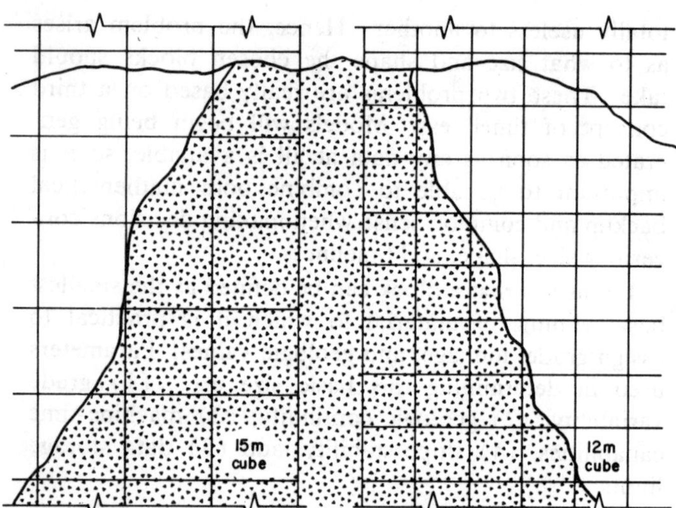

Fig. 1. A 15-m cube block and a 12-m cube block.

Required queuing and traffic area alone in this case could dictate that elemental blocks be at least 24 m wide. Capacities and cycle times also have their effect as to whether a small block is a viable time-related unit or a nuisance factor.

Stability of the final pit slope does not always play a remarkably drastic role in block size selection, but certain conveniences can be obtained if discretion is used here. Something as simple as adjusting the horizontal block size so that final bench offsets are even fractional multiples of the entire block size can be most helpful in personal visualization of plans. For instance, a final bench width of 4 m is an easy 0.2 multiple of a 20-m block size. If workable, simplifications like this one may avoid confusion in future layouts.

Finally, we must take into account the physical limitations of the computer with which we are working. Modeling a series of 5000 blocks is quite remote from modeling 50 000 blocks from the standpoint of time, money, and information derived. It is a known fact that accuracy of estimation decreases with decreasing block size in a porphyry deposit, so even though a small block offers a sense of security to the planning engineer, practicality must be considered in the final selection as to effect and cost.

Bear in mind that small blocks are flexible in that automated recombination can be used to form larger panels. The elemental block in Fig. 2a can be composited with neighboring blocks of similar value to establish patterns as in Fig. 2b.

These combining techniques and patterns will depend largely on the wishes of the person viewing them and the use for which they are intended. Always remember, however, no matter how many ways you have chosen

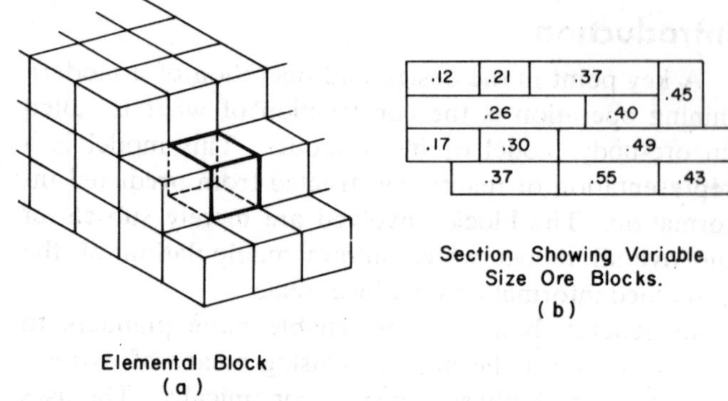

Fig. 2. Elemental block and section showing variable size ore blocks.

to allow recombining blocks, there will always be a person who needs to view the only method not programmed. Flexibility is the key.

Block shape is most easily handled when a cubic form is maintained. Polygonal shapes are best manipulated by hand and offer little advantage when viewed on an overall basis. Straight-line limits should be used to insure ease in data-file handling, since it is irrelevant to attempt estimation for blocks of continuously variable lateral limits. Surface contour information should be included so that blocks at the air/ground interface can be properly interpreted and flagged. Smooth pit contours, when desired, should be able to be produced through the use of optimizing techniques which are discussed later in this book. Simple smoothing may be readily obtained and checked by rotating grid orientation and interpolation by hand.

Understanding Model Applications

Regardless of the procedure used in estimating block values and assignment thereof to a mineral model, there are some basic fundamentals which should be applied to each effort. Each step when applied in sequence will help in building a successful and useful model. The logic applied is as follows: (1) define the problem for the size of operation; (2) list all available data; (3) specify the exact information desired; (4) choose the various parameterizing features necessary; (5) list potential data and information; and (6) evaluate benefits and analyze necessary equipment.

We must realize that each mine will exhibit different needs and capabilities for processing, but regardless of style, the most important step in the aforementioned process is problem definition. Here is where most engineers and management people miss the boat because there can never be a single model which is right for all. It does no one any good if processing is carried out with the wrong intentions.

A model which deals with block grades must consist of at least two different estimations. On one hand, we need a model that contains the best possible estimate of the average grade in each block. This will stand as the most logical basis from which planning will take place. On the other hand, we know that by definition, an estimation will contain inaccuracies or deviation from reality to some extent. If the estimation procedure we use yields information on estimation variances, and if those estimation variances for the elemental block compared to the grid spacing of drill holes are small, then our original model is good and will need no further investigation. If, however, the estimation variances are large this means that the possibility of considerably different pit designs might exist, and further simulation would be in order to investigate the range of possibilities.

The next step is to list what data will be available to us before model building begins. By knowing this, a more logical selection of estimation procedure can be made because some methods may need more or less data or data of different types and format. Also, the information desired in the final analysis will be directly affected by the amount and type of data used. These three concepts of data, method, and information must be considered in one thought and planning process in order to establish an effective model.

Very closely related to the previous items of concern is the idea of simulation. Consider the model as simulating reality by means of parameters which we insert and then showing the resulting consequences of those variables. This process is called conditional simulation and allows one to alter the facts enough that an adequate number of simulations can be viewed from data within the variance bracket. What makes the simulation conditional is the fact that the known sampling points are used with their respective known values in order to make the model close to reality. Fig. 3 shows the general structure of a model simulation program which brings up another important feature in block model construction. Parameterizing is a tool by which

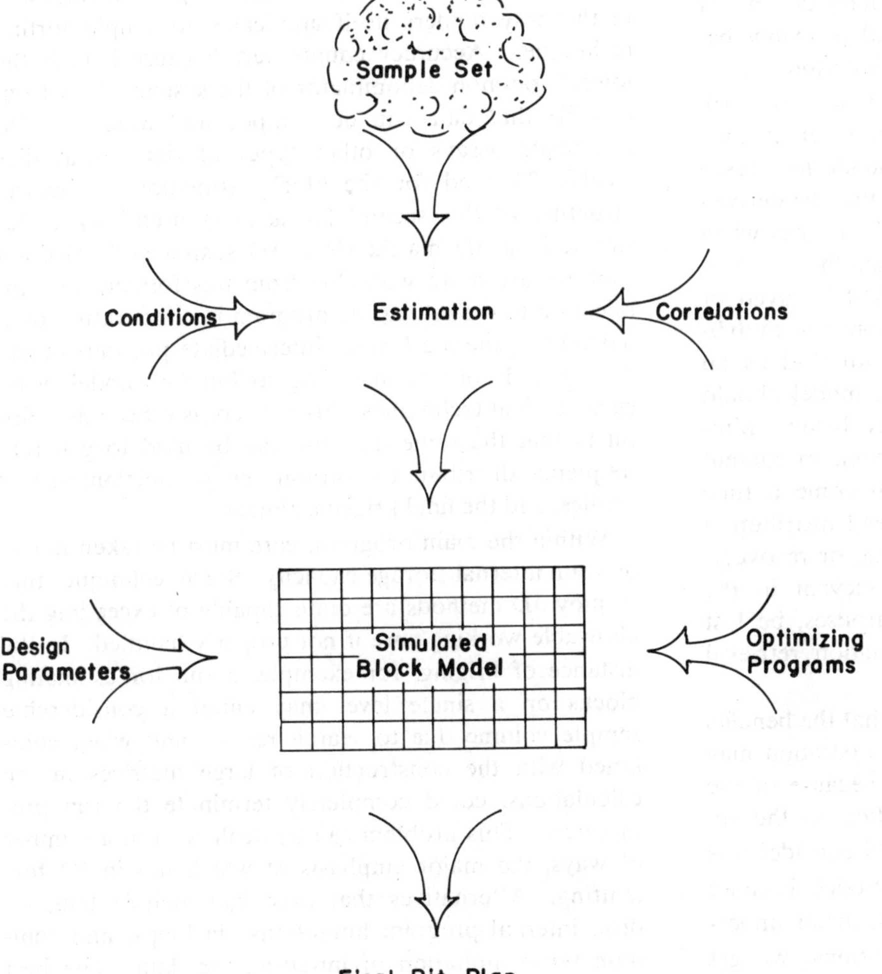

Fig. 3. General structure of a model simulation program.

one can alter the outcome of planning by specifying the parameters upon which the forecasts are made.

Final pit limits will be estimated by methods which are based upon two factors: (1) geologic features and (2) cost constraints. The geologic features are imperfect hypotheses upon which we must base our future calculations and plans, and this is why we choose to use estimation methods applicable to conditional simulations. Cost constraints are applied through what is called a parameterizing function which is based on both geometrical constraints and varying economic values. The profit value is what will ultimately dictate optimum pit design, and it is this profit value which is parameterized or characterized by building an algorithm corresponding to different economic variables. This is true in all cases, and no matter what estimation method and optimizing technique is chosen, we must remember that any resulting pit contour will require momentarily fixed economic, mining, and processing parameters. Each different set of conditions constitutes a separate simulation of possibly different results, and in order for these tools to be effective and useful for planning, the block model must be flexible enough to allow for these multiple investigations. All results are highly dependent on initial assumptions, all of which will probably become obsolete during the course of exploitation.

All too often when designs are first made on block models, an all-important feature is not given its due attention. A concerted effort must be made to foresee any potential data and information that may be derived from it. Many times data capacity or data types when initially chosen are lacking in one or another field of information or in the manner in which it is saved or input. For example, if an ore body contains molybdenum as its primary mineralization with lead as an uneconomical trace element, the block model should not ignore the lead concentrations as being unimportant. While it is true that the molybdenum content is the limiting design factor, there will come a time when the mill will need to know the lead distribution in order to plan for suppressing, leaching, or recovery. If this data had been regarded as irrelevant in the beginning, it could, for all practical purposes, be lost with respect to reasonably easy information retrieval in the future.

Finally, we must ask questions as to what the benefits are from such a program. A very large operation may necessitate highly mechanized methods because of the great amount of capital expenditure riding on the accuracy of the model. Small mines should consider less sophisticated means of fitting adequate models because the benefits derived may not warrant such an undertaking. Combined with these considerations, we get back to our original problem of capacities, time, and cost of available equipment. Ask whether or not the budget will be able to afford storage space, computer time, and also the lengths to which one must go to obtain the means to provide the required results, i.e., package programs, computer accessibility, and expertise.

Formatting

Correctly formatting a mineral model entails more than simple design of the final printed page. Indeed, there are at least three major concerns of format preparation: (1) input data, (2) internal core storage, and (3) output presentation.

The model must be able to quickly and efficiently manipulate files, turning data into data, and data into usable information. Data used for input is usually generated and input to files by the geology department. This raw data can be quite comprehensive in that it can contain assays, coordinates, and trace-element data as well as such general items as rock type, alteration, and fracture data. As can be seen in Table 1, a file of this type is easily built in a row-column type of format and in this way renders itself applicable to simple sorting techniques. Accuracy counts here because here is the lowest common denominator of the system. It is from this file that data can be stripped and used to build composite assays or other types of data input files (Table 2) used for the block estimation procedure. Structure of this second file is most useful when designed in a 3D matrix since 3D search and selection routines are more workable from this format, and application to the estimating program is direct rather than having to generate further intermediate portions of the data file. If one is intending to build a model using geostatistical techniques, this concept is especially helpful in that the same data file can be used to generate frequency distributions, variograms, proportional-effect studies, and the final kriged estimate.

Within the main program, care must be taken not to overrun internal storage capacity. Some techniques that employ 3D methods are quite capable of exceeding the allowable working core if not properly handled. In the instance of kriging, for example, a run for estimating blocks on a single level may entail a considerable sample volume due to search radius, and when combined with the construction of large matrices in the calculations, could completely terminate the run prematurely. This problem can be dealt with in a number of ways, the major emphasis of which lies in IO formatting. Alternatives that exist can include tape vs. disk, internal program limitations via loops, and common sense limitation of inputting the data. The best

Table 1. A Row-Column Type of Format Showing Assays, Coordinates, and Trace-Element Data

PAGE NO. 1
RUN DATE: 03/03/77
RUN TIME: 14:30:06

REPORT NO: 01

SITE	HOLE	TYPE	COLLAR	COORDINATES	ELEVATION	AZIMUTH	INCLINATION	INTERVAL	HOLE DEPTH	REMARKS
HU	0002	D	05010	05796	08054	090	+0	010	00500	NONE

SEQ NO	COORDINATES EAST	NORTH	ELEVATION	DISTANCE	AZI	INC	INT	MOS2	PERCENT WO3	MAX LNG	PC CR	RK QD	R C	RK TP	ALTERATION PSSACFT	MINERAL MFPFU
001	05015	05796	8054	5	90	+0	10	0.382	NA	0.4	37	54	6	U1	332405	0000
002	05025	05796	8054	15	90	+0	10	0.305	0.004	0.3	48	4	6	U1	332405	000R
003	05035	05796	8054	25	90	+0	10	0.246	0.002	0.2	59	0	5	U1	342406	0000
004	05045	05796	8054	35	90	+0	10	0.257	0.002	0.2	69	4	4	U1	232505	0000
005	05055	05796	8054	45	90	+0	10	0.229	0.002	0.4	60	4	5	U1	232505	0000
006	05065	05796	8054	55	90	+0	10	0.411	0.001	0.7	48	25	6	U1	132505	000R
007	05075	05796	8054	65	90	+0	10	0.277	0.004	0.7	38	18	7	U1	132304	P000
008	05085	05796	8054	75	90	+0	10	0.400	0.003	0.7	35	42	7	U1	132304	000R
009	05095	05796	8054	85	90	+0	10	0.287	0.001	0.5	42	12	6	U1	131304	000F
010	05104	05796	8053	95	90	−1	10	0.283	0.002	0.9	32	61	7	U1	132303	000R
011	05114	05796	8053	105	90	−1	10	0.290	NL	0.6	60	18	5	U1	122403	000F
012	05124	05796	8053	115	90	−1	10	0.504	NL	0.6	38	7	7	U1	143304	0000
013	05134	05796	8053	125	90	−1	10	0.286	0.002	0.9	26	76	7	U1	123306	0000
014	05144	05796	8053	135	90	−1	10	0.390	0.002	0.5	65	30	5	U1	122303	0000
015	05154	05796	8052	145	90	−1	10	0.545	0.002	0.8	44	42	6	U1	232204	000B
016	05164	05796	8052	155	90	−2	10	0.429	0.001	0.4	57	26	5	U1	232303	000F
017	05174	05796	8052	165	90	−2	10	0.346	NL	0.9	55	27	6	U1	123203	000R
018	05184	05796	8051	175	90	−2	10	0.253	NL	0.6	68	5	6	U1	122203	0000
019	05194	05796	8051	185	90	−2	10	0.374	NL	0.7	50	34	7	U1	122303	0000
020	05204	05796	8051	195	90	−2	10	0.248	NL	0.9	65	21	5	U1	332303	0000
021	05214	05796	8050	205	90	−2	10	0.483	NL	0.7	63	13	6	U1	253303	0000
022	05224	05796	8050	215	90	−2	10	0.315	NL	0.6	66	5	5	U1	243304	0H0R
023	05234	05796	8050	225	90	−2	10	0.702	NL	0.9	42	49	6	U1	252303	0000
024	05244	05796	8049	235	90	−2	10	0.347	NL	0.6	42	17	5	U1	243303	000F
025	05254	05796	8049	245	90	−2	10	0.332	0.005	0.7	39	55	6	U1	133303	0000
026	05264	05796	8049	255	90	−2	10	0.363	NL	0.7	56	30	5	U1	143303	0000
027	05274	05796	8048	265	90	−3	10	0.202	NL	0.6	52	18	5	U1	133403	00B0
028	05284	05796	8048	275	90	−3	10	0.323	0.002	0.6	54	13	5	U1	133403	000R
029	05294	05796	8047	285	90	−3	10	0.223	NL	0.5	33	16	4	U1	123302	00PB
030	05304	05796	8047	295	90	−3	10	0.320	NL	0.5	45	17	4	U1	322302	0000
031	05314	05796	8046	305	90	−3	10	0.476	NA	0.5	62	4	3	U1	132302	0000
032	05324	05796	8045	315	90	−3	10	0.301	0.004	0.7	51	31	6	U1	121302	000F
033	05334	05796	8045	325	90	−3	10	0.349	NL	0.7	42	38	5	U1	121302	0000
034	05344	05796	8044	335	90	−3	10	0.354	NL	0.5	54	20	5	U1	121203	0000
035	05354	05796	8044	345	90	−3	10	0.374	0.002	1.5	30	56	6	U1	121203	000F
036	05364	05796	8043	355	90	−3	10	0.500	0.001	0.7	32	56	6	U1	131302	0000
037	05374	05796	8043	365	90	−3	10	0.666	0.003	0.6	34	42	6	U1	121302	0000
038	05384	05796	8042	375	90	−3	10	0.701	0.001	0.8	76	22	2	U1	121303	0000
039	05394	05795	8042	385	91	−4	10	1.207	NA	1.4	35	54	4	U1	212502	0000
040	05404	05795	8041	395	91	−4	10	0.797	0.001	0.5	56	4	3	U1	111501	0000
041	05414	05795	8040	405	91	−4	10	0.988	NA	0.6	46	38	5	U1	111403	00ZF
042	05424	05795	8040	415	91	−4	10	0.465	0.002	0.6	60	17	5	U1	121303	0000
043	05434	05795	8039	425	91	−4	10	0.344	0.001	0.4	60	9	4	U1	221303	0000
044	05444	05794	8038	435	91	−4	10	0.400	NL	0.2	82	0	3	U1	141403	000R
045	05454	05794	8037	445	91	−4	10	0.885	NL	0.5	62	5	4	U1	121303	0000
046	05464	05794	8037	455	91	−4	10	0.643	0.002	0.4	60	0	4	U1	221303	0000
047	05474	05794	8036	465	91	−4	10	0.816	NL	0.5	55	13	5	U1	121303	0000
048	05484	05794	8035	475	91	−4	10	0.512	NL	0.6	33	25	4	U1	121303	000F
049	05494	05794	8035	485	91	−4	10	0.790	NA	0.8	45	24	5	U1	221303	0000
050	05504	05793	8034	495	91	−4	10	1.670		0.7	42	34	6	UU	0114030	00000

Table 2. Recomposited Data File

North	East	Elevation	Grade	Length	Variance
5025.00	4625.00	8050.00	0.36466646	60.00	0.00785232
5025.00	4675.00	7950.00	0.29124987	40.00	0.01989770
5025.00	4775.00	8000.00	0.24199992	50.00	0.00419311
5025.00	5175.00	7850.00	0.14599991	40.00	0.00869897
5025.00	5225.00	7900.00	0.21174991	40.00	0.00241017
5025.00	5375.00	8100.00	0.52699989	30.00	0.02594471
5025.00	5375.00	8150.00	0.60949934	40.00	0.00593096
5025.00	5425.00	7400.00	0.26224989	40.00	0.01512212
5025.00	5425.00	7450.00	0.28939992	50.00	0.02026063
5025.00	5425.00	7500.00	0.59199947	50.00	0.23509604

practice is to complete each individual block estimate one at a time, dumping the results after each loop. This relieves the core space somewhat for each new estimate. Specific formatting of program calculations is refinements applied to the method being used and should be considered on the merits for that method.

The output portion of the modeling program will enhance or severely limit its value. One can have the results written to tapes, disks, or printers. In all cases, the information is recorded sequentially which means that proper file building here is very important. Dumping to the printer eliminates any further file handling and can be done on the back of a two-part paper (thereby allowing pages to be pasted together and the darkened letters printed in any usual reproduction process). The problem with dumping is that once the information is printed, it can no longer be automatically reformatted into sections or specialized forms of reports. By writing results to an output file, a number of choices are available. Additional sorting programs can be employed, such as automatic optimizing techniques that rearrange the results into many usable forms. Different sections can be generated, unnecessary results removed, and statistical analyses are now possible on varying sets of data. Again, it is important here to have a 3D array type of storage file for flexibility.

Options

Once the model is constructed in the manner prescribed, it becomes available for active usage in the mine planning stage. Items of additional information can be inserted by addressing the XYZ indicators of any particular block or specifying a range of blocks in the driving program. A single driving program can be written to extract information either in sequence or by random access. Choices of information extracted are most easily built in using program options for most standard requirements. These requirements should be analyzed as to type of information, format, combinations thereof, etc. Options are not only used to view direct output but may also be used to build new data files for further program manipulation. In any case, the original model always stays intact, allowing many different analyses for different purposes.

Simulations and the Block Model

One must remember that simulations will continue to be made throughout the life of the mine, and each successive run will be made using the most recent data available. The original grade estimates will be replaced by drill-hole assays and mill-feed data as they become available and updated. It is known that sampling density and various other data will change characteristics, but it should be remembered that what we are attempting with the block model is the best unbiased representation of reality. Extremes of analyses will exist in any data set, so it is necessary to establish an information base that gives the average of all reasonable estimates. By accomplishing this task and making the resulting product as usable as possible for as many people as possible, the block model becomes a marvelous tool for accurate and meaningful mine planning.

References

David, M., 1975, *Geostatistical Ore Reserve Estimation*, Ecole Polytechnique de Montreal, PQ, Canada.

Francois-Bongarcon, D., and Marechal, A., 1977, "A New Method for Open Pit Design: Parametrization of the Final Pit Contour," *Proceedings*, 14th APCOM Symposium, AIME, New York.

Johnson, T. B., 1973, "A Comparative Study of Methods for Determining Ultimate Open Pit Mining Limits," *Proceedings*, 11th APCOM Symposium, Tucson, AZ.

Lemieux, M., 1977, "A Different Method of Modeling a Mineral Deposit for a Three-dimensional Open Pit Computer Design Application," *Proceedings*, 14th APCOM Symposium, AIME, New York.

5 Drill Hole Interpolation: Mineralized Interpolation Techniques

William E. Hughes
Ford, Bacon, and Davis

Roderick K. Davey
Kennecott Minerals Co.

William E. Hughes was formerly employed by Kennecott Copper Corp.'s Metal Mining Div. as a systems analyst. He holds B.S.E. and M.S. degrees in mechanical engineering from Brigham Young University and an M.S. in computer science from the University of Utah. He has many years of experience in the application of computer systems to engineering problems in the mining and metallurgical industries and other fields of scientific research.

Roderick K. Davey holds the position of technical superintendent at the Bingham mine of Kennecott Minerals Co. He obtained a B.S. degree in mining engineering at the University of Wisconsin and an M.S. degree in mining engineering from the Michigan Technological University. Mr. Davey is recognized throughout the industry for his expertise in the area of computer applications for the mineral industries.

INTRODUCTION

The objective of this chapter is to review and discuss interpolation techniques commonly in use in the mining industry today, including simple examples. This chapter will not attempt to compare the relative accuracy of each technique as this subject has adequately been covered by several authors (Knudsen and Kim, 1967; David, 1974; and Barnes, 1979). When appropriate, methods will be compared relative to the results of simple examples and computational ease. The examples presented are based on metallic deposits. The techniques, however, are not restricted to metallic deposits and can generally be applied to nonmetallic ones.

The purpose of interpolation, as applied to a mineralized deposit, is to extend the knowledge of the grade and geology of localized samples to an estimate of the grade and mineralogy of a larger block of ground or even an entire deposit. The word interpolation means to compute values between given values and is limited to that region between the known values. The distinction between interpolation and extrapolation is important and must be adhered to in the application of these techniques. In general, the interpolation must be limited to the area enclosed by verified samples.

The methods being discussed can be broadly broken into three categories: (1) geometrical methods, (2) distance weighting methods, and (3) geostatistical techniques.

GEOMETRICAL METHODS

Geometrical methods have traditionally been considered hand methods, but computer assistance works well for many situations. Advantages of most geometrical methods include the fact that sample points retain their identity throughout the interpolation, thereby simplifying checking and a reduction in the amount of mathematical number crunching characteristics of many other methods.

Polygon Methods

Polygon methods are the most popular geometrical methods and include as special cases: (1) square blocks, (2) rectangular uniform blocks, (3) triangular blocks, and (4) polygonal blocks.

This chapter will consider only the polygonal blocks which usually are generated by the mine planning engineer and geologist working in close conjunction to establish a set of interpolation rules for a specific deposit. Such a set of rules is shown in Table 1. These rules essentially conform to the mineralization controls as known. The examples described here apply to a horizontal section pierced by vertical drill holes, but the procedure will work equally well on nonhorizontal sections. For simplicity, these rules, and the subsequent discussion and examples of this chapter, were applied to a two-dimensional application. The procedures may readily be extended to three dimensions.

Considering that we plan on mining the deposit by open pit methods and with the rules in Table 1 in mind, the interpolation proceeds as follows:

1) Locations of drill holes and other samples are established for a specified level using available drill-hole survey data. Usually, the drill-hole location and assays of interest are depicted on a horizontal section.

2) Drill-hole interval assay data are composited to intervals consistent with bench height. The elevation of the sample is typically determined at the midpoint of the bench. Table 2 shows a typical drill-hole sample reporting form and corresponding composited grades (interval assay data).

3) Area of influence or radius of influence is established by geologic and mining experience.

4) Lines are drawn between drill holes that are within two times the radius of influence of each other.

Table 1. Example of Polygon Interpolation Rules*

1. Ultimate polygon shape is octagonal (eight-sided).
2. Radius of influence is R ft.
3. No polygon exceeds 2 x R ft from a sample point.
4. If drill holes are in excess of 5 x R ft apart, use a radius of R ft to show trend into undrilled area.
5. If holes are in excess of 4 x R ft, but less than 5 x R ft apart:
   a) Construct an R ft radius circle if assays are of unlike character, i.e., rock types, different mineralization, or one ore and the other waste.
   b) Use a 2 x 4 ft radius if assays are of like character to locate a point on a line between the holes; a line is then drawn to the point and tangent to the R ft diameter circle.
6. For holes less than 4 x R ft apart, construct a perpendicular bisector between the holes.
   a) If the holes are between 3 x R and 4 x R ft apart, use an R ft radius circle and connect the circles by drawing wings at a 30° angle from the center of the R ft radius circle to a 2 x R ft radius circle. The perpendicular bisector constructed above becomes the dividing line between 2 x R ft arc intersections.
   b) If the holes are less than 3 x R ft apart and a polygon cannot be constructed entirely from perpendicular lines from adjacent holes, then use an R ft radius circle and connect the circles with tangent lines. The dividing line is the perpendicular bisector between the holes.
7. After one pair of holes has been analyzed using these rules, another pair is evaluated, and this procedure is repeated until all combinations have been evaluated.
8. The assay value of the polygon will be the composited assay value of the drill hole that the polygon was constructed around.

*Metric equivalents: 1 ft × 0.3048 = m.

Table 2. Example of Drill-Hole Log and Composite Reporting Form

Drill-Hole ID = C-21A
Collar Location = 1800.0N 800.0E
Elevation = 5198.0
Azimuth = 0.0 Attitude = −90.0

Depth	Assay
5.	0.400
10.	0.560
15.	0.440
20.	0.480
25.	0.400
30.	0.380
35.	0.330
40.	0.590
45.	0.480
50.	0.600
55.	0.560
60.	0.320
65.	0.700
70.	0.210
75.	0.0
80.	0.080
85.	0.200
90.	0.070

Composites at 40 ft*

	E/W	N/S	Elevation	Assay
1)	800.00	1800.00	5179.00	0.4400
2)	800.00	1800.00	5140.00	0.3942
3)	800.00	1800.00	5114.00	0.1258

* Metric equivalent: 1 ft × 0.3048 = m.

This step may be altered by rules such as those in Table 1.

5) Perpendicular bisectors are constructed on each of these connecting lines.

6) Bisectors are extended until they intersect. If two lines run parallel or approximately parallel, and it is obvious that they will not intersect before the line closest to the drill hole intersects another line, the bisector that is closest to the drill hole is accepted as the polygon boundary.

7) In areas where drill holes are separated by distances greater than two times the radius of influence, an eight-sided polygon (octagon) form is drawn around the hole location, representing the maximum area of influence. This step may also be altered by rules such as those in Table 1.

8) Drill holes along the periphery of the ore body are extrapolated to the radius of influence and the octagonal form is drawn around the drill hole.

Example 1: As an example of hand-generated polygons, consider the composited samples and their locations on level 5140 of a theoretical open pit copper mine, shown in Fig. 1. Using a maximum radius of influence of 76.2 m (250 ft), a level map is generated, as shown in Fig. 2. Note that the polygon method degenerates to a rectangular method for regularly spaced drill holes that are closer to each other than two times the maximum radius of influence. Lightly shaded areas designate material projected to have a grade of ≥0.6%. By using a planimeter, and assuming a bench height of 12 m (40 ft), the projected tonnage of ≥0.6% Cu can be calculated to be 1 805 203 t (1,989,896 st) at an average grade of 0.93. [A tonnage factor of 1.88 t/m^3 (2.08 st per cu yd) was assumed. The procedure to calculate this example by hand after the composites had been calculated and plotted took approximately 1 hr.]

Polygons in Conjunction with a Geologic Model

Mineralization controls are easily introduced in the consideration of how far to extend the zone of influence of a particular sample; all of the rules described previously for the example of polygonal interpolation could also have been extended to include specific considerations concerning mineralization controls. For example, a fault located in a certain portion of the deposit is known to be postmineralization. Therefore, sample grades would not be projected from one side to the other, even though the maximum radius of influence had not been exceeded. The same condition would apply regarding different formations, favorable or non-favorable rock types. Also, metal grades of interest should not be projected into overburden. Rules should be determined to deal with assigning a metal grade to in-place material which is less than a full bench height thick, such as near the surface of the deposit.

Computer Applications of Polygon Methods

Using a computer, lists like Table 1 can automatically be considered as well as rules concerning mineralization controls for the specific deposit. In these procedures, we are always measuring or computing from the hole out to find polygon boundaries. However, it is very complicated to have a computer draw lines representing polygon boundaries in its memory and to assign area grades according to the procedures described previously. Very little accuracy is sacrificed by dividing the deposit into hypothetical blocks. The size of the block is dependent on sample density, objective of the project, and equipment to be used. Mineralization inventories

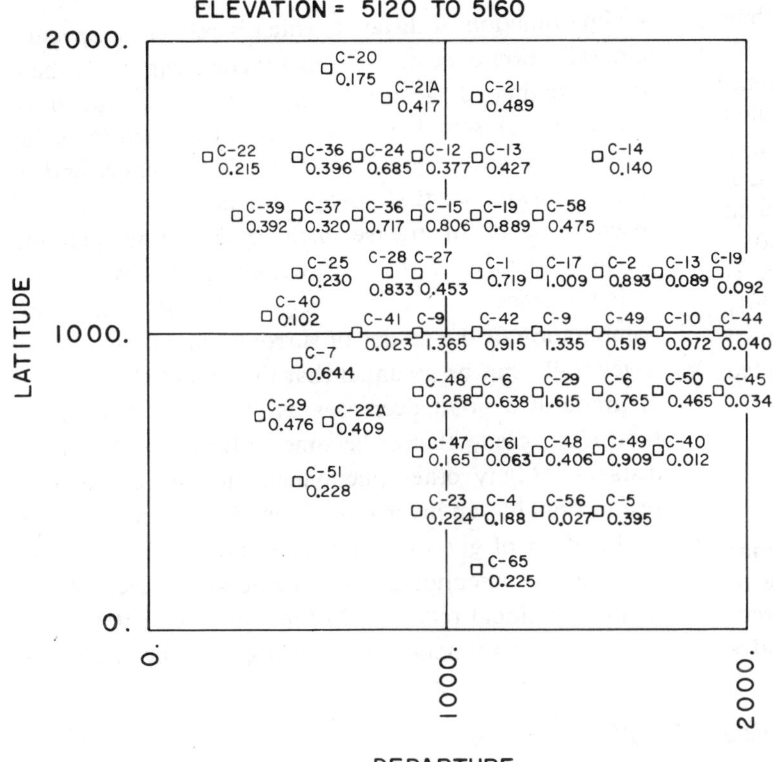

Fig. 1. Assay composites. C figures denote hole numbers and second figures are percent copper (composited values).

composed of hypothetical blocks can then be developed either manually or with the assistance of the computer. A typical procedure for generating the model is described as follows:

1) Generate a grid system dividing the deposit into

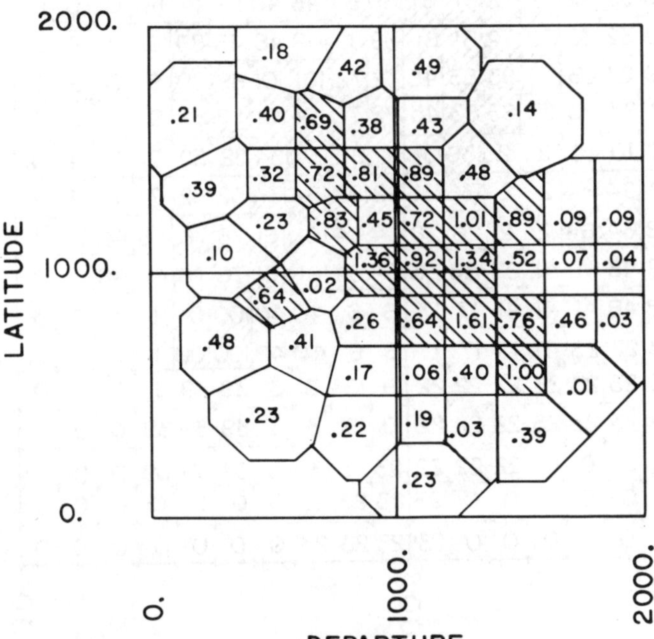

Fig. 2. Hand-generated polygons from composited samples in Fig. 1.

logical block units. The block size is based on operating practices, sample spacing, desired geologic resolution (dependent on sample spacing), and objective of the model. For example, it does not improve the accuracy of grade and tonnage projection to subdivide a model into 6×6 m (20×20 ft) by bench height blocks if sample spacing is 152 m (500 ft).

2) Geologic characteristics are assigned to each block of interest. This is usually accomplished by overlaying vertical geologic sections with a grid and assigning these characteristics to blocks halfway to the next section. This simplified procedure of generating a geologic model can be modified by using vertical sections in both directions and a set of appropriate rules.

3) Assign metal values and other characteristics of interest to each block by the following general procedure: (1) Compute the distance from the block center to all sample locations. (2) If no samples (or composites) are within the area of influence considerations (as described in Table 1), no grade is assigned to the block. (3) The nearest sample value is assigned to the block, provided the rock and mineral types of this block in the geological model are compatible with the rock and mineral types of the sample. In some cases, geologic compatibility may be disregarded if the sample is located in the same block being considered. In this situation, the value assigned to the block may be the

average of the multiple samples, if there is more than one.

Example 2: Fig. 3 shows the application of a computerized polygonal interpolation to the composited values shown as level 5140 in Fig. 1. The shaded area has been interpolated as mineralization $\geq$0.6% Cu. The mineralization model was divided into 30$\times$30 m (100$\times$100 ft) by bench height blocks before interpolation. Because the distance from block to composite is computed from the block center, results vary slightly from the defined polygons in Fig. 2. Accumulation of blocks with projected grades $\geq$0.6% Cu is calculated as 1 845 012 t (2,033,778 st) at an average grade of 0.92%.

DISTANCE WEIGHTING METHODS

Distance weighting methods became more popular when computer assistance became available because of the large number of repetitive calculations required. It normally is understood that the distribution of grades is some function of distance which is related to specific mineralization controls. If this function can be defined or approximated, accurate metal grades can be projected for nonsampled areas. All interpolation techniques assume that grade is related to distance, and it will be shown later that polygon methods are actually a special case of an inverse distance weighting method.

The objective of distance weighting methods is to assign a grade to a block or a point based on a linear combination of the grades of surrounding points. Since it generally can be assumed that the potential influence of grade at a point decreases as we move away from that point, grade change becomes a function of inverse distance. Many other phenomena in nature are also proportional to an inverse function of distance: change in the effect of gravity, magnetism, attenuation of light and sound, to mention a few. The questions are: "What is this functional relationship for a particular deposit?" and "Does this function change from deposit to deposit

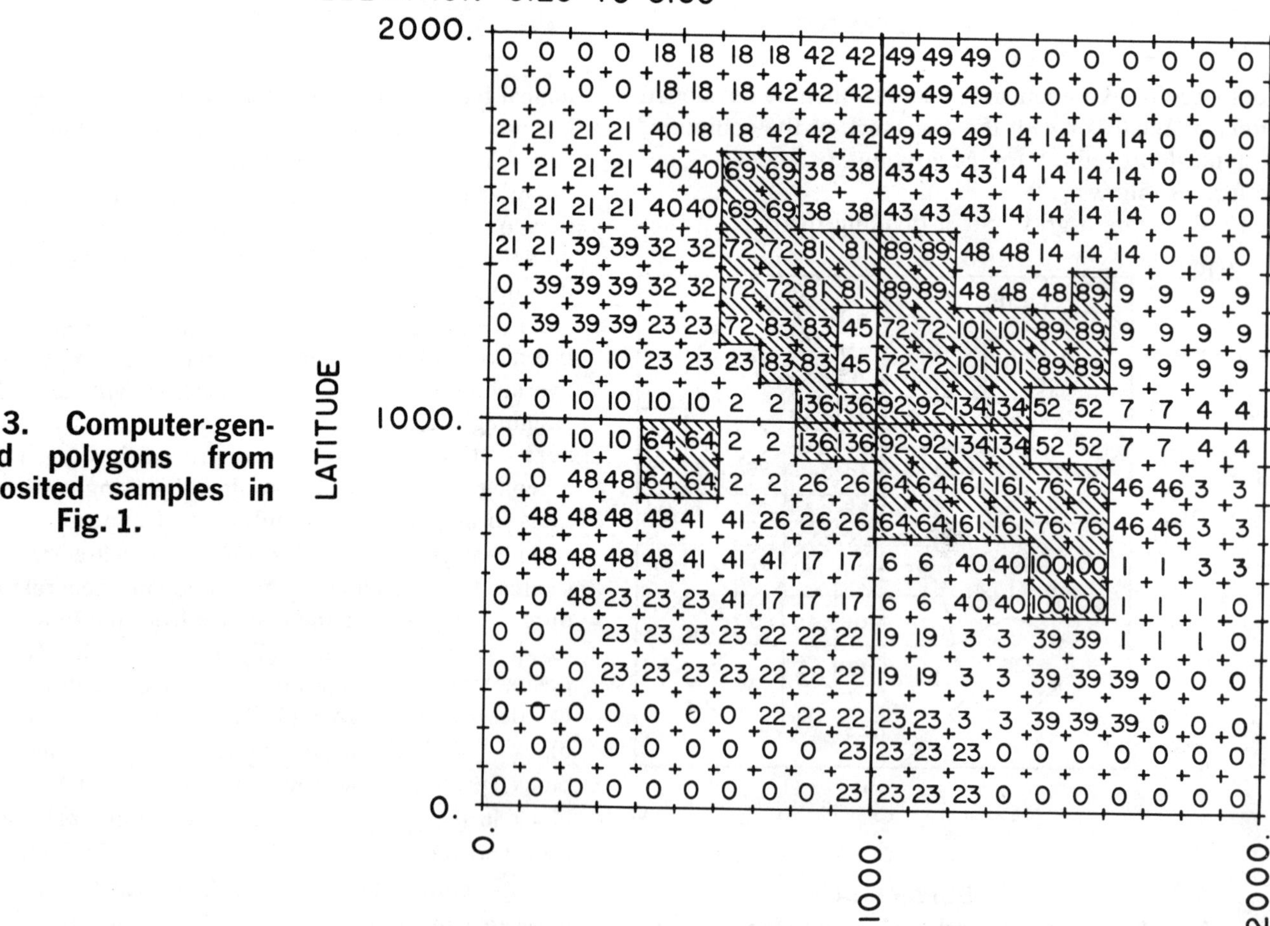

Fig. 3. Computer-generated polygons from composited samples in Fig. 1.

and between different geologic environments within a deposit?"

Determination of the Distance Exponent (Weighting Factor)

Two methods of determining the exponent power of distance in an inverse distance function are suggested.

Method 1 attempts to determine $f(d)$ in the relation-

$$G_2 = G_1/f(d) \quad (1)$$

ship where G_2 is an unknown grade influenced only by G_1, a known grade. G_2 and G_1 are separated by the distance d. Fig. 4 is a plot of the ratio of lesser grade to greater grade vs. distance separating samples for each drill hole compared to all other drill holes on level 5140 in Fig. 1. Also shown are fitted curves for several powers of d. The curve which best approximates this data is

$$G_2 = G_1/d^{.1} \quad (2)$$

which could be considered an appropriate model of the data. These data points are not independent and, therefore, each point within the ore body is probably influenced by several nearby sample points.

Because of geologic or mining experiences, the mine planning engineer may wish to weigh nearby samples more heavily than samples further away. He may choose a power for d quite different than the one suggested in Eq. 2. A common value chosen for these reasons is 2. Fig. 5 shows how the selection of m will cause the grade to vary between two known values. Note that when the exponent $m=\infty$, inverse distance interpolation degenerates to the polygon method, and if $m=1$, linear interpolation is performed.

Method 2 considers the fact that the interpolation model (mathematical function) chosen will be used to weigh nearby samples and compute a weighted average for the point being interpolated. This method uses nearby samples to compute the grade at the location of another sample. This computed grade then is compared to the sample grade at this location. Various powers for d are used to compute this grade and the power that best approximates all known locations would logically be selected for the weighting factor. In other words, each assay value in Fig. 1 was blanked out one at a time, and the grade at that point was computed from nearby samples using the relationship (model)

$$G = \frac{\sum_{i=1}^{n} G_i/d_i^m}{\sum_{i=1}^{n} 1/d_i^m} \quad (3)$$

where m is the power being considered. The variance or error between the computed grade and the grade assigned to that location was accumulated for all holes on that level and for each value of m investigated. The lowest error accumulation should be the best weighted average model for this deposit. Results of this test are shown in Table 3. These results indicate that 1.5 is the power m that should be used to interpolate this deposit.

The values in Table 3 fluctuate considerably and do not appear to follow a specific trend. A closer analysis of an actual deposit, making the increments of the power m smaller or continuous, would be in order.

The power of m that best fits the deposit may be a function of direction which should be considered in a

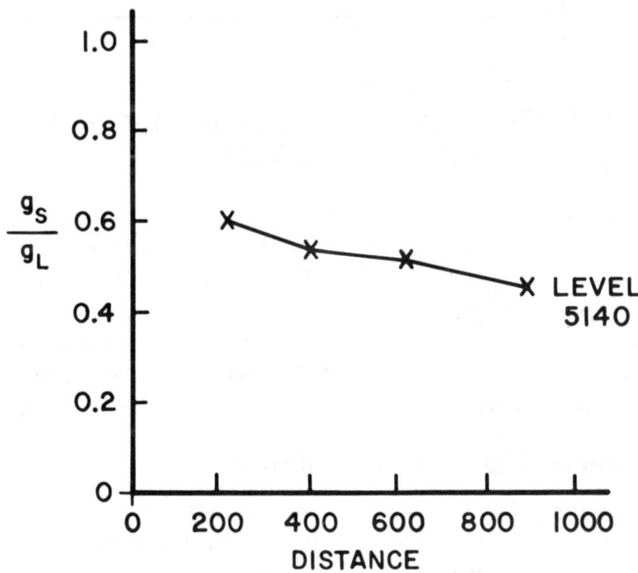

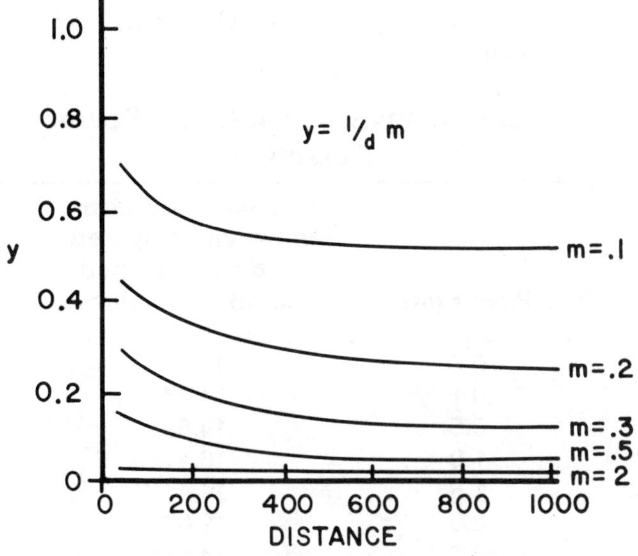

Fig. 4. Ratio of grade vs. distance.

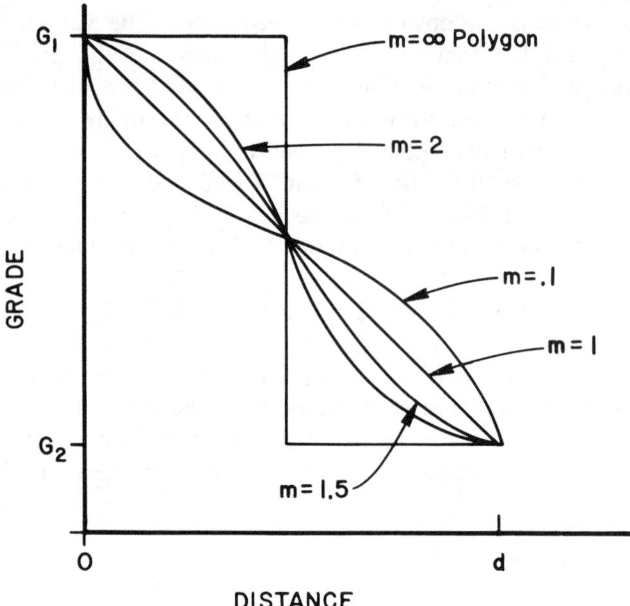

Fig. 5. Grade change vs. distance for various value m for $G_1/G_2 = 1/d^m$.

realistic analysis. For simplicity, the method herein described, and the examples herein, are restricted to a two-dimensional analysis. The methods of inverse distance weighting, however, lend themselves very well to a three-dimensional as well as directional extension, such as elliptical search patterns parallel to the structure and major mineralization controls of the deposit.

Inverse Distance Interpolation

A typical set of rules for inverse distance interpolation is shown in Table 4. With these rules in mind, the method of inverse distance weighting proceeds as follows:

1) Compute composites (bench height) and their locations on a level.

Table 3. Inverse Distance Power Selection

Power (m)	Accumulated error between computed grade & assigned composite value
0.0	17.36
0.1	14.47
0.5	14.85
1.0	16.41
1.5	14.44
2.0	19.01
2.5	14.57

2) Determine composited samples within the radius of influence of the point to be evaluated.

3) Exclude assay values that violate an angular limitation. An angular limitation may be needed to exclude interaction of nearby holes and to reduce the possibility of overweighting along a trend line. In other words, if two samples are within the angular limitation, they should really be considered as one sample and Eq. 3 would be overweighted if one of the samples was not eliminated.

4) Evaluate Eq. 3.

5) Assign the computed grades to the block or point.

Note that in Eq. 3 the term d^m in the numerator is the weighting factor and the term $1/d^m$ in the denominator is a normalizing factor which reduces the weighting factor to a linear combination of assay samples. Because of these properties, Eq. 3 can be represented in the following form:

$$G = \sum_{i=1}^{n} a_i G_i \tag{4}$$

where a_i are the normalized weighting coefficients. These weighting coefficients have the following properties:

$$\sum_{i=1}^{n} a_i = 1 \tag{5}$$

$$0 \leq a_i \leq 1 \tag{6}$$

Because of these properties, it is important to be cautious of accidentally slipping into a mode where interpolation becomes extrapolation. For example, this can occur if samples are eliminated because of one or more rules like the ones in Table 4, leaving two samples separated by a little more than the exclusion angle but both on the same side of the point being evaluated. This might occur along the edge of a deposit where cutoff drilling had not been completed. Because of the characteristics of weighting coefficients as shown in Eqs. 5 and 6, the interpolated grade will be somewhere between the two known grades which is more than likely not a good choice. For this reason, it is suggested that any set of rules using inverse distance or any other weighted average method also include an inclusion angle which would abort the interpolation of a point if samples were not represented over a range of at least 160° around the point.

Other methods of dealing with the extrapolation problem include (1) assigning pseudosamples in unsampled areas, and (2) limiting application of weighting techniques to a periphery described by known sample locations.

Example 3: This example is adapted from Weiss and

Table 4. Example of Inverse Distance Interpolation Rules*

1. Develop rock type distance factors. These factors are sets of A, B, and C coefficients for equations of the form $A \cdot X^2 + B \cdot X + C = Y$, where Y is the average standard deviation between grades and X is the distance between the sample points. This is done for all combinations of formations plus within each formation.
2. Develop geologic model with rock type code for each block being evaluated. Rock type codes are assigned to each composite value.
3. Radius of influence equals R ft and angle of exclusion equals A°.
4. The block must pass one of the following in order to be assigned any grades:
   a) The block must be within R ft of a composite.
   b) The block is within R ft of a line connecting two composites which are within 3 x R ft of each other.
   c) The block is within R ft of a line connecting two composites that are within 3 x R ft of a third composite.
   d) The block is inside a triangle formed by three composites, any two legs of which are equal to or less than 3 x R ft long.
5. Collect all assay composites for the level that is within 5 x R ft of the center of the block.
6. Count the number of composites having the same rock type as the block. A rock type the same as the block is defined as:
   a) The rock type of the composite matches the rock type of the block.
   b) The rock type of the block is unknown or undefined.
7. If no composites are found to match the rock type of the block, extend the radius of search outward by R ft increments until one or more composites are found within an increment. Add these composites to the list of ones affecting the block grade assignment.
8. Compute distances from the block to each composite having a different rock type than the block, such that the new distance would be equivalent to the two points being in the same rock type. If the equivalent distance is less than the original distance, use the original distance. The original distance rather than the equivalent distance will be used by the minimum angle screening.
9. Compute the azimuths from the block to each composite influencing the assay assignment.
10. For each assay of the mineralization model, compute the angle between each pair of composites having data for the assay. Check to see if the angle is less than A°. If the angle is less than A°:
    a) And the rock type of the composite farther away matches the rock type of the block, the nearer composite is rejected.
    b) And both composites match the rock type of the block and only two composites match the rock type of the block, both composites are retained.
    c) And the rock type of the closer composite matches the rock type of the block, the further composite is rejected.
    d) And the rock type of neither composite matches the rock type of the block, the furthest composite is rejected.
11. The grade assignment for the block is computed as:

$$G = \sum_i (G_i/D_i^2) / \sum_i (1/D_i^2)$$

    where G_i is the sample assay value and D_i is the equivalent distance to the ith composite.
    a) Unless there is a nonzero composite value within or on the boundary of the block, in which case that composite will be used directly.
    b) Unless there is only one composite, in which case the closest composite from the reject list having the same rock type is included. If no second composite can be found with the same rock type, the closest composite from the reject list is included.
12. If the resulting grade assignnment is zero, it will be increased to the smallest nonzero number which can be represented in the model.

* Metric equivalent: 1 ft × 0.3048 = m.

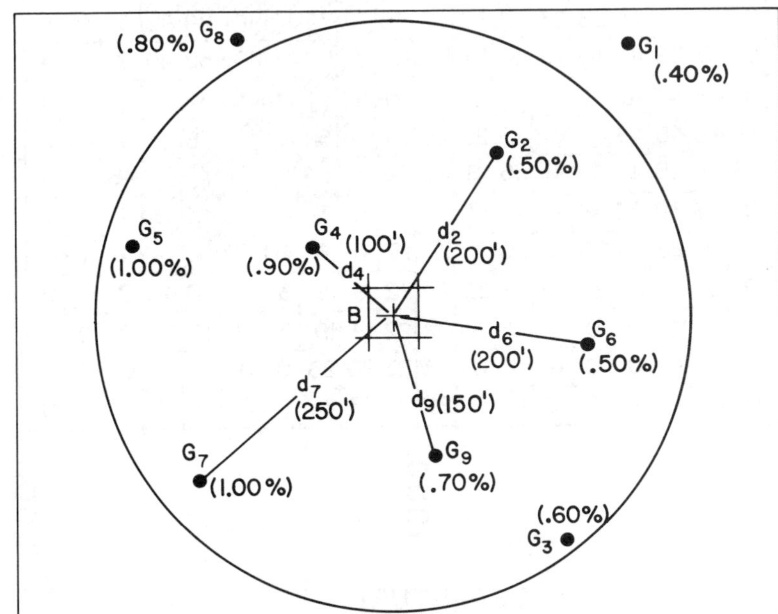

Fig. 6. A hypothetical block calculation from neighboring samples. G is assay composite value, d is distance, and B is block assay.

$$B = \frac{G_2 \times \frac{1}{(d_2)^2} + G_6 \times \frac{1}{(d_6)^2} + G_9 \times \frac{1}{(d_9)^2} + G_7 \times \frac{1}{(d_7)^2} + G_4 \times \frac{1}{(d_4)^2}}{\frac{1}{(d_2)^2} + \frac{1}{(d_6)^2} + \frac{1}{(d_9)^2} + \frac{1}{(d_7)^2} + \frac{1}{(d_4)^2}}$$

$$B = \frac{0.5 \times \frac{1}{(200)^2} + 0.5 \times \frac{1}{(200)^2} + 0.7 \times \frac{1}{(150)^2} + 1.0 \times \frac{1}{(250)^2} + 0.9 \times \frac{1}{(100)^2}}{\frac{1}{(200)^2} + \frac{1}{(200)^2} + \frac{1}{(150)^2} + \frac{1}{(250)^2} + \frac{1}{(100)^2}} = 0.77\%$$

O'Brian (1967) and is shown in Fig. 6. The radius of influence is shown as the circle around the point to be interpolated in Fig. 6. Rules include: (1) an angular exclusion of 18° (excludes G_3 and G_5), (2) maximum of seven nearest holes (excludes G_1 and G_8), and (3) power $m=2$.

Example 4: Fig. 7 shows an inverse distance computer evaluation of level 5140 shown in Fig. 1. Rules for this interpolation include: (1) 76.2 m (250 ft) maximum radius of influence, (2) power $m=2$, and (3) angular exclusion=18°. Accumulation of blocks ≥0.6% Cu is calculated as 1 817 057 t (2,002,963 st) at an average grade of 0.91% Cu.

The rules in Table 4 also exemplify one way of treating a deposit containing different geological environments, i.e., by generating an equivalent distance between different geological zones based on data of grade change in each independent zone.

Equivalent distances are created by developing variance of known samples vs. distance curves within each individual geological environment, as well as between each zone. The primary environment is selected as a base and distances in other zones or between zones are expressed as an equivalent distance in the base zone. This is accomplished by ascertaining the variance in the geological zone in question from the known distance between samples and determining the distance in the base geological environment at this same variance.

Another way of handling this problem is to exclude holes of incompatible geology even though they may fall in the area of influence, in a manner similar to the way the problem was handled using polygon methods.

GEOSTATISTICAL TECHNIQUES

The theory of regionalized variables developed by Matheron (1971) forms the basis of geostatistics. The sample value is a regionalized variable, i.e., it displays an apparent structural relationship to other samples within the region of study but simultaneously retains a random characteristic. Such regionalized variables cannot be represented by some derivable function de-

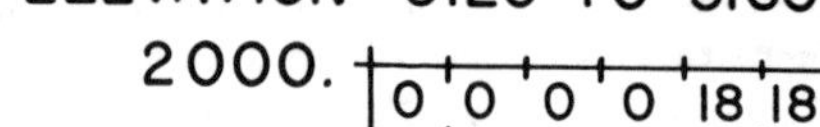

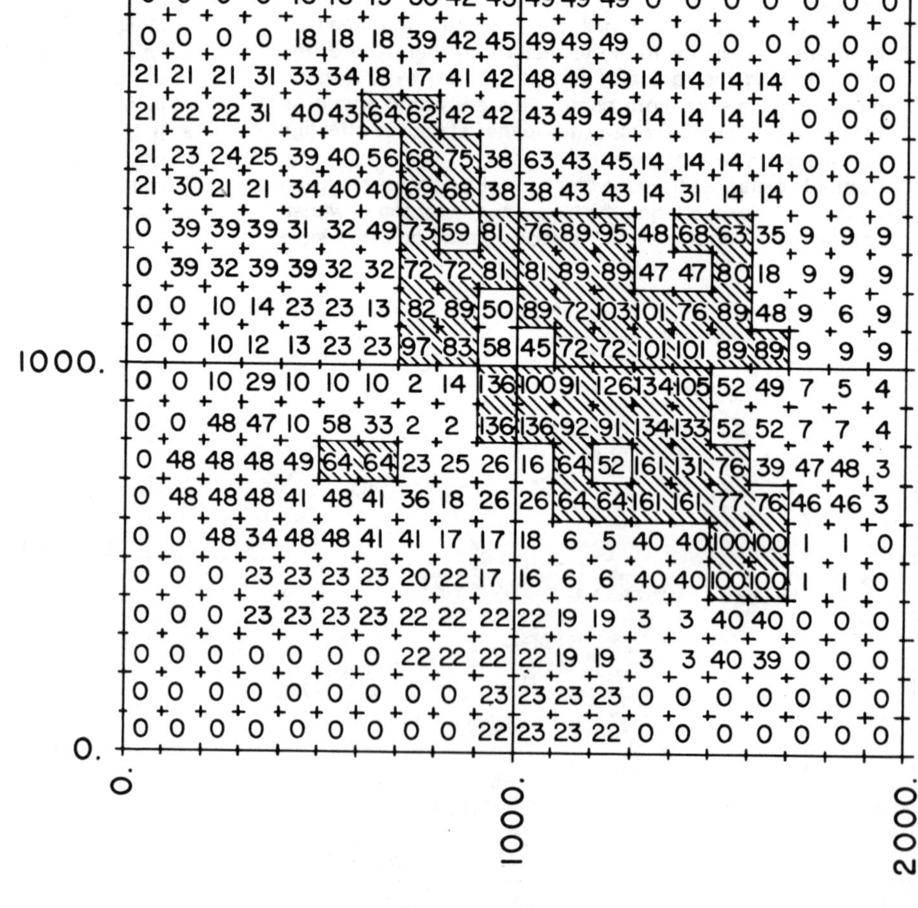

Fig. 7. Inverse distance interpolation from composited samples in Fig. 1.

scribing the underlying phenomenon. In contrast, the stress in a beam can be represented by a function derived from knowledge of the strength of materials.

Variograms

Regionalized variables can, however, be studied by using a statistical analysis of variances and using a graphical tool called a variogram (David, 1974, and Matheron, 1971). A variogram is a plot showing how grade in a deposit varies versus distance. The function of ore grade variance is defined as γ (gamma) and is computed according to the following working relationship:

$$2\gamma = \sum_{x=1}^{n} (G_x - G_{x+h})^2/n \qquad (7)$$

where n is the number of samples, G_x is the sample grade at point x, G_{x+h} is the sample grade at a point h distance from G_x, and h is the distance between samples.

In practical application, a variogram is developed in the following manner:

1) The distance between a sample and all other samples within a region of analysis is measured or computed.

2) The term $(G_x - G_{x+h})^2$ is computed for each different distance; for example, 7.6, 15.2, 22.8, 30.4 m (25, 60, 75, 100 ft), etc. Often the distance differences may be lumped together for small distance changes in order to develop a satisfactory number of samples, $n(n \geq 30$ has been suggested).

3) Repeat the two previous steps for all samples.

4) Accumulate all of the $(G_x - G_{x+h})^2$ terms for the same distance h or for lumped accumulations of similar h's.

5) When all holes have been analyzed in this manner, compute γ by dividing the accumulated $(G_x - G_{x+h})^2$ terms by two times the number of pairs accumulated for each Δh.

6) Plot γ vs. h.

A typical variogram is shown in Fig. 8. A mathematical model can be fit to the data points of the variogram. For classical variograms of the form shown in Fig. 8, a spherical model is generally fit to the data. The spherical model is as follows (David, 1974, and Matheron, 1971):

$$\gamma = C_1\left(\frac{3h}{2a_r} - \frac{h^3}{2a_r^3}\right) + C_o \qquad h \leq a \qquad (8)$$

$$\gamma = C_1 + C_o \qquad h > a$$

where $C_1 + C_o$ is the sill value in Fig. 8 and represents the general variance of the deposit, C_o is the nugget effect and represents the error of a point sample, a_r is the range of the data and represents the distance at which a sample has no further effect but blends into the general variance of the deposit, and h is the distance between samples.

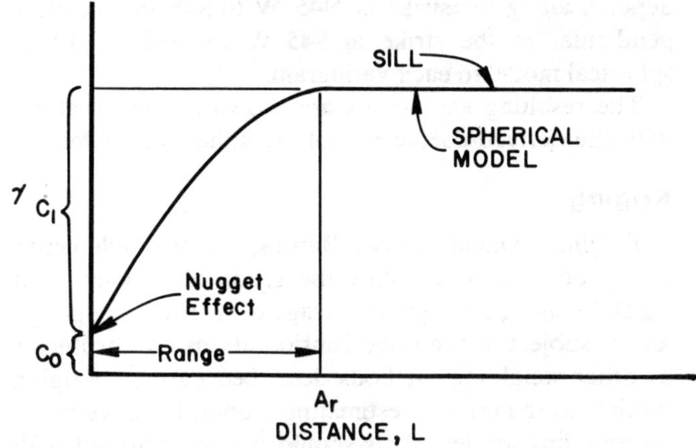

Fig. 8. Classical variogram, fit with spherical model.

Possibly the greatest contribution of the variogram is the analytical computation of a_r, the range or radius of influence, which can now be defined as a function of sample data in the ore body.

For each deposit, specific mineralization controls result in inherent characteristics concerning metal distribution. Variations, as a function of direction, are called anisotropies. Anisotropies can be ascertained by developing variograms at various locations and three-dimensional directions in the deposit. Usually significant contrast can be defined by taking variograms along the perpendicular axis of the deposit, i.e., along the strike and perpendicular to the strike or down, dip. The anisotropies show primary differences in the range a_r and can be modeled by adjusting a_r by a multiplying factor as angle is varied.

The art of defining the experimental variograms for significant mineral populations is a most critical part of geostatistical analysis and requires considerable knowledge and experience and assistance from those most familiar with the deposit: geologists and engineers. One should be aware of the limitations of modeling, curve fitting, and the validity of the data being used. For example, often drilling patterns are such that no two holes are close enough to accurately define the variogram at distances closer than, say, 152 m (500 ft). If this data is modeled and used to predict grades in 15-m (50-ft) blocks, a risk is taken because the variogram has been extrapolated to small distances.

Example 5: Compute variograms for the drill-hole data shown in Fig. 1. Compute variograms for the entire deposit, along the strike at N45°W to S45°E, and perpendicular to the strike at S45°W to N45°E. Fit a spherical model to each variogram.

The resulting variograms are shown in Fig. 9, along with the derived coefficients of the spherical model.

Kriging

Kriging (David, 1974; Barnes, 1979; Lellement) is a procedure to estimate the grade of a point or a block by using a weighted average of surrounding points and is subject to the same considerations and problems as other weighting methods described earlier. Kriging minimizes the error of estimating a point from weighted surrounding grades. This should not be confused with sampling, assay, or other errors associated with overall deposit analysis. The estimation error is merely the statistical error associated with using a weighted average and, in most cases, is small in comparison with the overall error of deposit modeling. The estimation error can be expressed in terms of the variogram:

$$\sigma_k{}^2 = \sigma_z{}^2 - 2\sum_{i=1}^{n} a_i\, \sigma_{zx_i} + \sum_{i=1}^{n}\sum_{j=1}^{n} \sigma_{x_ix_j} \tag{9}$$

where $\sigma_z{}^2$ is the variance between points similar to the one being considered, σ_{zx_i} is the covariance between the point being considered and the sample point x_i, a_i is the weighting coefficient, and $\sigma_{x_ix_j}$ is the covariance between samples x_i and x_j. In order to krige a single point, this error function can be minimized by differentiating the function with respect to the unknown a_i and setting the result equal to zero. This produces a set of n simultaneous equations to be solved for the unknown weighting coefficients a_i. This can be expressed in a matrix-vector equation:

$$\bar{A} = [E]^{-1}\bar{D} \tag{10}$$

A detailed description of the elements of Eq. 10 can be found in David (1974, pp. 202-204). Matrix $[E]$ and vector D are made up of the covariance between samples and the covariance between the point being considered and all samples being considered. The covariances between samples i and j can be evaluated directly from the variogram because

$$\sigma_{ij} = C_o + C_1 - \gamma(h)$$

Various schemes have been incorporated to krige from point samples to a block in a mineralization model. It has been shown (Parker, 1976) that only a small error can be associated with using a point estimate from punctual kriging to represent the grade of an entire block.

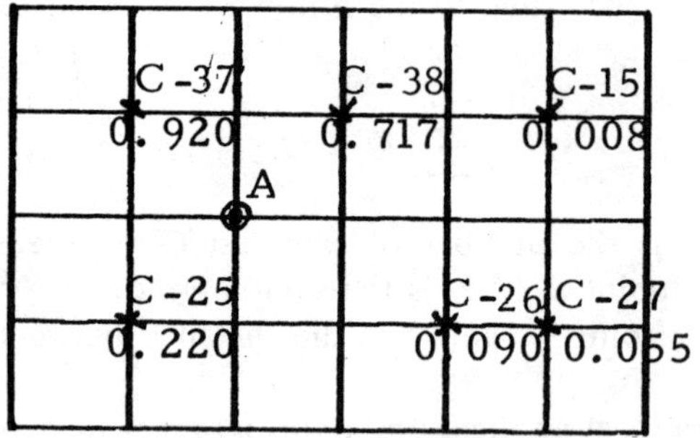

Example 6: Estimate the value at point A using kriging and the variogram model in Fig. 9 for the entire level. [Each block is 30×30 m (100×100 ft) by bench height.]

1) Evaluate σ_{ij} for all sample values.
For example:

$$\sigma_{37,\,38} = \sigma_{38,\,37} = \sigma_{38,\,15} = \sigma_{15,\,38} = \sigma_{25,\,37} = \sigma_{37,\,25} = \sigma_{27,\,15} = \sigma_{15,\,27} = C_1 + C_o - \gamma(200) = 0.02 + 0.16 - 0.10 = 0.08$$

2) Evaluate $\sigma_{A,\,x_i}$ for all sample values.
For example:

$$\sigma_{A,\,38} = C_1 + C_o - \gamma(141) = 0.12$$

3) Exclude sample points separated by $d > a$. $\sigma_{25,\,15}$ and $\sigma_{27,\,37}$ are excluded because d>137.1 m (d>450 ft).

4) $\sigma_{ii} = C_1 + C_o = 0.18$

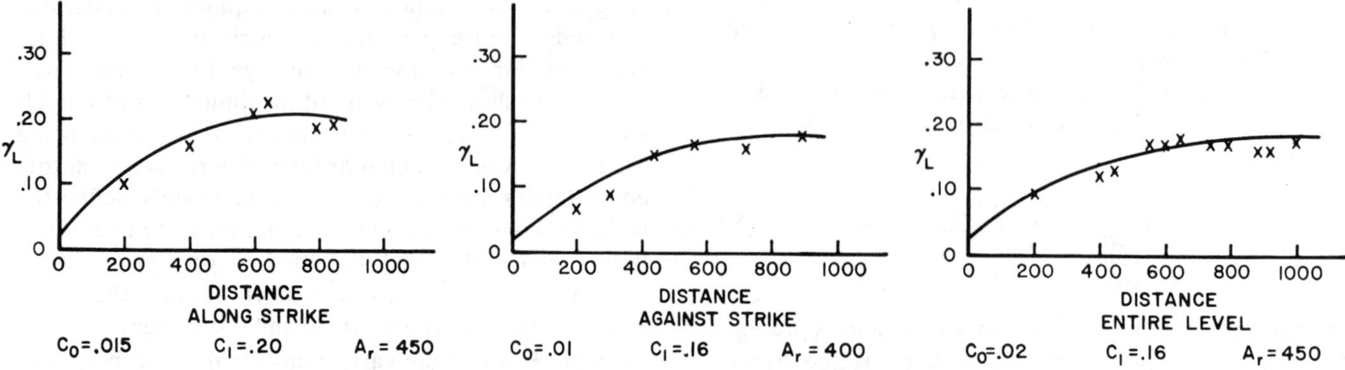

Fig. 9. Variograms of sample data in Fig. 2.

5) Invert the matrix E. This may require a computer program.

6) Multiply E^{-1} by $\bar{D}$. (Matrix-vector multiplication.)

7) The resulting weighting coefficients are:

$$a_{37}=0.042 \; a_{38}=0.553 \; a_{15}=0.017$$
$$a_{25}=0.097 \; a_{26}=0.289 \; a_{27}=0.002$$

8) The resulting grade at point A is 0.687%. (Note that $\sum a_i=1$ and $0 \leq a_i \leq 1$.)

Example 7: Fig. 10 shows a kriging evaluation of level 5140 shown in Fig. 1. Rules for this interpolation include: (1) 76.2 m (250 ft) radius of influence, (2) $C_o=0.015$ and $C_1=0.20$ along the strike, and (3) $C_o=0.01$ and $C_1=0.16$ against the strike. Accumulation of blocks $\geq 0.6\%$ Cu grade is calculated as 1 900-921 t (2,095,407 st) at an average grade of 0.86%.

Common problems associated with use of the kriging technique include the following:

1) Variograms do not accurately represent the mineralized zone because of inadequate data.

2) Mathematical models do not accurately fit the variogram data, or variograms have been improperly interpreted.

3) Kriging is insensitive to variogram coefficients.

4) Computational problems and expense are associated with repeatedly inverting large matrices.

5) The matrix form of [E], i.e., constants along the diagonal and a mirror reflection of the same order of magnitude values across the diagonal, tends to be ill-conditioned. Ill-conditioned is a mathematical term meaning that sometimes the inversion of matrix [E] will not produce correct answers, no matter what level of precision is maintained. This means that under certain geometries, kriging doesn't work well.

6) There are problems associated with weighting coefficients. These problems were discussed earlier under the heading "Inverse Distance Methods."

Additional information about geostatistical methods and their application can be found in Knudsen and

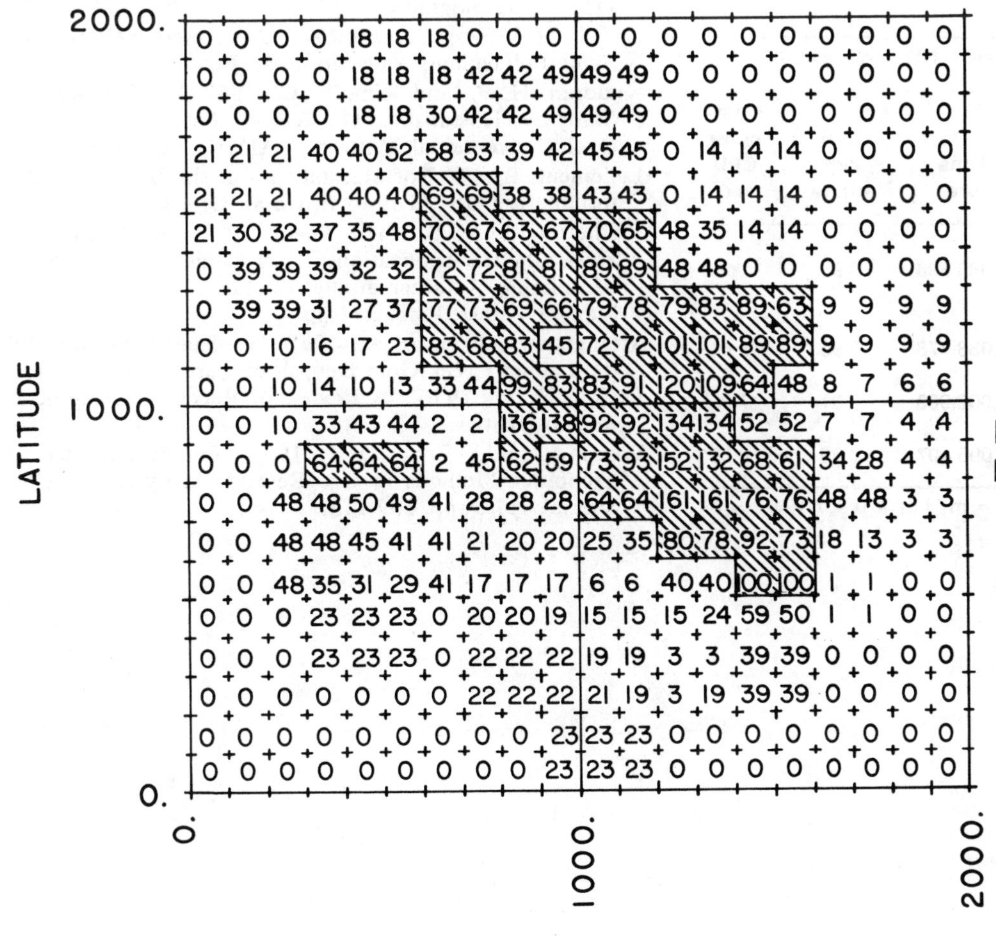

Fig. 10. Kriging interpolation from composited samples in Fig. 1.

Kim, 1967; David, 1974; Matheron, 1971; Barnes, 1979; Montgomery, et al., 1977; Lellement; and Parker, 1976. Many of the references describe methods to improve geostatistical applications considering what types of deposit are being investigated, the available data, and desired results. Several of the references also indicate complex mathematical methods to alleviate some of the problems noted herein.

CONCLUSIONS

Table 5 shows a comparison of some of the methods described in this chapter. The decision as to which method to apply to any particular deposit evaluation is left to the user, and would certainly depend on the deposit, the data available, sample density, the type of results required, the required accuracy, and the amount of time, money, and energy that one is willing to expend on the evaluation of a specific deposit.

Table 5. Comparison of Methods Applied to the Composite Level Data in Fig. 1*

	Block size = 100 x 100 x 40 ft.			
	Ore blocks	Tons ore	Ore avg. grade	Computer CPU sec.
Example 1				
Hand polygon		1,989,896	0.93	
Example 2				
Computer polygon	66	2,033,778	0.92	6.61
Example 4				
Inverse distance	65	2,002,963	0.91	7.08
Example 7				
Kriging	68	2,095,407	0.86	19.55

* Metric equivalents: 1 ft × 0.3048 = m; 1 st × 0.907 184 7 = t.

There are pros and cons of each method. Many innovative designers have combined what they consider the best of various techniques to their particular application. The astute observer will note that the inverse distance weighting rules described in Table 4 actually are a combination of inverse distance and geostatistical methods. Although the discussion herein has been restricted to two-dimensional examples, three-dimensional applications, especially with the aid of computers, is quite practical and generally results in a better interpolation of the deposit.

There also is no question that the incorporation of geologic data into the grade assignment process is essential for generation of the most accurate model of the deposit.

REFERENCES

Barnes, M. P., 1979, "Mineral Inventory vs. Production Planning, Case Study—Sacation Mine, Arizona," *Computer Methods for the 80's,* A. Weiss, ed., AIME, New York, to be published.

Brookes, P. I., "Block Estimation at Various Stages of Deposit Development."

David, M., 1975, *Geostatistical Ore Reserve Estimation,* Ecole Polytechnique de Montreal, PQ, Canada.

Knudsen, H. P., and Kim, Y. C., 1967, "A Comparative Study by Geostatistical Ore Reserve Estimation Methods Over Conventional Methods," AIME, New York.

Lellement, B., "Use of Geostatistics at the BRGM to Determine the Best Way to Prove an Ore Body."

Matheron, G., 1971, "The Theory of Regionalized Variables and Its Applications," Les Cahiers de Centre de Morphologic Mathematique, Booklet No. 5, Ecole Nat. Sup. des Mines, Paris, 211 pp.

Montgomery, J. H., et al., 1977, "Geostatistical Study of the Ladner Creek Gold Deposit of Caroline Mines Ltd."

Parker, H. M., 1976, "A Review of Recent Developments in Geostatistics."

Weiss, A., and O'Brian, D. T., 1967, "Practical Aspects of Computer Methods in Ore Reserve Analysis," CIMM, Special Vol. 9, pp. 109-113.

6 Drill-Hole Interpolation: Estimating Mineral Inventory

Marvin P. Barnes
M. Barnes Associates

Marvin P. Barnes graduated from the University of Utah in 1955 with a B.S. degree in geological engineering, Cum Laude and Tau Beta Pi. He received his Ph.D. in geological engineering in 1970.

He was appointed chief computer geologist for the exploration dept. of ASARCO, Inc. and shortly afterward was given the title of regional exploration manager for the Salt Lake division. He has made computer-assisted appraisals of many mineral deposits throughout the world. He worked in both positions until his resignation from ASARCO, Inc. in 1976 when he organized the consulting firm of M. Barnes & Associates in partnership with T. T. Farr.

Definition

The term mineral inventory is a relatively new designation, born of the computer age and grown to common usage with the development of computer techniques for extending drill-hole sample values to regularly gridded data blocks representing in situ mineral values. The term has been generally accepted by geologists and engineers because it has a precise meaning which avoids many of the ambiguities of ore reserve.

A mineral inventory is an estimated inventory of mineral in place and is usually achieved by dividing the deposit into regularly spaced gridded blocks to which are assigned estimated values using various extension techniques. A mineral inventory makes no presumptions about the minability of the blocks or their cutoff grade. A mineral inventory estimate may be good or poor according to how much is known about the deposit, how extensively it has been sampled, and what techniques have been employed to assign estimated grade to the blocks. Such an inventory, however, does not change with time, economic conditions, or mining technology, but only with new and better data and methods. Thus, the mineral inventory is an excellent designation for describing mineral value distribution within a deposit's geological setting.

Objectives of Mineral Inventory

Mineral inventories are computed for the purpose of determining the quantity, the quality or value, and the spatial distribution of potentially economic minerals. Such computations are made at all stages in the life of a mineral extraction process, from the early exploration phase to the final year in the life of a mine. Popoff (1966) states, "They are the most responsible and irreplaceable tasks in the valuation of a mineral deposit. Efficiency in extraction and productiveness is impossible without accurate reserve computations."

The accuracy of estimate required for each of the three parameters of quantity, value, and location is dependent upon the objective for which the inventory estimate is being made. The objective may be somewhat different for each exploration, development, and production stage, with need of increasing accuracy with each new investment requirement. During the development and production stages, accurate predictions of spatially distributed values are particularly important. In the latter stages, however, modern mathematical techniques made possible by the computer play an increasingly important role.

Conventional Extension Functions (Principles of Calculating Mineral Reserves)

An extension function may be defined as a technique or mathematical function used to extend sample values to estimate the value of surrounding volumes of mineral. Any method of calculating mineral inventories from sample data can properly be called an extension function.

Various methods have been applied over the years for computing mineral reserves. Popoff (1966) describes the principles and conventional methods for computing reserves of mineral deposits as employing three main principles: (1) the rule of gradual changes, (2) the rule of nearest points or equal influence, and (3) the rule of generalization.

The rule of gradual changes or law of linear functions implies that all sample elements of a mineral body change gradually and continuously as a linear function along a straight line connecting two adjacent sample points. The rule can be applied to other parameters of a mineral body besides grade and weight factors, i.e., areas, volumes, and tonnages. The principle of gradual changes is employed in the conventional triangles method for calculating mineral inventories. Fig. 1 illustrates the triangle or triangular prism method in which all drill holes or sample points are connected by straight lines into a system of triangles. Each triangle represents the base area of an imaginary prism of some thickness. The average grade of each prism is usually calculated as the arithmetic mean of the three samples or as the thickness-weighted mean of the three samples.

Fig. 2 illustrates another much used geometric method for calculating reserves based upon the gradual changes principle.

The standard cross section method is often used when fences of drill holes extend across an irregularly shaped mineral deposit. Each internal block is defined by two sections and each end block by a single section.

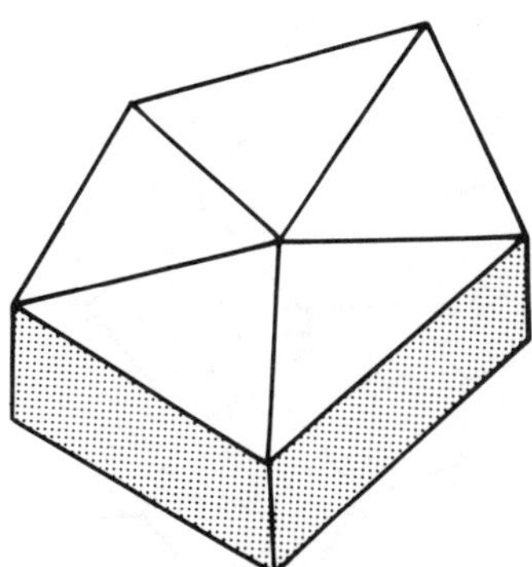

Fig. 1. Triangular prisms.

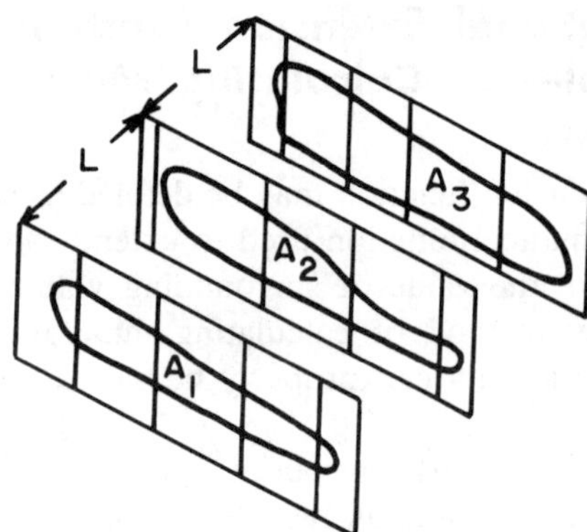

Fig. 2. Cross sections.

The usual procedure for calculating mineral reserves by this method is as follows:

1) Determine the mineral area in all sections: planimeter irregular outlines and other geometric schemes.

2) Calculate average values for each section by using length-weighted sample value average, using area-weighted sample value average, or using the arithmetic average of sample values.

3) Compute volume for each block. The simplest and most used formula for volume between two partial sections of area A_1 and A_2 and perpendicular distance, L is:

$$\text{Vol} = \frac{A_1 \; A_2}{2} L \text{ and t (tons)} = \frac{A_1 \; A_2}{2} LF^*$$

where $F^* = \text{t/m}^3$ (st per cu ft).

4) Sum the results of all blocks and compute average value for entire mineral body (volume-tonnage weighted value).

The rule of nearest points or equal sphere of influence implies that the value of any point between two samples is constant and equal to the value of the nearest sample. The rule assumes that the value of a sample extends halfway to any adjacent sample. The well-known polygonal method is based upon this rule or principle (Fig. 3) as is the rectangular block method for uniform sample spacing, etc. See Fig. 4.

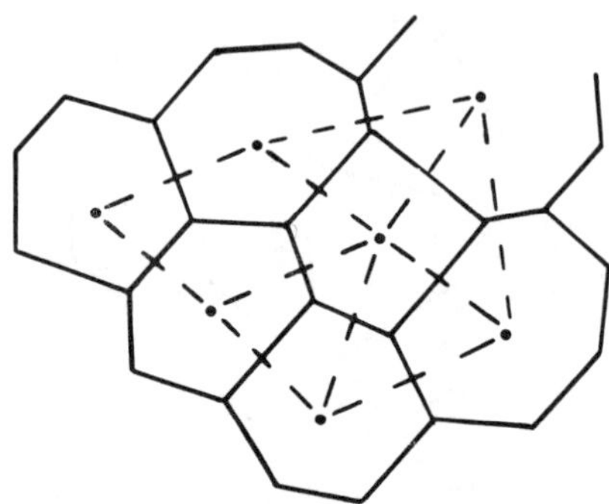

Fig. 3. Method of polygons.

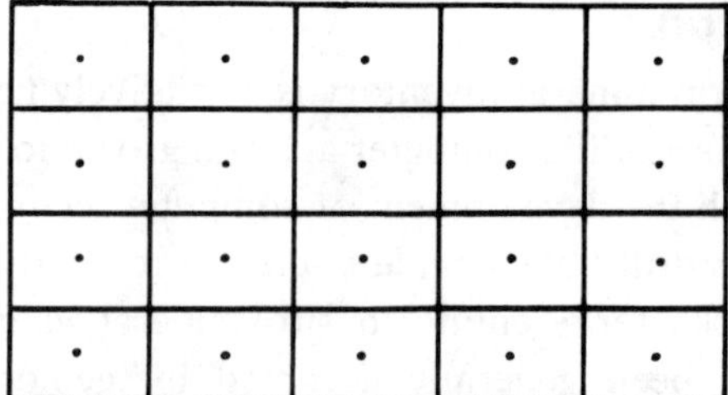

Fig. 4. Rectangular blocks.

The polygons of the polygonal method are constructed by drawing perpendicular bisectors to lines connecting all sample points. The area of influence for the outside perimenter of the mineral zone is limited by using a standard mean radius of influence or by making the polygon equidimensional around the sample point.

The polygonal mineral reserve is computed by measuring the area of each polygon (usually with planimeter), multiplying the area by thickness at each sample point, applying the proper tonnage factor [m^3/t (cu ft per st)] and assigning to the polygonal prism the grade of the sample. The grade and tonnage for the total deposit is the sum of all tonnage and the volume-weighted average grade of all the samples. The rectangular block method is a simplified rule of nearest point technique for ideal uniformly spaced samples.

The rule of generalization may also be stated as the empirical method or even the rule of thumb. The rule of generalization is really no rule at all and is usually arbitrarily applied as a matter of judgment reflecting past experience and opinions. Adapting a definite weighting factor for reserve computations for application to one mineral deposit based upon determinations from another similar deposit is probably the most common example.

In some cases, the use of the rule of generalization is justifiable and necessary. Classifying reserves by categories for certain types of mineral deposit and assuming factor values for reserves based upon production data rather than directly from sparse drilling data of erratic and doubtful values are generalizations. Projecting continuity of mineralization beyond the outermost workings along strike or at depth and fixing arbitrary boundaries for computations are other examples. The problems associated with generalized techniques are related to the subjectivity of the methods and the difficulty of duplicating results by another examiner.

Conventional extension functions may be summarized as methods for calculating mineral reserves based upon geometric relationships among samples. The methods are functions of geometry and distance between samples which simplify the calculations of volume and grade. They are not functions of the mineralization characteristics which they propose to measure.

Computerized Extension Functions Using Conventional Principles

Mineral inventories prepared manually usually consist of irregularly shaped blocks or panels that are related to the sample density and proposed mining units. The computer is most efficient in creating inventories made up of usually much smaller, regularly shaped blocks over the extent of the mineral zone. The choice of block size is based upon several factors and requires careful consideration by the analyst.

The extension function is designed to assign values to all such blocks via the computer. These values are estimated by using one or more variations of the three conventional principles: gradual changes, nearest points, and generalization.

The computer was first employed in speeding up conventional inventory calculations similar to manual techniques. Its power and speed were then recognized for making possible the use of other mathematical functions which were not time-economical without the computer. The variation in techniques is limited only by the ability of the analyst and geologist to conceive mathematical functions which can simulate the natural mineralogical structuring of the deposit.

The Computer Polygonal Method

The least complicated computer technique for assigning estimated values to mineral inventory blocks is based upon the nearest point rule. Each inventory block is defined for the computer in *xyz* coordinates, as are also the sample points within the deposit. The procedure for assigning the value to the block requires only a simple distance calculation between the center of the block and the sample points to find the closest sample, the value of which is then given to the block. The technique is often referred to as the computerized polygonal method, and some consideration must be given to limiting the extension of values to blocks outside the sampling perimeter.

The Inverse Distance Squared Method

Another widely accepted computerized extension function uses the principle of gradual changes for making value estimates and is generally referred to as the inverse distance squared method (IDS). The inverse distance interpolation technique uses straightforward mathematics for weighting the influence of all surrounding samples upon the block being evaluated. However, it is this assumption that often causes difficulty in the reliability of computer inventories unless

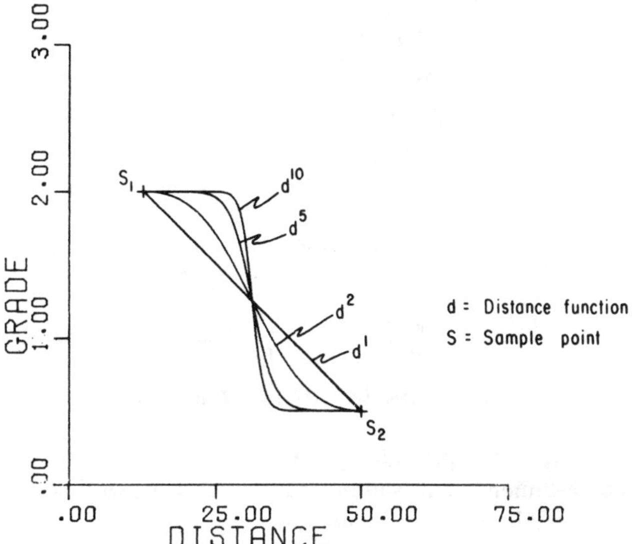

Fig. 5. Effect of exponent upon weighting inverse distance method.

the program can take into account mineral trends and population boundaries. The capacity of the computer program to use only those samples of the proper mineralogical population, even though others may be in close proximity to the block, is very important.

Eq. 1 represents a linear distance-weighting function for computing the value of a block from surrounding samples:

$$B = \frac{(V_1/d_1) + (V_2/d_2) + \ .\ .\ .\ .\ + (V_n/d_n)}{1/d_1 + 1/d_2\ .\ .\ .\ .\ + 1/d_n} \qquad (1)$$

where B is estimated value of block, V is value of sample, and d is distance of sample to center of block.

More weight can be assigned to the nearest sample by squaring the distance between the sample and block, resulting in an exponential distance function. The higher the exponent, the more weight is given to the nearest sample until an exponent of 5 closely approximates the nearest point rule. See Fig. 5.

It has been found that the inverse distance raised to the square is generally the most acceptable distance interpolation function for mineral deposits. The computer is usually programmed to select sample points within a limited distance from the block being estimated. This search radius may be programmed to vary according to sample point density.

The IDS method can also be modified to reflect anisotropism within the deposit and vary the distance weighting function according to direction. Fig. 6 illustrates the computational process for an isotropic two-dimensional IDS block value estimation. When the trend and anisotropic factors are known, a computational process such as illustrated in Fig. 7 may be

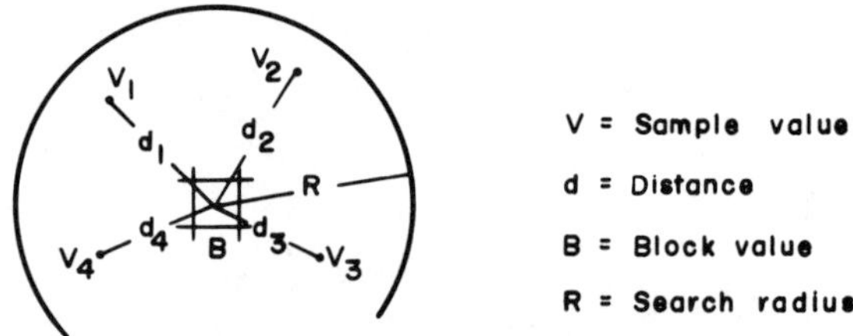

Fig. 6. Block estimate-inverse distance squared method (IDS).

$$B=\frac{(V_1/d_1^2)+(V_2/d_2^2)+(V_3/d_3^2)+(V_4/d_4^2)}{1/d_1^2+1/d_2^2+1/d_3^2+1/d_4^2}$$

Isotropic Mineral Structure

employed to provide greater estimation accuracy. Three-dimensional sample search and weighting are also used in the same manner.

It should be stressed that the inverse distance interpolation techniques are also functions of geometry and distance between sample points and are not functions of mineralogical structuring which they attempt to simulate.

Geostatistical Mineral Reserve Estimation Method

One of the newest and most powerful techniques for estimating mineral inventories falls under the heading of geostatistics, defined as the application of the theory of regionalized variables* to the study of mineralized volumes of rock. The theory was advanced by George Matheron in 1962 and has been widely recognized throughout the world as a superior method for estimating the grade of in situ mineralization because it provides a sound theoretical and practical basis for quantifying the geological concepts of (1) area of influence of a sample, (2) the continuity or lack of continuity of mineralization within the ore body, and (3) the lateral changes in mineralization according to the trend direction of an ore body and its orthogonal components, or in other words, a measure of the anisotropy of the deposit.

Royle (1971) has defined the objectives of geostatistics as being: (1) to estimate the most likely value of blocks of ore or the values of the whole deposit, and (2) to estimate the errors of such estimates. This latter is important as, in addition to providing a check on unwarranted optimism, it shows where more valuation work may be needed.

The fundamental tool of geostatistical analysis, which permits the quantification of the geologic parameters mentioned previously, is called the semivariogram or simply the variogram. The variation that exists among samples some distance apart within a continuous mineral deposit is a measure of their spatial correlation. Experience and common sense tell us that the nearer two samples are taken to each other, the more alike they should be (assuming that variation in sampling and measuring techniques is negligible). As the distance between samples increases, their difference in value will, on the average, become greater.

The variogram is an arithmetically simple graph which plots the average differences between sample values at specified distances or lags apart. Because the difference between two paired values may be either plus or minus, it is necessary to apply a technique of classical statistics and square the difference, sum them, and divide by twice the number of pairs found.

The result is called the geostatistical variance $\gamma(h)$ and is represented by the formula:

$$\gamma(h)=\Sigma|f(x+h)-f(x)|^2/2n(\vec{h}) \qquad (2)$$

where $f(x)$ is the grade at sample point x, $f(x+h)$ is the grade at a point $x+h$ m (ft) away, $\vec{h}$ is a distance vector function (directional), and $n(h)$ is the number of data pairs counted along directional lag (h).

As with other statistical parameters, the geostatistical variance or gamma function $\gamma(h)$ must have a large enough number (n) pairs for each lag (h) to be statistically significant. Fig. 8 illustrates a simple case of n samples regularly distributed along a line h m (ft) apart. Thus we have $(n-1)$ pairs to compute $\gamma(h)$, $(n-2)$ pairs to compute $\gamma(2h)$, $(n-3)$ to compute $\gamma(3h)$, and so forth.

* Regionalized variable is a variable whose magnitude depends on neighboring values which are distributed in two- or three-dimensional space.

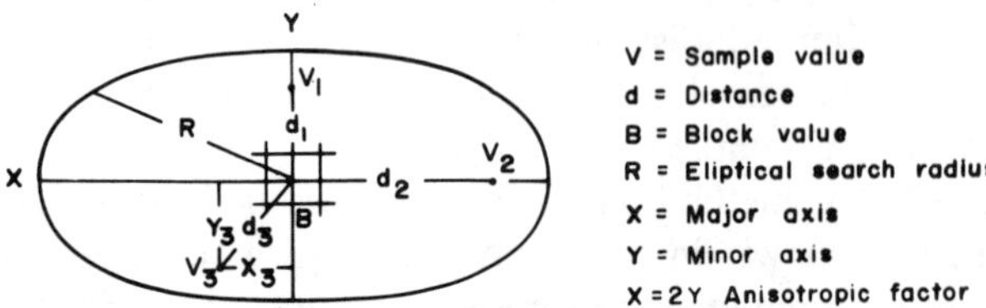

Fig. 7. Block estimate-anisotropic IDS method.

$$B=\frac{(V_1/2d_1^2)+(V_2/d_2^2)+(V_3/x_3^2+2y_3^2)}{1/2d_1^2+1/d_2^2+1/(x_3^2+2y_3^2)}$$

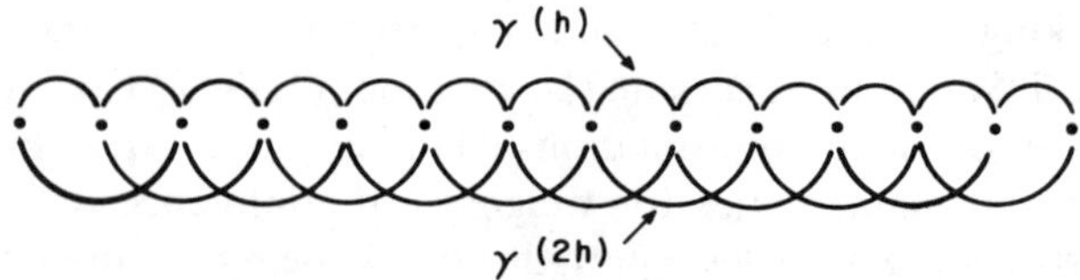

Fig. 8. Variogram computation scheme using sample pairs a given distance apart.

The theoretical variogram should be properly expressed by the following continuous function.

$$2\gamma(\vec{h}) = \frac{1}{V^3} \int_V \int_V \int_V |f(x+h) - f(x)|^2 \, d_x d_y d_z \quad (3)$$

where V is the volume of the deposit, and $\vec{h}$ is the vector function that may vary according to direction.

In practice, the theoretical variogram is never realized, and the gamma function must be estimated by formula 2 and is called the experimental variogram. The computations necessary for an adequate variogram are very laborious and are best done with the help of a computer.

Properties of the Variogram

Inherent in the variogram is the quantitization of spatial mineralogical characteristics that in the past could only be estimated by the geologist and were almost totally subjective in nature. Experience with one mineral deposit sometimes led the geologist or engineer to apply the same spatial characteristics to another deposit, solely for lack of better criteria. The variogram effects a solution to this problem by providing the following information:

A Measure of Continuity of the Mineralization: A rate of increase of $\gamma(h)$ near the origin and for small values of h reflects the rate at which the influence of a sample decreases with increasing distance from the sample site. The growth curve demonstrates the regionalized element of the sample, and its smooth steady increase is indicative of the degree of continuity of mineralization.

The intersection of the curve with the origin provides a positive measure of the nugget effect of the samples from which the variogram has been generated and indicates the magnitude of the random element of the samples.

Figs. 9, 10, and 11 are real experimental variograms generated from three different types of deposits.

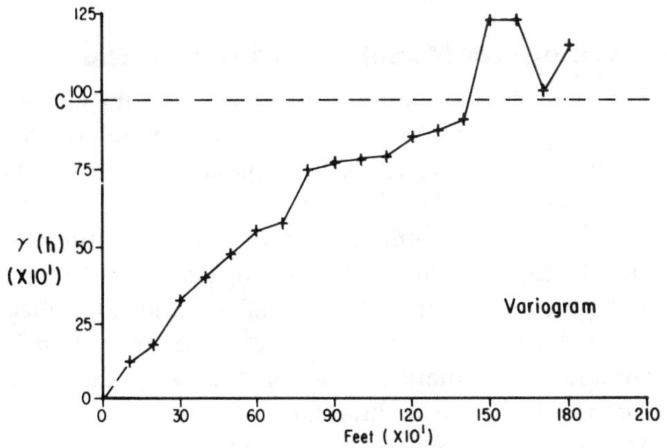

Fig. 9. Stratabound deposit.

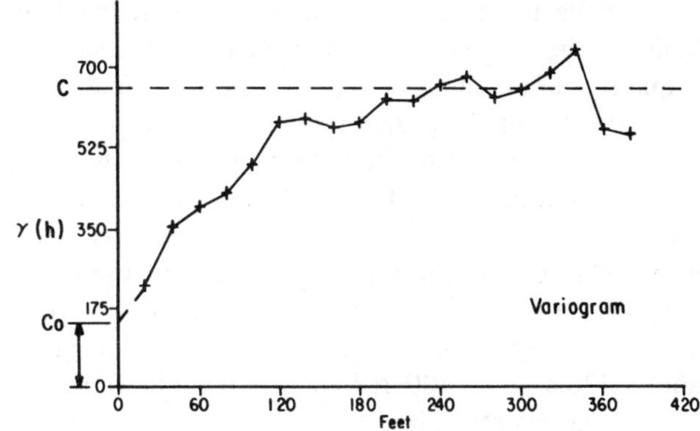

Fig. 10. Porphyry-copper deposit.

The slow steady growth of $\gamma(h)$ from zero in Fig. 9 is characteristic of many stratigraphic and stratiform deposits with fairly uniform mineralization having a high degree of continuity.

Fig. 10 was generated from porphyry-copper deposit data where mineral veinlets, changes in structural intensity, and other discontinuous features created a significant nugget effect due to changes over very short distances. Beyond the short-range effects, however, $\gamma(h)$ shows a fairly uniform growth curve and reaches a plateau at the sill of the variogram that is the overall variance of all the samples.

The experimental variogram of Fig. 11 is illustrative of a total random effect found in some gold deposits. The mineral continuity is nonexistent, and the samples appear to be completely independent no matter what the distance between them. Geostatistical ore reserve estimation techniques cannot make any contribution toward evaluating the deposit having a pure nugget effect since no regionalized element is present. The application of classical statistics and random variables is appropriate for evaluation of such a deposit.

A Measure of the Area of Influence of a Sample: The zone of influence of a sample is the distance or range in any direction over which the regionalized element is in effect. When samples reach a point far enough apart

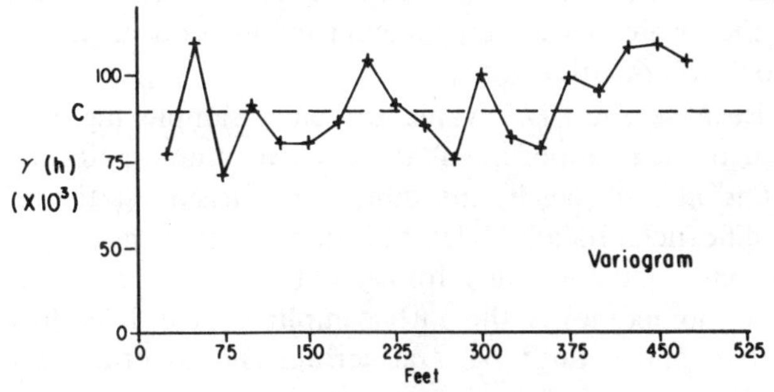

Fig. 11. Gold deposit.

so as to have no influence upon each other, we have established the range or zone of influence of the sample. The quantification of the range or zone of influence in various directions has important applications in the design and spacing of development drill holes within a deposit. The total zone of influence is indicated by the point at which the $\gamma(h)$ growth curve reaches a plateau, referred to in the Matheron or spherical scheme as the sill. In Fig. 9 it is apparent that the range of the stratiform copper deposit in the direction of the variogram is about 426.7 m (1400 ft). The gold mineralization of Fig. 11 has no range because no regionalized element can be recognized from the variogram.

A Measure of Mineral Trend or Mineral Anisotropics of the Deposit: The fact of mineral anisotropism in various types of deposits has long been recognized. The range of influence of a sample is greater along the strike or trend of the deposit than it is normal to trend. Most of the time, another anisotropism is evident in the vertical dimension. Prior to the variogram, there was no satisfactory way of determining the three-dimensional area of influence of a sample. By the simple process of computing variograms in different directions as well as vertically, one can readily determine not only the mineralogical trend but the magnitude of the directional changes in the zone of influence. Knowing quantitatively the mineralogical range in three dimensions, it is relatively simple to assign directional anisotropic factors that will give proper weighting to samples relative to their location from the point or block being evaluated. For example, if the range of influence along the trend is twice as great as the range normal to trend, we can multiply the distance in the normal direction by a factor of two to restore geometric isotropy in terms of the major trend direction.

Practical Aspects in Computing Variograms

The selection of a sample composite value representative of the desired mineralogical zone dictates the support† to be used for variogram computation. The experimental variogram will then be fitted by a mathematical model or the intrinsic function that will be used in the kriging extension function for mineral inventory block computations.

Because the $\gamma(h)$ value for each sample lag is a statistic, it is important that sufficient sample pairs are found at each lag in any direction to assure statistical significance. Ideally at least 30 such pairs are necessary to compute the variance for each lag in any given direction. Sometimes in the early sampling stage, it is difficult to find enough pairs at certain lags to produce a viable variogram point, and lesser numbers may be used. A variogram program that will output a different symbol when plotting the $\gamma(h)$ value for all lags having less than 30 pairs is useful for quick recognition of less reliable points.

Although each lag is thought of as a specific distance, in practice, the lag distance usually represents the mean of a distance class interval. In other words, the lag distance of 15 m (50 ft) may represent all pairs of samples falling between 11½ and 19½ m (37½ and 62½ ft) apart. Such a practice is necessitated by the uneven spacing of most samples, especially when computing directional variograms that are not parallel or normal to a roughly rectangular sampling pattern.

The importance of equal support lengths is stressed in making variogram computations. However, in the case of most vein and stratigraphic deposits where composited values are derived across a nonuniform total thickness, a simple technique, often employed by gold miners, can be used which yields valid variograms and lends itself very well to geostatistical analysis. Instead of computing $\gamma(h)$ as a grade variable only, a new variable is created by multiplying the grade by the thickness, i.e., feet by percent or inch by dead weight ton, etc. This metal quantity has been termed the accumulation and can be operated upon geostatistically by representing the tabular ore zone as a plane and the accumulation as a value at a point on the plane. The variogram will be computed using the accumulation variable. Since both the thickness and the grade are regionalized variables, so also is their accumulation.

In order to determine the grade of an ore block estimated by the accumulation variable, it is necessary to also compute a variogram for thickness and estimate the thickness of each ore block. Thickness may be estimated using techniques other than geostatistics. The estimated grade for each ore block can then be obtained by dividing the estimated accumulation for the block by its estimated thickness.

The Variogram Model or Intrinsic Function

The experimental variogram computed from samples or composited samples taken from a mineral deposit represents one realization of the spatial behavior of the regionalized variable. In order for the experimental variogram to be useful as a tool for estimating the grade of blocks within a deposit, a generalized model must be formulated that theoretically will fit all realizations of the spatial behavior. Such a model will be a continuous mathematical function that will adequately encompass the experimental data.

The mathematical model is called the intrinsic function because it describes the spatial behavior of the

† Sample size or length.

regionalized variable within the ore body and is an intrinsic feature of such regionalization. The application of intrinsic theory requires that certain assumptions be made about the spatial behavior, mainly that the regionalized variable exhibit at least a weak or quasi-stationarity. The word stationarity, as defined in geostatistics, implies that similar mineral distributions occur throughout the mineral zone. This does not imply that similar mineral values will occur, but that the increment of change (Δ value) is the same throughout the deposit or is stationary. The true variogram of the deposit, which in practicality can never be obtained, is none other than the intrinsic function of the spatial behavior of the regionalized variable.

Intrinsic Schemes

Several intrinsic schemes or mathematical models have been proposed over the years to represent a probability function that describes the behavior of various experimental variograms. Each of these intrinsic schemes or mathematical models has its own variogram: (1) the linear scheme (Fig. 12) $\gamma(h)=A(h)$, (2) the De Wijsian scheme (Fig. 13) $\gamma(h)=3 \log (h)$, and (3) the spherical scheme (Fig. 14)

$$\gamma(h)=C(3/2\, h/a-1/2\, h^3/a^3)+Co \quad \text{when } h \leq a$$
$$\gamma(h)=C+Co \quad \text{when } h>a$$

where a is range or zone of influence.

Much of Matheron's early work employed the De Wijsian scheme, which was commonly associated with hydrothermal deposits. Both the linear scheme and the De Wijsian scheme, which will produce a straight line when the lag (h) is plotted to log scale, imply that $\gamma(h)$ increases infinitely with increasing distances. Experience has shown that both models often accurately fit experimental variogram data near the origin, but break down when h becomes large.

The spherical model or Matheron model, as it is sometimes called, is one in which the variogram reaches a finite value as h increases indefinitely. This finite value, referred to as the sill of the spherical variogram,

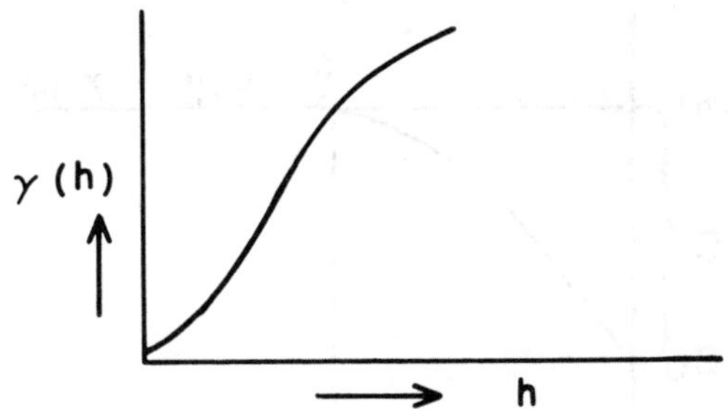

Fig. 13. De Wijsian scheme.

is the overall variance of the deposit and is reached when the grades are far enough apart to become independent of each other. The spherical model has become the most important intrinsic scheme, and many practicing geostatisticians have adopted it as an almost universal model. The model has been found to adequately represent such diverse deposits as iron ore bodies, porphyry-copper deposits, stratibound lead-zinc deposits, bauxite and lateritic nickel, as well as uranium and phosphate deposits. This model will be the only scheme discussed in this text.

The spherical scheme is defined by the formula (Fig. 15):

$$\gamma(h)=C(3/2\, h/a-1/2\, h^3/a^3)+Co \quad \text{when } h \leq a$$
$$\gamma(h)=C+Co \quad \text{when } h>a$$

where $C+Co=\gamma(\infty)$ and is called the sill, Co is the nugget effect (usually present), and a is the range or maximum zone of influence.

The fitting of the spherical model to the experimental variogram is not a difficult procedure. Experience has shown that a visual fit is usually sufficient. A practical technique for determining the model is to draw a best fit straight line through the first few points of the experimental variogram from the origin to the sill, $C+Co$ (the computed overall variance). The line will intersect the sill at $2/3a$, which will define the range. The intersection of the line with the origin defines Co, the nugget effect.

Because the experimental variogram will usually be derived with the help of a computer, it is not difficult

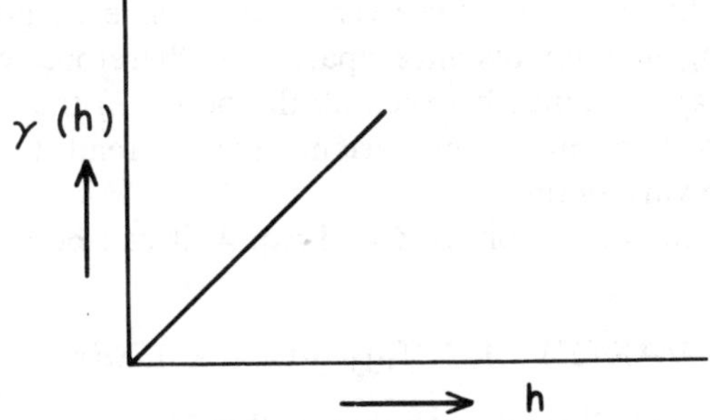

Fig. 12. Linear scheme.

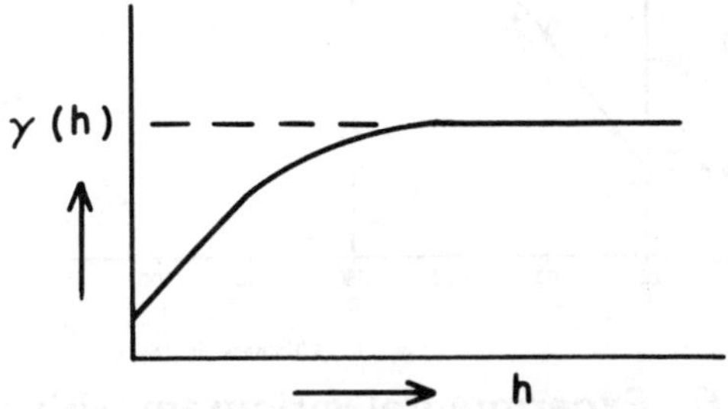

Fig. 14. Spherical scheme.

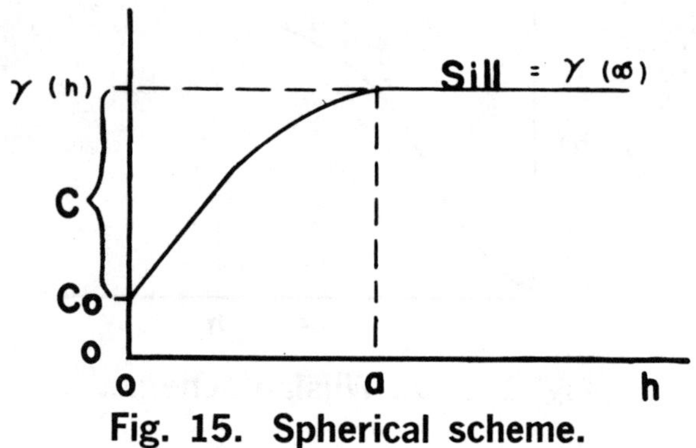

Fig. 15. Spherical scheme.

to write a program that will superimpose a spherical model upon the experimental variogram data. A program by which the observer can interactively change the range and nugget effect of the spherical model and see the fit to the experimental data on a cathode ray tube has been found to be tremendously helpful in plotting out modeling variograms. Fig. 16 illustrates a model fitted to an experimental variogram plot.

Experience has shown that for purposes of grade estimation, the spherical model is not sensitive to changes in the range (value of a). Significant error in fitting the range does not produce corresponding errors in grade estimation. The exactness of fit that may be possible through using a least-squares fitting program is not justified in matching models to experimental variogram points. It is more important that care be exercised in determining the nugget effect, Co. The estimate of error associated with estimating grade for ore blocks is sensitive to change in the nugget effect.

The Estimation Variance

The estimation variance may be defined as the variance of the error made in estimating the grade of a

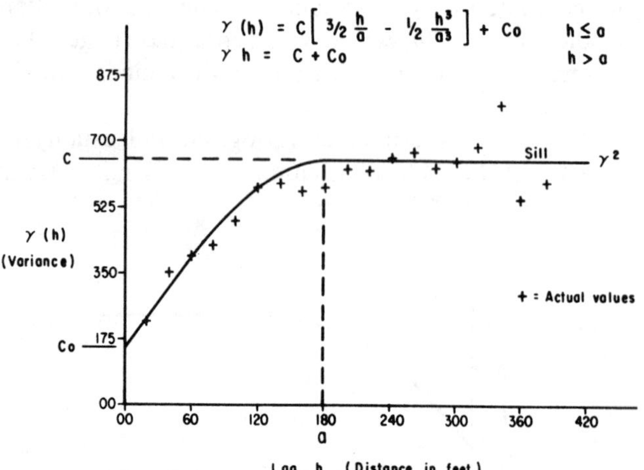

Fig. 16. Experimental variogram with spherical best fit model superimposed.

panel or block of ore by assigning to it the value of the samples lying in and around it. It is apparent from the variogram that the relative locations of the samples to the block will have an influence upon the weighting that can be assigned to each sample value.

In order to understand the variance of the error of estimation, suppose we visualize a block of ore of volume V with the variable Z as its true unknown grade with Z being defined as the probabilistic mean of the random function $f(x)$.

$$Z=\frac{1}{V}\int_V f(x)dx. \tag{4}$$

Furthermore, suppose that n samples have been taken in and near V at points such as X_i. Let Z^* represent the estimated grade of V as determined from weighting the known sample grades according to their relative positions.

$$Z^*=a_iX_i. \tag{5}$$

Then Z^*-Z will be the error made in assuming that the value Z^* extends over V when the true value is Z and VAR $(Z-Z^*)$ is the variance of this error. Stated another way

$$\text{VAR (Error)}=\text{VAR }(Z-Z^*) \tag{6}$$

$$=\text{VAR } Z-2\text{ COV}|(Z)(Z^*)|+\text{VAR } Z^* \tag{7}$$

thence by substitution of identity 5 for Z^*

$$\text{VAR (Error)}=\overset{(1)}{\text{VAR } Z}-\overset{(2)}{2\Sigma a_i\text{ COV}|(Z)(X_i)|}+\overset{(3)}{\sum_i\sum_j a_ia_j\text{COV }(X_iX_j)}. \tag{8}$$

Let us consider each of the VAR and COV terms of Eq. 8 in order to better understand their meaning in the physical sense, and how their value can be determined from the variogram, which is a variance function.

1) VAR(Z) is the variance of the grade within the block. The internal variance of the block grade will depend upon the average difference in grade that exists between any two points within the block. The variogram is by definition the average difference of grades according to their distance apart, and therefore, when the average distance between all the points in the block has been determined, the variance can be read directly from the variogram.

From the definition of Z and Eq. 4, it can be shown that:

$$\text{VAR}(Z)=1/V^2\int_V\int_V \sigma(X-X')dxdx' \tag{9}$$

where X and X' are point values within V.

2) COV$|(Z)(X_i)|$ is the covariance of the grade of

the block and the grade of the sample. The covariance is a measure of the correlation between the grade of the entire block and the grade of a given sample, or it is the average variance which exists between any point in the block and a point in the sample. Again such a variance is expressed in the variogram function and can be readily obtained when the average distance has been determined.

The covariance term 2 can be expressed thus:

$$2\Sigma a_i \text{ COV}|(Z)(X_i)| = 2/nV \sum_{i=1}^{n} a_i \int_V \sigma(X - X_i)dx \quad (10)$$

where X is value of any point in V and X_i is value of sample.

3) COV (X_iX_j) is the covariance of the grades of two samples. The correlation or lack of correlation that exists between two samples is simply a special case of COV $(Z)(X_i)$. The variance between any two samples can be read directly from the variogram with the distance between already being known.

$$\sum_i \sum_j a_i a_j \text{COV}(X_i X_j) = 1/n^2 \sum_{i=1}^{n} a_i \sum_{j=1}^{n} a_j \sigma(X_i - X_j). \quad (11)$$

Recombining the variance and covariance terms of Eq. 8 by their equivalents in Eqs. 9, 10, and 11, we now have the fundamental formula for determining the estimation variance of the estimated grade of a block of volume V as determined by n punctual (point) samples.

$$\begin{aligned} \text{VAR}(Z - Z^*) = {} & 1/V^2 \int_V \int_V \sigma(X - X')\,dxdx' \\ & - 2/nv \sum_{i=1}^{n} a_i \int_V \sigma(X - X_i)\,dx \\ & + 1/n^2 \sum_{i=1}^{n} a_i \sum_{j=1}^{n} a_j\, \sigma(X_i - X_j). \quad (12) \end{aligned}$$

Having looked at the estimation variance in terms of relative positions of samples and blocks for punctual samples, i.e., samples representing a value at a point, we now must consider the relative volume relationships. All samples represent some volume and we have mentioned the concept of sample support and sample composites previously.

Let $\sigma^2(o/V_b)$ be the variance of punctual samples in a block of volume V_b, and $\sigma^2(o/v)$ the variance of punctual samples in samples of volume v. The variance of the samples v in the block V_b is then given by:

$$\sigma^2(v/V_b) = \sigma^2(o/V_b) - \sigma^2(o/v). \quad (13)$$

If we next consider the blocks of volume V_b situated in a total deposit of volume V_d, we have:

$$\sigma^2(V_b/V_d) = \sigma^2(v/V_d) - \sigma^2(v/V_b) \quad (14)$$

and

$$\sigma^2(v/V_d) = \sigma^2(v/V_b) + \sigma^2(V_b/V_d). \quad (15)$$

Eq. 15 thus states that the variance of samples of volume v in the total deposit of volume V_d is equal to the variance of samples in the block plus the variance of blocks within the deposit. This relationship was noted in the experimental work of D. G. Krige and is known as the krige relationship.

Matheron (1962) has shown that the variance of the error of estimation $Z_V - Z_v$ which is committed when assigning to V the grade of v after defining Z_V and Z_v as

$$Z_V = 1/V \int_V f(x)\; dx$$

$$Z_{v'} = 1/v' \int_{v'} f(x)\; dx$$

is as follows:

$$\begin{aligned} \text{VAR}(Z_V - Z_{v'}) = {} & 2/Vv' \int_V \int_{v'} \sigma(x - x')\,dxdx' \\ & - 1/V^2 \int_V \int_V \sigma(x_1 - x_2)dx_1 dx_2 \\ & - 1/V'^2 \int_{v'} \int_{v'} \sigma(x'_1 - x'_2)dx'_1 dx'_2. \quad (16) \end{aligned}$$

It will be observed that Eqs. 12 and 16 are essentially the same expressions: Eq. 16 is an expression of the generalized fundamental formula and is the theoretical basis for determining the magnitude of error involved in an estimation procedure when $\sigma(h)$ is known. Eq. 12 is the specific case of discrete samples, which is used in practice. Since all values of x are defined in 3D space, the integrals are sextuple ones. Prior to the ready availability of the computer, the solution of the fundamental formula in three dimensions using various intrinsic schemes was not a trivial problem. However, with the help of the computer, adequate solutions to the fundamental formula can be quickly determined.

Confidence Limits From the Estimation Variance

In the study of statistical inference for independent variables, it has been shown that one can determine the possible error of the estimation of the mean for certain confidence levels. For example, assuming a normal population, we can say that there is 95% probability that the true mean lies within two standard deviations either side of the estimated mean. By determining the standard error of the mean $\sigma_e = \sigma/\sqrt{n}$) and using the student t table, one can establish the confidence interval of the mean at any given level of confidence.

In the case of nonindependent and spatially correlated variables such as exist in most mineral deposits, the same concept of confidence intervals can be applied. It has been shown in the preceding section that the variance of the error can be determined by application of the fundamental formula 12, if the variogram is known and there is a program to estimate the integrals. The square root of the estimation variance is the standard error of the mean and can be used in confidence interval calculations exactly as the normalized standard error in the preceding paragraph. It will be noted that the number of samples (n) is taken into consideration in the solution of the fundamental formula and should not enter into the calculation a second time.

Geostatistical Extension Function—Kriging

Kriging is the name given by Matheron to designate a best linear unbiased estimator for assigning values to mining blocks using geostatistical techniques. The technique involves assigning weighting factors to sample values, which depend upon the geostatistical parameters of the deposit and the geometry of the sample points relative to the block being estimated.

The best linear unbiased estimator makes three significant implications:

1) Unbiased means that the computed value should be, on the average, equal to the real value and not systematically higher or lower.

Consider again a block B having a true unknown grade of Z_b and estimated grade Z_b^* from n samples of known value $Z(X_i)$ $(i=1, \ldots n)$ which will make the weighted average

$$Z^*_b=\sum_{i=1}^{n} a_i Z(X_i).$$

Then to meet the unbiased condition, since

$$E(Z^*_b)=m$$

then

$$E(\sum_{i=1}^{n} a_i Z(X_i))=m$$

also

$$E(Z(X_i))=m;$$

therefore

$$\sum_{i=1}^{n} a_i=1.$$

2) In regression analysis, the best fit is the equation of a line or surface which minimizes the error between the surface and the sample point. The fundamental equation of the estimation variance or the variance of the error of estimation is expressed as $\mathrm{VAR}(Z^*-Z)$ or:

$$\sigma_e^2=\sigma_V^2-2\sum_{i=1}^{n} a_i\sigma_V X_i+\sum_i\sum_j a_i a_j \sigma X_i X_j \qquad (16)$$

where σ_V^2 is the variance of the grade of block of volume V, $\sigma_V X_i$ is the covariance of the grade of block V and sample X_i, and $\sigma X_i X_j$ is the covariance of the samples X_i and X_j.

Each of the variance and covariance values can be determined from the variogram function ($\sigma=\gamma$ of the variogram), thus making it possible to minimize the variance of the error of estimation of the kriging function σ_K^2 by differentiating Eq. 16 with respect to the a_i's and setting the derivatives equal to zero. However, when there is a constraint ($c=0$), the Lagrange principle says that $F=Q+2\mu C$ should be minimized, where μ is the new unknown, the Lagrange multiplier. Thus, for our case, we should take the derivative of:

$$F=\sigma_K^2+2\mu(\sum_i a_i-1)$$

or

$$F=\sigma_V^2-2\sum_{i=1}^{n} a_i\sigma_V x_i+\sum_i\sum_j a_i a_j \sigma X_i X_j+2\mu(\sum_i a_i-1)$$

the derivative of which is:

$$\frac{\partial F}{\partial a_i}=-2\sigma_V x_i+2\sum_j a_j\sigma x_i x_j+2\mu=0, \; V(i=1, \ldots n)$$

$$\frac{\partial F}{\partial \mu}=\sum a_i-1=0.$$

3) It is now obvious that the derivatives are a linear system of $n+1$ equations with $n+1$ unknowns which can be written in the form:

$$\sum_j^n a_j \sigma x_i x_j+\mu=\sigma V x_i \quad (i=1, \ldots n)$$

$$\sum_i^n a_i=1. \qquad (17)$$

The only efficient method to solve such a system of equation is to convert them to matrix form of the order $|\Sigma|\;|A|=|D|$ where Σ, A, and D are:

$$\Sigma=\begin{vmatrix} \sigma_{11}\sigma_{12}\cdots\cdots\sigma_{1n} & 1 \\ \sigma_{21}\sigma_{22}\cdots\cdots\sigma_{2n} & 1 \\ \cdot & \\ \cdot & \\ \cdot & \\ \cdot & \\ \cdot & \\ \sigma_{n1}\sigma_{n2}\cdots\cdots\sigma_{nn} & 1 \\ 1\;\;1\;\cdots\cdots\;1 & 0 \end{vmatrix} \quad A=\begin{vmatrix} a_1 \\ a_2 \\ \cdot \\ \cdot \\ \cdot \\ \cdot \\ \cdot \\ a_n \\ \mu \end{vmatrix} \quad D=\begin{vmatrix} \sigma_{Vx1} \\ \sigma_{Vx2} \\ \cdot \\ \cdot \\ \cdot \\ \cdot \\ \cdot \\ \sigma_{Vxn} \\ 1 \end{vmatrix}$$

Observe that $|\Sigma|$ is a symmetrical matrix of the covariance between all sample values, and $|D|$ is a matrix of covariance between the block being estimated and the sample values. All variances and covariances can be derived from the variogram, and the values for weighting coefficients are obtained by solving the linear system for $|A|$ according to the form

$$|A|=|\Sigma|^{-1}|D|. \qquad (18)$$

There are many computer algorithms written to solve large systems of simultaneous equations. The reader is left to determine which program can most efficiently meet this need. The computer cost involved in kriging is strongly influenced by this choice.

Example of Point Kriging

In order to illustrate the mathematics of kriging, let us take an example of samples in a two-dimensional setting. Instead of estimating the grade of a block, we will first simply estimate the grade of one particular point, or, we will use surrounding samples to estimate the expected grade of a sample taken at point Xo. By estimating a point value, it is not necessary to solve the covariance integral between the samples and the block, and the coefficients can be determined directly from the spherical variogram model when the distances from samples to the point and between samples have been determined.

Consider an experimental variogram for a vein-type silver deposit fitted by a spherical variogram model having a range $a=76$ m (250 ft), $Co=17$, and $C=66$. Fig. 17 shows the relationship of four samples in the vein designated S_1, S_2, S_3, and S_4, and X_o marks the point at which we want the best grade estimate possible. Uniform vein thickness is implied over the sample area.

According to Eq. 17, we have only to set up a system of simultaneous equations to solve for the weighting coefficients for each sample value, after being able to determine the covariance coefficients between sample points from the variogram. For the case illustrated in Fig. 17 we have:

$$\begin{aligned} a_1\gamma_{11}+a_2\gamma_{12}+a_3\gamma_{13}+a_4\gamma_{14}+\mu&=\gamma_{01}\\ a_1\gamma_{21}+a_2\gamma_{22}+a_3\gamma_{23}+a_4\gamma_{24}+\mu&=\gamma_{02}\\ a_1\gamma_{31}+a_2\gamma_{32}+a_3\gamma_{33}+a_4\gamma_{34}+\mu&=\gamma_{03}\\ a_1\gamma_{41}+a_2\gamma_{42}+a_3\gamma_{43}+a_4\gamma_{44}+\mu&=\gamma_{04}\\ a_1\ +\ a_2\ +\ a_3\ +\ a_4\ +0&=1 \end{aligned}$$

where the diagonals γ_{11}, γ_{22}, γ_{33}, γ_{44} are all equal and are the variance of the sample with itself. Applying the spherical model defined by the formula

$$\begin{aligned} \gamma(h)&=C|3/2\ h/a-1/2\ h^3/a^3|+Co \quad \text{when } h\leq a\\ \gamma(h)&=C+Co \quad \text{when } h>a \end{aligned} \tag{19}$$

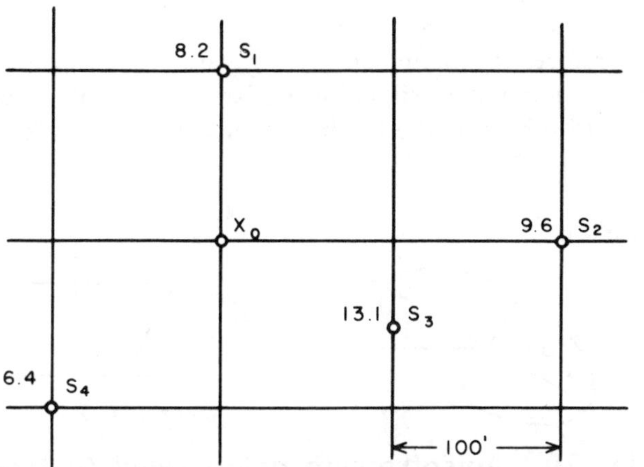

Fig. 17. Sample locations silver vein deposit silver value in ounces per ton. Metric equivalents: 1 oz×0.028 349 5=kg; 1 st×0.907 184 7=t.

it is apparent that the variance of a sample with itself ($h=0$) is zero. The covariance between other samples and the point to be estimated are as follows:

$$\begin{aligned} \gamma_{12}&=\gamma_{21}=\gamma_{14}=\gamma_{41}=\gamma(\sqrt{200^2+100^2})\\ &=66\left|3/2\ \frac{\sqrt{200^2+100^2}}{250}-1/2\frac{\sqrt{200^2+100^2}^{\,3}}{250^3}\right|\\ &\quad+17=81.94 \end{aligned}$$

$$\begin{aligned} \gamma_{13}=\gamma_{31}&=\gamma(\sqrt{100^2+150^2})=76.02\\ \gamma_{23}=\gamma_{32}=\gamma_{30}=\gamma_{03}&=\gamma(\sqrt{100^2+50^2})\ =58.32\\ \gamma_{24}=\gamma_{42}&=\gamma(\sqrt{300^2+100^2})=83.00\ (h>250)\\ \gamma_{34}=\gamma_{43}&=\gamma(\sqrt{200^2+50^2})\ =80.13\\ \gamma_{10}=\gamma_{01}&=\gamma(100)\ =54.48\\ \gamma_{20}=\gamma_{02}&=\gamma(200)\ =79.30\\ \gamma_{40}=\gamma_{04}&=\gamma(\sqrt{100^2+100^2})=67.02. \end{aligned}$$

At this point, we must turn to the computer to solve the linear system of equations by plugging the covariance values into Eq. 19, inverting the matrices to the form of Eq. 18, and solving for the a_i's, which have the following values:

$$\begin{aligned} a_1&=0.393\\ a_2&=0.022\\ a_3&=0.329\\ a_4&=0.256. \end{aligned}$$

To complete the example, we have only to multiply the sample value by its weighting coefficient to determine the estimated grade for point Xo:

$$\begin{aligned} G_x&=a_1\,S_1+a_2\,S_2+a_3\,S_3+a_4\,S_4\\ &=0.393(8.2)+0.022(9.6)+0.329(13.1)\\ &\quad+0.256(6.4)\\ &=9.38 \text{ oz per st.} \end{aligned}$$

It is appropriate to note here that because of the need for program efficiency when solving large linear equation systems that David (1975) and others use a sigma function for defining all the covariance relationships. The sigma function can be thought of as the complement to gamma of the variogram where $\sigma_o=C+Co$, and

$$\sigma_{ij}=(C+Co)-\gamma_{ij}.$$

The proper solution of the linear system using either the variogram gamma function or its complement, as defined, will produce the same solution for sample weighting coefficients.

Block Kriging

To move from point kriging to block kriging involves only one additional step for determining the covariance between the sample and the block rather than the samples and a point. Note that the covariance between samples remains the same, and the sample covariance matrix need only be computed once.

In order to determine the covariance between a block V and sample X, it is necessary to evaluate the integral given by the following general formula:

$$\sigma VX_i = C|1/V \int\int_v \sigma(\sqrt{(X_i - X_o)^2 + (Y_i - Y_o)^2}\, d_x d_y| + Co$$

where V is the area of block V and X_o and Y_o are coordinates of sample X.

For all practical purposes, the evaluation of the integral is best achieved by approximating the continuous function with a series of discrete points taken within the block and then averaging the results. In other words, instead of determining the covariance of sample and block, we estimate the average covariance between the sample and a number of mesh points within the block as in Fig. 18. It is apparent that the greater number of mesh points used, the more nearly will the average covariance approach the continuous function. However, great precision is irrelevant due to uncertainties in the variogram model, and a practical trade-off must be made between precision and the computation time involved.

To proceed from two-dimensional to three-dimensional block kriging is to consider the triple integral in $x\ y\ z$ space and estimate the solution using sample and mesh points in a three-dimensional mesh grid.

Block Kriging and Anisotropism

Knowing that anisotropism exists within a mineral deposit, how then can such anisotropisms be adequately accounted for when estimating block values using kriging techniques? Since the variogram constitutes the intrinsic function used in kriging mathematics, it should be apparent that the anisotropic factors established by

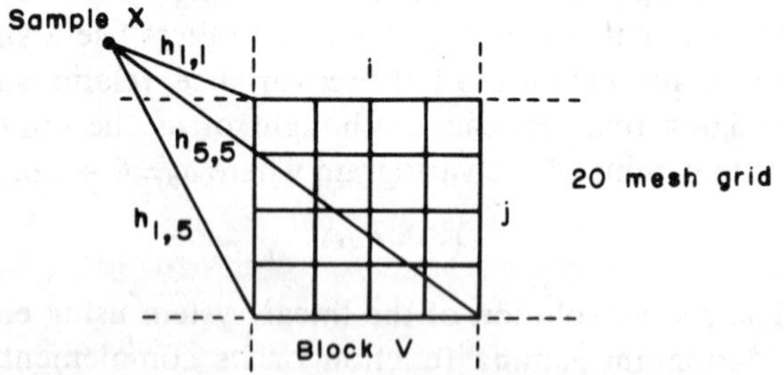

Fig. 18. Method of computing covariance between sample and block by averaging covariance between discrete points within block.

$$\begin{array}{c} \sigma(h_{1,1}) \\ \sigma(h_{1,2}) \\ \vdots \\ \sigma(h_{i,j}) \\ \hline 1/n\sum_{1}^{n}\sigma(h_{i,j}) \simeq \sigma_{VX_i} \end{array}$$

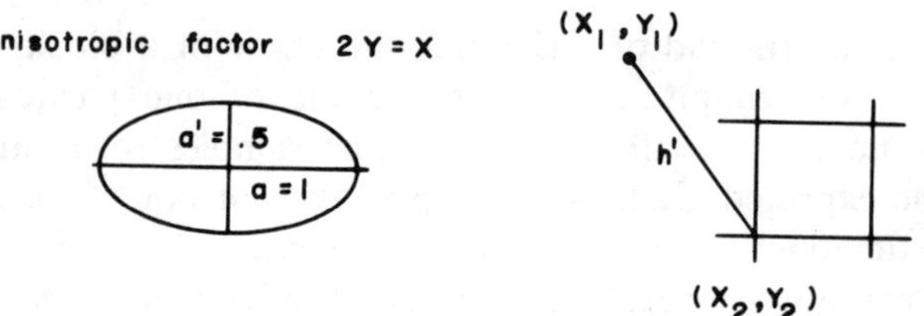

Fig. 19. Anisotropic parallel to block coordinate axes.

directional experimental variograms can be applied directly in the kriging estimation process.

Assume, for example, that range of a spherical variogram along trend is double the range perpendicular to trend in a deposit exhibiting geometrical anisotropism. The sill and nugget effect will remain the same, and the range will vary. If the coordinate axes of the block being estimated are parallel to the trend (anisotropic axes) of the deposit as in Fig. 19, the modified distance h' as a function of the coordinates of the two end points (X_1, Y_1), (X_2, Y_2) is:

$$h' = \sqrt{(X_1 - X_2)^2 + K^2(Y_1 - Y_2)^2}$$

where $K = 2$.

If the coordinate axes are not parallel to the anisotropic axes, then it is necessary to make a coordinate transformation through the angle Q as illustrated in Fig. 20.

Kriging Estimation Variance or Estimation of Error

It will be recalled that the geostatistical estimation variance, or the estimation of error, is expressed by the fundamental formula:

$$\sigma_e^2 = \sigma_V^2 - 2\sum_i a_i \sigma_{VX_i} + \sum_i\sum_j a_i a_j \sigma_{ij}. \qquad (8)$$

Therefore, to obtain the kriging variance, designed to be the minimum estimation variance, it is only necessary to substitute the kriging coefficient for the a_i's into

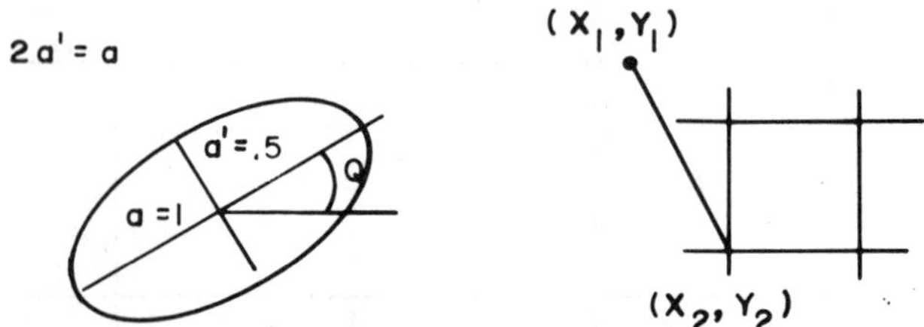

Fig. 20. Anisotropism not parallel to block coordinate axes.

$$h' = \sqrt{|(X_1 - X_2)\mathrm{Cos}Q + (Y_1 - Y_2)\mathrm{Sin}Q|^2} + \sqrt{K^2|(X_1 - X_2)\mathrm{Sin}Q + (Y_1 - Y_2)\mathrm{Cos}Q|^2}$$

where $K = 2$

Eq. 8. However, it is possible to take advantage of a simplification of the fundamental formula by using terms which can also be used in computing kriging coefficients.

Consider the equation of the kriging system

$$\sum_j a_j \sigma X_i X_j + \mu = \sigma_{VX_i} \tag{17}$$

multiplying each side of the equation by a_i and summing gives

$$\sum_i \sum_j a_i a_j \sigma X_i X_j + \sum a_i \mu = \sum_i a_i \, \sigma_{VX_i},$$

but since $\Sigma a_i = 1$, we have

$$\sum_i \sum_j a_i a_j \, \sigma X_i X_j + \mu = \sum a_i \, \sigma_{VX_i}.$$

Then, by substituting the identity back into the fundamental formula 8, it can be observed that the minimum estimation variance, now called the kriging variance, is equal to:

$$\sigma_e^2 = \sigma_K^2 = \sigma_v^2 - \sum_i a_i \, \sigma_{VX_i} + \mu. \tag{20}$$

The terms μ and $\Sigma a_i \, \sigma_{VX_i}$ are also computed for determining kriging coefficients, and the first term, σ_V^2 is the variance of a block and is the average difference in value which exists between any two points in the block. If all the blocks in the deposit or subset of the deposit are uniform, it is necessary to compute the block variance only once.

Computer Programs for Kriging

Geostatistics and kriging had their origin prior to the widespread industrial use of digital computers, particularly in the mineral industry. However, it should now be apparent that without the power of the computer, the practice of geostatistics can be very difficult and time-consuming. Such, indeed, was the case in the early 1960's.

Many kriging programs have now been written and are available throughout the mineral industry. It is extremely difficult for any one program to do all things for all applications. If the principles are thoroughly understood, it is possible to modify kriging algorithms to accomplish nearly all mineral estimation problems within reasonable costs to the user.

David (1975) has outlined the basic structure of a kriging program as follows: one has on hand a file of samples with their grades and coordinates, a file of blocks to be estimated from the first file, and a variogram function.

Now for each block the following process should be repeated: (1) the sample file is searched for samples having an influence on the block; (2) the covariances between these samples should be computed; (3) the covariance of these samples and the block should be computed; (4) these covariances should be arranged in a linear system form; (5) the linear system of equations should be solved; (6) the solution is the set of weights which one requires; (7) the grade is obtained by multiplying the set of weights by the grade of samples retained; and (8) the precision on the grade estimation is computed after formula 20.

Careful analysis and optimization of each of these steps for the particular problem at hand can make possible an economic geostatistical estimation for nearly every mineral deposit. A study of the rapidly growing published literature in this field will yield examples and solutions for nearly every problem encountered today.

The Mineral Population Boundary Problem

Mineral deposits rarely are completely homogeneous and often consist of many subsets separated by geological and mineralogical boundaries. These boundaries may include: changes in mineral suites, lithological horizons, host rock types, faulted and offset segments, zones of enrichment or depletion, and several other geological and mineralogical changes that affect the mineralogical structuring and distribution within the deposit.

When employing conventional methods for assigning estimated values to irregularly shaped inventory blocks, it is only necessary for the geologist or engineer to confine the block limits to the known geologic boundaries. Thus, if a fault interrupted the mineral zone, the trace of the fault can be used as a block boundary. Intelligent decisions can always be made as to which blocks to include in each mineral population. However, with the advent of the computer which permits the computation of mineral inventory blocks using better mathematical estimation techniques, geologists are faced with the boundary problem. Unless the computer can be programmed to recognize population differences and geologic boundaries, greater estimation errors often will be found in computer-generated mineral inventories than in less sophisticated conventional inventories. Much disillusionment has been expressed over the years by experienced mineral appraisers as a result of their study of computer-generated inventories that failed to adequately account for geologic vagaries. Many, thus disillusioned, have concluded that computer techniques are not satisfactory or have only limited application to simple deposits.

Mineral population boundaries can be grouped into three broad classifications: (1) physical or topographic boundaries which include surface topography and subsurface topography; (2) mineralogical boundaries which consist of gradational boundaries; mineral suite zonation boundaries; oxidation, leached, and enrichment boundaries; and mineralogical changes due to different host rock environment; and (3) geological boundaries comprising lithological and/or stratigraphic boundaries, wall rock boundaries, fault boundaries, and intrusive boundaries (postmineral).

The solution of the mineral boundary problem for adequate and accurate computer-generated mineral reserves and inventories has two facies: (1) getting the boundary resolved into a digital x y z matrix that can be entered into the computer and by which various mineral population subsets can be defined and (2) coding the various programs that contribute to the mineral inventory in such a way that they recognize mineral boundaries and functions within the proper geological-mineralogical constraints.

Step one involves defining the bounding surface or surfaces at selected grid intervals (the same grid interval used for computation of mineral inventory blocks). In the case of surface topography, there is more than one way to digitize, i.e., assign elevations to specific grid points on a topographic map. The most common method is to use an electronic digitizer to record the Z elevation for the selected x and y coordinates.

Subsurface boundaries, however, cannot be observed directly and must be interpolated from the point values intercepted within the random drill holes. If enough points are known, a fairly accurate subsurface map can be defined. Surface topographic mapping is based upon the assumption of linear gradient between selected points. It is logical, therefore, that a linear function be used for interpolating between known points of a subsurface boundary with the objective of establishing an absolute surface that will represent exactly the known elevation at each drill-hole intercept. A computer interpolation program based on a linear gradient function between known data points best serves this purpose.

Having generated an x y z boundary matrix for each mineral population, we now have a method whereby the computer can be programmed to check the coordinates of each mineral inventory block against the boundary matrix and tag the block as to which population it belongs. Such a block tag can be recorded on only one to four machine bytes, and a total 3D block matrix of even a very large deposit can be stored in relatively few machine words of core. It is this compactly stored tag array that will be interrogated by the mineral inventory program to determine whether or not to use the sample taken from a certain population when computing an estimated value for the mineral block. The mineral population boundary problem and its solution can be summarized briefly as follows: (1) determine the various mineral population subsets within the total deposit by using geological and mineralogical differences and geostatistical analysis, (2) separate and composite the samples according to the population they represent, (3) generate digital mineral boundaries between the population subsets by means of electronic digitizer (surface topography) and a linear interpolation program using only real data (drill-hole intercepts) and using both real and pseudodata of geological interpretations, (4) generate gridded 3D block tag arrays that designate the population to which each block belongs within the limits of the entire deposit, and (5) compute the estimated values for all of the mineral inventory blocks in the deposit according to the population they represent and using only samples from that population.

References

David, M., 1975, *Geostatistical Ore Reserve Estimation,* Ecole Polytechnique de Montreal, PQ, Canada.

Matheron, G., 1971, "The Theory of Regionalized Variables and Its Applications," Les Cahiers du Centre de Morphologic Mathematique, Booklet No. 5, Ecole Nat. Sup. des Mines, Paris, 211 pp.

Popoff, C. C., 1966, "Computing Reserves of Mineral Deposits: Principles and Conventional Methods," Information Circular 8283, US Bureau of Mines.

Royle, A. G., 1971, *A Practical Introduction to Geostatistics,* Dept. of Mining and Mineral Science, Leeds, England.

C. D. Broadbent, editor

Contents

INTRODUCTION TO SUBSECTION 2C

Monetary and physical values go hand in hand through all phases of mine exploration, planning, development, and operation. The topics grouped in this subsection focus more on dollar values as the primary objective, whereas in the other areas, physical objectives are of primary importance to the task. The four authors address the methods for valuing ore in the ground, tracking operating costs, assessing future market prices (for planning), and finally valuing the combined mineral operations. All of these topics are extensively interrelated and interdependent—each supporting the other in some part of the overall planning process.

R. K. Davey summarizes input requirements for an ore reserve block model. He follows with examples of break-even grade calculations with and without leaching credits, a review of both marginal analysis and variable cutoff strategies to optimize results, and looks at the cost impact of alternative haulage configurations. R. F. Winkle reminds us that "A detailed breakdown of mining costs . . . is mandatory for a controlled and efficient mining operation," and further notes "Cost awareness intensifies the daily search for better and cheaper methods and promotes improved scheduling." This latter challenge is critical in the "slim picking" environment in which we find ourselves. His review of records and record forms should become a valuable reference.

A. C. Noble tackles the unforgiving chore of Price Projection. He cautions that "The assumptions used for estimating price should be clearly stated for use in cost forecasts." Four principal methods of forecasting are described and the pitfalls of each are noted. The value of a sensitivity analysis is explained because as Mr. Noble notes, "The final pit slope . . . will be determined by a future price which may be drastically different from the best estimate." A detailed example of trend line and confidence band calculations concludes his contribution. F. J. Stermole rounds out the group of "Economics" papers with a presentation on economic evaluation of open pit mines, an iterative process that enters the picture at most every planning step—from initial exploration, through preliminary planning, to final decision analysis. A particularly useful discussion on inflation, escalation, and constant dollars is included, as are two complete cash flow calculation examples—one with and one without other taxable income. The subject is well highlighted in his examples.

A common thought in all papers in the group is that input data must be detailed, valid, and well-qualified if the results are to be worthy of consideration in the planning process. Quantities derived from inadequate records or values derived from unknown assumptions could be pitfalls that cloud and compound errors in many following results. Collectively, the authors imply that one cannot be too cautious in these analyses.

7 Mineral Block Evaluation Criteria

Roderick K. Davey
Kennecott Minerals Co.

Roderick K. Davey is technical superintendent at the Bingham Mine of Kennecott Minerals Co. His responsibilities include preparation of department budgets, and planning and supervision of personnel and the computing resources provided to the department to assist operating divisions with mine planning, maintenance planning, production reporting, metals accounting, and financial analysis.

Davey graduated from the University of Wisconsin, Madison, with a B.S. in mining engineering and from Michigan Technological University with an M.S. in mining engineering. His career with Kennecott includes service as a junior mining engineer with Chino Mines Div. and as a systems analyst and senior systems analyst with the Computing Center.

Introduction

In any business, it is essential that we select those alternatives which are not only technically feasible, but will be the most profitable to the business in terms of corporate objectives.

Some of the studies and activities requiring support of financial analysis to select the alternatives which most closely satisfy the management objectives and technical considerations are as follows:

1) Cutoff grade determination.
2) Ore reserve estimation.
3) Production forecasts (long- and short-term) for an operating mine, development of a new mine, or expansion of an existing mine.
4) Evaluation of haulage alternatives.
5) Evaluation of equipment.

Effective analysis requires the best information available concerning:

Mine Planning and Operational Aspects: Drilling data, mining plans, metallurgical characteristics, equipment performance, availability of water, and general considerations concerning the shape of the deposit, continuity, and strength of the materials (as this concerns cavability, cost of development, and ultimate pit slope ∡).

Market Potential: Projected supply and demand of the resource involved and associated prices of products.

Environmental Considerations: The projected capital and operating costs associated with meeting environmental regulations regarding air, water, and reclamation.

Labor: The availability of skilled labor, particularly craftsmen and miners.

Transportation: Included here would be developing roads and railroads, construction of special loading facilities such as docks, etc.

Governmental: Included here might be consideration for taxes, governmental stability, governmental participation in ownership, etc.

Royalty: Will a royalty have to be paid to another party, private or governmental, and what are the alternatives?

Energy: Availability and cost.

Availability and Cost of Capital: as well as inflation.

Infrastructure: of the company concerning smelting, refining, fabrication, and marketing of the product.

The financial data can be derived from cost sheets, depreciation schedules, and other financial documents. The accurate forecasting of metal prices is, at best, difficult. It is not uncommon to project prices based on the current profit margin and historical cost trends. With today's market and prices for many products, this would not be a satisfactory approach for estimating ore reserves (ultimate pits). However, some estimate of the average profit margin that might prevail throughout the projected life of the operation is required.

The metallurgical characteristics, which are just as critical as estimated price, costs, and assays, usually are based on operating experience.

Each financial analysis may involve some, or all, of the items mentioned and perhaps even some that have not been mentioned.

Above all, experience of the project personnel is a key factor for establishing the data base for any evaluation and performing the evaluation.

The following examples of financial analyses will be presented in this paper:

1) Determination of cutoff grade. Several examples will be presented including: (a) calculations considering average costs; (b) calculations considering variable haulage costs; (c) calculations considering the potential for mill/flotation vs. leaching to recover the resource; and (d) an example of marginal analysis, as applied to a deposit to be mined by underground methods.

2) Comparison of costs associated with alternative haulage configurations. For example: (a) different truck sizes (performance characteristics), and (b) different haulage profiles.

The financial and operational criteria shown in these examples are not the only financial or business criteria to be considered in selection of the most favorable alternative for investing capital.

A basic premise for all financial analyses is that there is an alternative which is competing for the money to be invested.

This chapter will not attempt to explain the theory behind the techniques applied. The theory is amply covered by other authors, some of which are noted in the references. Also, a glossary of terms has been included.

Cutoff Grade Calculations

When developing ore reserves, or mining plans for a potential mine or production forecast for an operating property, it is essential to determine the grade at which the mineral resource can no longer be processed at a profit, the break-even grade. In reality, this is not a simple matter because costs, metallurgical characteristics, and stripping (open pit) vary throughout the deposit. For example, we have a case where the break-

Glossary of Terms

Annual Return on Original Investment

Formula is:

$$\frac{\dfrac{\text{Annual earnings after tax}}{\text{Original investment}}}{2} \times 100$$

Capital

Sum of long-term debt and shareholder's equity.

Capital Expenditure Evaluation

Method of providing the tools for allocating capital resources and measuring the anticipated future returns that can be expected from the incremental capital invested.

Cash Flow

Sum of net earnings plus depreciation.

Cash Outflows

Represent expenditures or costs necessary to support a capital project.

Cost of Capital

The sum of debt cost or interest and costs attributable to equity financing which a company must pay or earn to satisfy all investors both on a short- and long-term basis.

Cost of Goods Sold

Material, labor, and overhead attributable to the completion of a finished product which is available for sale.

Declining Balance Method

An accelerated method of depreciation which uses a constant factor which can be no higher than double the straight-line method but is computed on the undepreciated balance each year.

Depreciation

A simple way of allocating the investment cost in an effort to insure investment recovery.

Discounted Cash Flow

Deals principally in the future in measuring the time value of money.

Discounted Payback Method

Introduces the interest factor relating to the cost of money which is a variation of the payback method using discounting cash flow techniques.

Discounting

The reverse of compounding. The value of money to be received in the future is shifted back to the present.

Straight-Line Depreciation

Charges to earnings are spread evenly over the life of the depreciable asset.

Useful Life

Estimated time in years and months that a fixed asset may be expected to be replaced.

even grade for an open pit copper mine has been determined to be 0.35% copper, based on average costs and metallurgical characteristics. However, a particular portion of the deposit might be projected as high grade but have a low metal recovery. Thus, if we were to process the material, a loss would be incurred.

In these examples, the cutoff grade and break-even grade will be considered equal. It should be noted that the cutoff grade may be set at a point other than the break-even grade.

Also, operating costs vary, depending on length of haul, type of material, and age and type of equipment used. Thus, even though the grade of the portion of the deposit being considered is higher than the average cutoff grade, it may not be classified as ore. The *obvious* reason for this situation is that the cutoff grade was determined using an average haulage cost, based on haulage from the centroid of the deposit to the centroid of the dumps and/or crusher.

The physical and operating circumstances of each mine vary considerably, and the potential spectrum of financial analyses is almost infinite; therefore, in this chapter, only a few of the techniques that might be applied will be addressed.

Examples of different techniques and circumstances for calculating a projected cutoff grade are as follows:

1) Computation of the projected cutoff grade, including byproduct values and using average costs and metallurgical characteristics.

2) Computation of the projected cutoff grade, including byproduct values and variable mine haulage costs.

3) Computation of the projected cutoff grade, in-

cluding consideration for leaching (dump or vat) and variable mine haulage cost.

4) Computation of the projected cutoff grade by marginal analysis.

Computation of the Projected Cutoff Grade—Average Costs and Metallurgy

The economic model for the entire mine, based on averages, can be expressed as an equation. The equation defines the relationship of the grade of the material to the net value. If the net value is divided by the cost/ton of stripping, we have an "allowable break-even stripping ratio equation." In this case, the projected cutoff grade is determined by solving the equation for the grade which generates a zero net value—break-even.

Net Value: Compute the net value of one ton of copper ore having a grade of 0.55% copper, a mill recovery of 80% (of the copper contained), and a concentrate grade of 20% copper. In this example, the mining and processing costs used will be the average for the deposit (or a portion of the deposit) as will the metallurgical characteristics. Also, a singular metal price has been projected and taxes are not included. The derivation of the grade vs. net value equation follows:

1) The pounds * of salable copper are computed as follows:

Pounds of copper recovered (mill) per ton of ore $=0.0055\times2000\times0.80=8.8$

The ratio of concentration =

$$2000\times\frac{0.20}{8.8}=45.45$$

Smelting loss per ton of ore @ 10 lb per ton of concentrate $10/45.45=0.22$ lb per ton of ore

Pounds of blister copper per ton ore $=8.8-0.22=8.58$

Refining loss @ 5 lb per ton of blister copper $=5/(2000/8.58)=0.02$ lb ton of ore

Net copper per ton ore $=8.58-0.02=8.56$ lb

The recoverable quantities of byproducts, if any, would be calculated much in the same way as the recoverable copper was. Basic data required includes estimated quantities of the value in the raw material and estimated percent recovery for milling, smelting, and refining and/or any other processes involved.

* One pound (lb) × 0.454 = kilogram (kg).

2) The net value per ton of ore is computed as follows:

Production Cost, Excluding Stripping—The ton of mineralized material is envisioned to be on the surface.

	\$ per ton of ore
Mining	0.42
Ore haulage	0.25
Milling	1.60
General	0.35
Amortization & depreciation	0.56
Subtotal	3.18
Treatment Costs—	
Freight @ \$1.40/ton concentrate	0.03
Smelting @ \$50.00/ton concentrate	1.10
Freight @ \$50.00/ton blister copper	0.21
Refining @ \$130.00/ton copper	0.56
Selling & delivery @ \$0.01/lb copper	0.09
General plant @ \$0.07/lb copper	0.60
Total treatment cost	2.59
Total Production Cost—	
Less byproduct credits	0.65
Total cost	5.12
Value of Salable Copper @ \$0.65/lb—	5.56
Net Value—	
Value salable copper	5.56
Less total cost	5.12
Net value	0.44

Relationship Net Value to Grade: After determining the net value for a minimum of two different grades of ore, the relationship (equation) between the net value and the grade can be derived.

Cutoff Grade: The cutoff (break-even) grade is determined by solving the equation for the grade which produces a zero net value.

Calculating Net Value: It should be noted here that this example assumes a linear relationship between the net value and the grade of the material. In the situation where the metallurgical characteristics vary with the grade, the net value would be calculated for several different grades and a curve fit to the values. The function which describes the curve would then be solved for the grade at which the net value equals zero, the cutoff grade. Remember that we have considered the cutoff grade to be the break-even grade in this case. The curve generated, using the example, is shown in Fig. 1. Using the curve, we then can determine that at a grade of 0.55% copper, we could strip 1.76 tons (max.) at \$0.25/ton.

Thus, if the average stripping ratio was 1:1 (waste to ore), then the break-even grade would be approximately 0.48% copper.

In some cases, it may be desirable to consider a minimum profit. This can be treated as a cost. The end effect is the cutoff grade (break-even) is shifted to a higher grade.

Also, we have presented this example from the perspective of the entire deposit. The same procedure can be used for any portion of a deposit, including shovel cuts or on a block-by-block basis. A block may be regular or irregular, but it represents a specific portion of the deposit.

Cost Factors: The following remarks concern factors which should be taken into consideration when generating the costs to be used in any of these computations:

Mining Cost—

1) Drilling and blasting costs may vary because of different rock types and/or drill sizes.

2) Loading costs may vary because of different loading conditions, age of equipment, and/or type of equipment.

3) Roads and dump costs are typically considered constant.

4) Water pumping costs should be included, if pertinent to the operation.

5) Mine haulage costs can be determined as follows:

The tons per operating hour that can be hauled by proposed equipment over each proposed route are calculated using mining plans. The average tons hauled per hour for the entire operation is calculated by totaling the projected tonnage (mining plan) over specified profiles and dividing this total by the total number of hours required to haul all material.

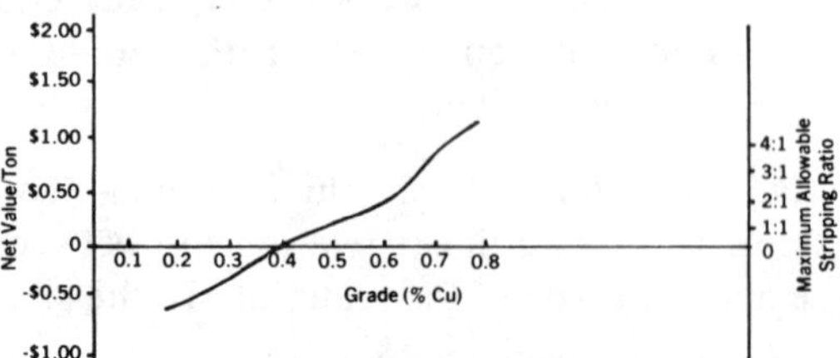

Fig. 1. Net value/ton vs. grade percent copper and maximum allowable stripping ratio vs. grade percent copper.

The average cost per operating hour is determined for each type of haulage equipment. This projection includes the cost of haul roads, labor (fringe benefits included), maintenance and repairs, fuel, supervisors, and depreciation. The average cost per operating hour is determined by multiplying the operating cost per hour for each piece of equipment (or different type) times the hours it will haul. This is repeated for each piece of equipment and then the product of cost per hour times the hours are totaled and divided by the total hours to arrive at the projected average cost per operating hour.

The projected average mine haulage cost per ton is computed by dividing the projected average operating cost per hour by the projected average tons hauled per operating hour.

The foregoing computation is greatly simplified if there is only one type of equipment being used, and there are no particular differences in the haulage practices.

6) General plant costs may be broken into: mine general and maintenance general. These costs and their allocations against ore and waste tonnages depend on local accounting practices.

7) Ore reloading costs should be included, if appropriate.

8) Ore haulage cost typically varies by the distance between the mine and concentrator, as well as the haulage method used.

9) Concentrating costs may vary because of geologic and metallurgical characteristics of the ore.

10) Items 1 through 10 are calculated on a per ton ore or waste basis.

11) Concentrate delivery costs vary between operations because of haulage distances involved between concentrator and smelter. This cost is stated in terms of cost per ton of concentrate.

12) Smelting cost is stated per ton of concentrate. Environmental costs may be considered as pertaining to acid production.

13) Freight costs of smelter product to refinery.

14) Refinery costs.

15) Selling and delivery costs.

These cost definitions and calculations will vary according to the resource being considered and its particular processing, fabricating, and marketing circumstances and economics.

Computation of the Projected Cutoff Grade—Variable Mine Haulage Costs

In the example just shown, the mine haulage was treated as though it were the same for estimating the cutoff grade for all portions of the deposit regardless of its location and the potential haul, or the haulage equipment. The haulage cost which is a portion of the mining cost is dependent on the size, type, and condition of the equipment to be used and the haulage profiles involved. Specifically, we are interested in the length and lift of the profile and the type of equipment to be used, as well as variation in the operating practices.

In this next example, we will compute the projected profitability of a specific portion of the deposit. We also will calculate the anticipated mine haulage cost for this material. The projected total copper content of the material is 0.55%, the mill/flotation recovery 80%, and the copper in the concentrate 20%.

The material in question has the same characteristics of material which currently is being mined; the same drilling and loading equipment will be used; therefore, it is anticipated that the drilling and blasting as well as the loading costs will remain constant. However, the length of the haul will be increased from 10,000 ft † one way (haulage profile A) to 11,400 ft one way (haulage profile B), and the lift will increase from 180 ft (profile A) to 280 ft (profile B). The elevation of the loader for profile A is 1500 ft and 1400 ft for profile B.

A 150-ton ‡ truck will haul a projected 296 stph over haulage profile B, 11,400 ft one way and 280 ft of lift. This assumes a decrease of 45 stph compared to profile A. The cost of operating the 150-ton truck is estimated to be $42 per operating hour. Therefore, the projected cost per ton is $42/341 = $0.12 for profile A, and $42/296 = $0.14 for profile B. The projected rate of change in the cost/ton is $0.02/100 ft of increased lift or $0.0002/ft of increased lift.

Thus, if our current mining cost/ton including drilling and blasting, loading, supervision, roads and dumps, mine haulage (labor, maintenance and repair, fuel and tires), and depreciation is $0.42/ton, we can establish the following equation, for calculating the mining cost:

Estimated Mining Cost/Ton of Material
= $0.42 + $0.0002/ft below 1500-ft elevation.

Pounds of Salable Copper:

Pounds of copper recovered (milling) per ton of ore, 8.8

The ratio of concentration (2000 × 0.2)/8.8 = 45.45

Smelting loss per ton of ore @ 10 lb per ton of concentrate, 0.22 lb

Pounds of blister copper per ton ore, 8.58

Refining loss @ 5 lb per ton of blister copper, 0.02 lb

Net copper per ton of ore, 8.56 lb

Computation of Net Value per Ton of Ore:

	$ per ton of ore
Production Cost, Excluding Stripping—	
Mining (elev. 1400 ft) $0.42 + $0.0002 × (1500 − 1400)	0.44
Ore haulage	0.25
Milling	1.60
General	0.35
Amortization & depreciation	0.56
Subtotal	3.20
Treatment Costs—	
Freight @ $1.40/ton concentrate	0.03
Smelting @ $50.00/ton concentrate	1.10
Freight @ $50.00/ton blister copper	0.21
Refining @ $130.00/ton copper	0.56
Selling & delivery @ $0.01/lb copper	0.09
General plant @ $0.07/lb copper	0.60
Total treatment cost	2.59
Total Production Cost—	
Less byproduct credits	0.65
Total cost	5.14
Value of Salable Copper @ $0.65/lb—	5.56
Net Value—	
Value of salable copper	5.56
Less total cost	5.14
Net value per ton ore	0.42

This same procedure can be followed to develop an equation that considers varied haulage cycle times resulting from increased haul distances because of added lift, longer hauls to dumps, lateral movement in the pit on the same level, etc.

Calculation of Projected Cutoff Grade—Considering Dump Leaching as an Alternative for Recovery of Copper

In this example, a ton of material will be evaluated as to whether it is more profitable to recover the copper by conventional milling/flotation or leaching.

Leaching parameters including costs and metallurgical parameters are added to those presented in the previous example for conventional milling/flotation.

In this example, experience has demonstrated that the mineralization being evaluated is amenable to dump leaching. The leaching characteristics of the material considered are as follows:

† One foot (ft) × 0.3048 = meter (m).
‡ One short ton (st) × 0.907 = metric ton (t).

Mineralization type	Projected % Cu recovered @ 0.4% total Cu	Fixed tails, % Cu
1	50	0.20

Examples of how these parameters are used follows:

1) If we have one ton of material designated as type 1, with a total copper value of 0.45%, we can calculate the copper recovered as follows:

0.50×0.0045 * 2000=4.5 lb of copper are projected to be recovered

2) If we have one ton of material designated as type 1 with a total copper grade projected to be 0.30%, how much copper would we expect to recover? The estimated pounds of recoverable copper per ton of the No. 1 type material=

$$\frac{0.30-0.20}{0.30}0.0030\times2000=0.33\times6=2 \text{ lb Cu}$$

recovered/ton of material leached.

The "fixed tails" is defined as that grade below which no copper recovery can be expected.

The foregoing parameters are typically derived by experience and/or test results, preferably a test dump.

To determine which method, leaching or milling/flotation, would produce the largest projected profit, we will expand example No. 2 to include leaching cost and metallurgical parameters.

We previously assumed that the break-even (cutoff grade) milling/flotation cutoff grade for a particular ton of material is 0.40% Cu. Also, the material is classified as type No. 1 with a total copper value of 0.55%. It also has been established that material with a projected copper grade ≥ 0.65% total copper will only be evaluated as mill ore.

The financial evaluation considering leaching is as follows:

Compute the net value of one ton of copper ore having a projected grade of 0.55% Cu, a projected leach recovery of 50%, and a precipitate grade of 85% Cu.

Computation of Pounds of Salable Copper:

Pounds of copper recovered per ton= 0.0055×2000×0.50=5.5

The ratio of concentration (in terms of leaching)=

$$2000\times\frac{0.85}{5.5}=309.09$$

Smelting loss per ton of leach material @ 30 lb per ton of precipitate, 0.096 lb

Pounds of blister copper per ton leach material, 5.404

Refining loss at 5 lb per ton of blister copper, 0.014 lb

Net copper per ton leach material, 5.390 lb

Computation of Net Value of Salable Copper per Ton of Leach Material:

Production Costs (Excluding Mining Costs)—	$ per ton of leach material
Precipitation cost @ $0.20/lb copper (through to smelting)	1.10
Treatment Costs—	
Smelting (@ $50.00/ton of precipitates)	0.16
Freight (@ $50.00/ton blister copper)	0.14
Refining (@ $130.00/ton copper)	0.35
Selling & delivery (@ $0.01/lb copper)	0.05
General plant (@ $0.07/lb copper)	0.38
Subtotal	1.08
Total Production Cost—	2.18
Value of Salable Copper @ $0.65/lb—	3.50
Net Value—Value of salable copper	3.50
Less total cost	2.18
Net value	1.32

However, if the ton of material were shipped to the concentrator, the total cost of producing 8.56 lb of salable copper (through the mill, smelter, refinery) would be $5.14 (with $0.65 credits deducted).

Also, the net value of this copper, recovered by milling/flotation, at a selling price of $0.65/lb of copper would be $0.42 as compared to $1.32 for leaching.

If this evaluation is used for estimating ore reserves, management must decide whether or not it would mine a portion of the deposit that is submill grade, solely for dump or vat leaching. That is, the material being removed and placed in the dumps to be leached does not involve any mill/flotation material, nor will its removal uncover, strip, any significant amount of milling material.

The foregoing example could be applied to other situations where there is potentially more than one process available for recovering a particular resource.

Another point of debate on this subject is that we might expect to recover an average 50% of the copper from a ton of type No. 1 material, but it is estimated that it will take five years. This is in comparison to milling the material and recovery of the copper by

Table 1. Copper Recovered by Leaching, Five-Year Period

Year	Estimated % recovered annually	Estimated lb.* recovered @ 0.55% Cu	Values of recovered copper discounted @ 8% annually, $
1	20	2.20 × 0.936	2.20 × 0.936 × 0.65 = 1.33
2	15	1.65 × 0.857	1.65 × 0.857 × 0.65 = 0.92
3	7	0.77 × 0.794	0.77 × 0.794 × 0.65 = 0.40
4	5	0.55 × 0.735	0.55 × 0.735 × 0.65 = 0.26
5	3	0.33 × 0.681	0.33 × 0.681 × 0.65 = 0.14
Total	50	5.50	

* Metric equivalent: 1 lb × 0.453 592 4 = kg.

flotation, all of which takes place in a matter of weeks or months.

Table 1 shows the percent of the total copper projected to be recovered by leaching each of the five years.

Thus, if we were to discount the value of the copper recovered by the rate of 8%, we would reduce the value of the copper recovered from $3.54 to $3.05.

In actual day-to-day operation, implementation of the foregoing procedure to establish which material will go to the mill or the dump requires that someone with knowledge of leach types and geology be available to evaluate the changing situation in the mine. In a highly variable deposit where the grades and geology of mining faces change rapidly and sometimes unpredictably, this can be difficult; also, it makes it difficult to meet production goals. However, when mining plans are being made, the procedure should be considered for broader areas, for example, a shovel cut.

As to whether or not the value of copper recovered by leaching should be discounted or not is a matter for mine management. Also, the byproducts associated with many copper deposits, such as molybdenum, gold, and silver, will not be recovered if the material is leached.

The concepts described in this example might be applied to situations such as a choice of milling or leaching uranium or gold ores in addition to copper.

In conclusion, all of the foregoing calculations can be performed manually. However, if there are a large number of blocks, shovel cuts, etc., involved, a computer can be of considerable assistance. It is essential to perform some of the calculations manually to verify the computer's efforts.

By performing the evaluation of each block with the computer, we are in a position to ascertain whether projected tonnages, grades, and metallurgical characteristics of a small portion of the deposit warrant being included in the ore reserves or are the most favorable economically and operationally, considering the company's business plans. Thus, averages don't have to be used, simply because we have some assistance with the calculations.

Marginal Analysis

This is a procedure for estimating the cutoff grade of a deposit and it is especially useful when evaluating a new deposit and determining plant capacity. The technique is also applicable to estimating the cutoff grade for an operating mine. Vickers (1961) presents an excellent description of the technique; thus in the example presented here we will only repeat definition of the key variables and the basic steps.

Definition for Marginal Analysis Example (Vickers, 1961)

Average Revenue per Pound of Metal: Average revenue is the total price received for a product divided by the numbers of pounds sold.

Average Total Cost per Pound of Metal: Average total cost is the total operating cost, i.e., variable cost plus fixed cost, incurred in producing a product divided by the total number of pounds produced.

Variable Cost: Variable cost is that total cost incurred in producing a product which tends to vary with the level of output, divided by the total units produced.

Fixed Cost: Fixed costs are total costs for that outlay of factors which do not vary with the level of output, remaining the same regardless of the level of recovery.

Marginal Revenue: Marginal revenue is the additional revenue obtained when one additional unit is sold. If the total output increases by one unit, the resulting change in total revenue is marginal revenue.

Marginal Cost: Marginal cost is the total additional cost that is incurred by the production of one additional unit.

Level of Recovery: Level of recovery, as used in this discussion, is the total number of pounds of metal

Table 2. Grade Distribution

Grade interval % equiv. Pb	Grade midpoint % equiv. Pb	Frequency distribution	Tonnage distribution × 10^3	Cumulative tonnage × 10^3
16–15	15.5	0.002	20	20
7– 6	6.5	0.065	733	753
6– 5	5.5	0.467	5243	5,996
5– 4	4.5	0.158	1773	7,769
4– 3	3.5	0.137	1533	9,302
3– 2	2.5	0.171	1920	11,222

actually extracted from a given body of ore; thus, it is a function of the grade.

Rate of Recovery: Rate of recovery is the annual production rate of the mining operation expressed in tons.

In this example, the business objective is to determine the lowest grade material that can be extracted and processed which maximizes the net present value of the profits available from mining the hypothetical deposit. The alternative objective of maximizing the total profit could be accomplished easily by not proceeding to the discounting step as required for estimation of the net present value.

The assumptions which are made for this computation are as follows:

The marginal and average revenue per pound were the same and constant at 2000 stpd, regardless of the recovery. The average cost per pound of product will vary with haulage distance and mining and development costs will vary, depending on continuity of ore. The continuity becomes more of an adverse factor for the higher grade intervals.

The hypothetical deposit is a lead-zinc ore body to be mined by underground methods—trackless room-and-pillar. The first step will be to complete a tonnage-grade distribution for the deposit. This can be done by determining the range of metal values possible and dividing the range into reasonable increments. For example, in this case, where the equivalent lead grade ranges from 0.1-16.00%, we have subdivided the range into 1.00% increments. Also, there was no material with a projected grade between 7 and 15%; therefore, nothing will be shown in the following computations for the 7-15% interval.

The zinc values were included by equivalencing zinc to lead. Then the number of tons of material were tabulated according to the projected grade of the sample volume (Table 2). The sample volume can be blocks, regular or irregular-shaped, polygons, etc. One additional suggestion is that the tonnages for each increment coincide with a mining plan, even though it may be just a "first cut" plan. Thus, there is some consideration given to mining sequence and basic operational practices.

The second step is to compile a table (No. 3) of the tonnages and recoverable pounds of lead equivalents for each grade increment.

Table 3. Metal Distribution

Grade interval % equiv. Pb.	Grade midpoint % equiv. Pb.	Cumulative tonnage† × 10^3	Tonnage† distribution × 10^3	Metal* distribution × 10^3	Cumulative* lb.† × 10^3	Average* equiv. grade % Pb
16–15	15.5	20	20	5,744	5,744	14.36
7– 6	6.5	753	733	92,984	98,728	6.56
6– 5	5.5	5,996	5,243	537,885	636,613	5.30
5– 4	4.5	7,769	1,773	145,578	782,191	5.04
4– 3	3.5	9,302	1,533	96,242	878,433	4.72
3– 2	2.5	11,222	1,920	94,019	972,452	4.33

* Considering recovery and dilution.
† Metric equivalents: 1 st × 0.907 184 7 = t; 1 lb × 0.453 592 4 = kg.

Table 4. Operating Analysis

Average grade	Tons§ × 10³	Equiv.* lb.§ Pb. × 10³	Revenue × 10³, $	Total cost × 10³, $	Net before tax × 10³, $	Federal income† tax @ 48% × 10³, $	Net after tax × 10³, $	Years‡ life	Total cost × 10³, $	Present value × 10³, 10% Disc., $
14.36	20	5,744	862	180	681	327	354	0.04	507	354
6.56	753	98,728	14,809	6,779	8,030	3,854	4,176	1.51	10,634	3,738
5.30	5,996	636,613	95,492	53,965	41,528	19,933	21,595	11.99	73,898	12,272
5.04	7,769	782,191	117,329	69,923	47,406	22,755	24,651	15.54	92,678	12,262
4.72	9,302	878,433	131,765	83,722	48,043	23,061	24,982	18.61	106,783	
4.33	11,222	972,452	145,868	101,000	44,868	21,537	23,331	22.45	122,536	

* Consideration has been made for recovery and dilution.
† Cost and percentage depletion allowances were not considered in this example.
‡ Years life calculated for 2000-tpd operations, which will operate an estimated 250 days per year.
§ Metric equivalents: 1 st × 0.907 184 7 = t; 1 lb × 0.453 592 4 = kg.

The third step is to use Table 3 plus the mining costs and projected lead prices to generate Table 4. Profits after taxes increase until an average grade of 4.72% equivalent lead is reached.

The cutoff grade which is directly related to the average grade of 4.72% is determined by developing Table 5. The data in Table 4 is used to generate Table 5, the marginal analysis of the hypothetical deposit. Projected reserves were arranged in the same order as in Table 4, descending grade. The tonnage and average grade of each increment is depicted along with the accumulated average grade and tonnage after each increment. The incremental equivalent pounds of lead are calculated.

The total cost, including operating costs, depreciation, and taxes, are derived from Table 4. The marginal cost of the material in the increment equals the incremental total cost divided by the incremental pounds of equivalent lead. The average cost is calculated by dividing the total accumulated cost to this point by the total accumulated pounds of equivalent lead. A similar procedure would be followed to determine the marginal and average revenues.

However, in this example, marginal and average revenue were held constant.

An analysis of the results indicates that the marginal cost and revenue are equal between the grade midpoints 4.5 and 2.5% equivalent lead. Total profits have also reached a maximum. Observing Fig. 2, we can see that the marginal cost and marginal revenue are equal for an average equivalent lead grade of 4.8%. The cutoff falls in the range of 4.5 to 3.5% equivalent lead. A more definitive answer concerning the cutoff grade would require that the tonnage and grade distribution be analyzed in greater detail. Since it was our objective to determine the cutoff grade at the maximum net present value, we have added an additional column to Table 4. The net present value was calculated at 10% discount rate, and plotted on Fig. 2. The net present value is a maximum for an average grade (reserves) of

Table 5. Marginal Analysis

Avg. equiv. grade	Reserves, tons* ×10³	Equiv. grade interval	Equiv. Pb × 10³	Total Revenue × 10³, $	Total cost × 10³, $	Incremental cost × 10³, $	Incremental revenue × 10³, $	Avg. cost/ lb*	Marginal cost/ lb*	Average and marginal revenue per lb*
14.36	20	16–15	5,744	862	507			0.9¢		15¢
	733	7– 6	92,984			10,127	13,948		10.9¢	
6.56	753		98,728	14,809	10,634			10.8¢		15¢
	5,243	6– 5	537,885			63,264	80,683		11.8¢	
5.30	5,996		636,613	95,492	73,898			11.6¢		15¢
	1,773	4– 5	145,578			18,780	21,836		12.9¢	
5.04	7,769		782,191	117,329	92,678			11.8¢		15¢
	1,533	3– 4	96,242			14,105	14,436		14.7¢	
4.72	9,302		878,433	131,765	106,783			12.2¢		15¢
	1,920	2– 3	94,019			15,754	14,103		16.8¢	
4.33	11,222		972,452	145,868	122,536					

* Metric equivalents: 1 st × 0.907 184 7 = t; 1 lb × 0.453 592 4 = kg.

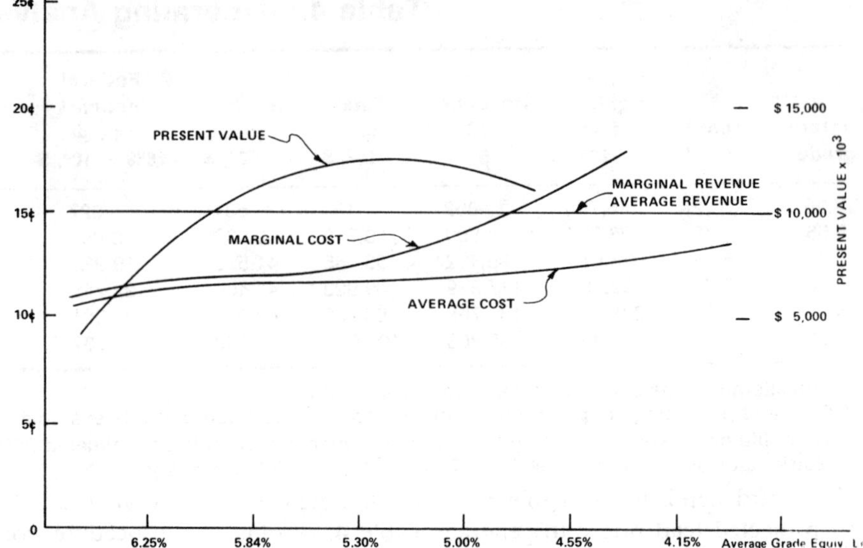

Fig. 2. Marginal analysis.

approximately 5.2% equivalent lead. The cutoff grade falls in the range of 4.5 to 5.5% equivalent lead. A more detailed analysis of the tonnage vs. grade function in this range would permit us to more precisely estimate the cutoff grade which would maximize the net present value.

The foregoing analysis could be repeated for various rates of recovery (mining rates) and mining sequences. Subsequently, net present values could be compiled for each rate and level of recovery to determine the mining rate and cutoff grade which would provide the maximum present value. Also, this procedure could be followed for each element (a couple of years' production, for example) of a mining sequence.

Variable Cutoff Grade Analysis

Lillico (1973) presents a marginal analysis technique for developing a variable cutoff grade strategy which would generate the highest net present value.

Lillico states, "when mining a variable grade ore body, it is common practice to apply a fixed cutoff grade. This can result in simplified mill control and provide an even flow of profit. In general, this fixed cutoff grade policy does not consider the earning power of money."

"The correct variable (cutoff grade) policy will produce more return than a fixed cutoff—the amount of improvement depending on the ore body under consideration."

He also discusses the development of this variable cutoff grade strategy considering variable capacity and mining sequence. In general, this approach stresses mining the material with the highest net value per ton first. Many of the techniques employed in the marginal analysis are used by Lillico in his example, specific consideration being given to the time value of money.

Comparison of Costs Associated with Alternative Haulage Configuration

In many situations, we are required to select the most potentially profitable method of operation and/or equipment, but because the circumstances or equipment are different, we must first equate them to a common datum. Two examples of evaluation techniques for comparing equipment and/or operating practices were prepared by an engineering department of an operating mine. They are as follows:

Selection of Most Cost Effective Haulage Unit

Develop a unit of measure which will adequately compare the cost effectiveness of different truck haulage operating configurations. The unit of measure should incorporate variables such as road grades, haulage distance, weather, road conditions, etc., and could be used instead of the more detailed cycle time analysis.

Solution:

Haulage cost per unit of work

Work = Tonnage hauled x [feet lift + (horizontal feet hauled x % rolling resistance)]

One unit of work is one foot-ton (meter-ton).

Example:

	Operation A	Operation B
Haulage profile		
Avg. horizontal distance	10,000 ft	6,000 ft

Avg. lift	250 ft	150 ft
Rolling resistance	5%	4%
Tonnage hauled	1,000,000	1,000,000
Haulage cost	$200,000	$150,000
Truck type	A	B
Cost/ton	$0.20	$0.15
Cost/foot-ton	$0.00027	$0.00038

Truck haulage for operation A is seen to be more efficient for the work being performed although the cost per ton would not indicate this conclusion. It should be noted that all of the computations shown here do not supplant good input and experience.

Determination of Most Cost Effective Haulage Profile

Find the point (in distance) where advance of a lower waste dump has greater haulage cost than starting a new dump at a higher elevation. Both dumps have a common starting point and a road exists to the higher dump from the common starting point. Use truck speeds in the analysis and assume that one truck operating minute costs the same on a level haul as on a grade.

Solution:

Definition of terms:

L = incremental height between higher and lower dump (feet)

SLG = speed loaded on upgrade (mph)§

SLO = speed loaded on level road (mph)

SEG = speed empty return on downgrade (mph)

SEO = speed empty return on level road (mph)

G = grade of road to higher dump

$F = \dfrac{5280 \times 60 \text{ (fph)}}{3600 \text{ sec/hr}} = 88$ fps @ 60 mph

SDG = slope distance to high dump (feet)

TTG = round trip travel time on grade (min)

TTO = round trip travel time on level road (min)

HDO = economic distance from common starting point for advance of lower dump (TTO=TTG)

Equations:

$$SDG = \sqrt{L^2 + \left(\frac{L}{G}\right)^2}$$

$$TTG = \frac{SDG \times \left(\frac{1}{SLG} + \frac{1}{SEG}\right)}{F}$$

$$HDO = \frac{TTG \times F}{\left(\frac{1}{SLO} + \frac{1}{SEO}\right)}$$

§ One mile per hour (mph) × 2.237 = meter per second (m/s).

$$HDO = \frac{\sqrt{\left(L^2 + \left(\frac{L}{G}\right)^2\right) \times \left(\frac{1}{SLG} + \frac{1}{SEG}\right)}}{\left(\frac{1}{SLO} + \frac{1}{SEO}\right)}$$

Example:

Given $L=80$, $SLG=9$, $SLO=28$, $SEG=23$, $SEO=28$, and $G=7\%$

$$HDO = \frac{\sqrt{\left(80^2 + \left(\frac{80}{0.07}\right)^2\right) \times \left(\frac{1}{9} + \frac{1}{23}\right)}}{\left(\frac{1}{28} + \frac{1}{28}\right)}$$

$HDO = 2479$ ft.

Each circumstance of evaluating the most cost effective equipment and operating practice will require that the engineer utilize actual cost and production statistics. If projections are required, caution must be exercised to make certain that the basic data and factors being used fit the situation being proposed.

Also, the more complex the situation becomes, the more advantageous it is to consider the use of a posed operation and any alternatives.

References

Collins, T., 1975, "Mining's Renaissance on Wall Street," *Mining Engineering,* September.

Davis, G. O., 1977, "How to Make the Correct Economic Decision on Spare Equipment," *Chemical Engineering,* Nov. 21.

Dran, J. J., Jr., and McCarl, H. N., 1975, "An Examination of Interest Rates and Their Effect on Valuation of Mineral Deposits," SME Preprint 75H342, SME Fall Meeting, Salt Lake City, September.

Epstein, B. S., 1975, "Financing the Acquisition of a Going Coal Mine," *Mining Engineering,* September.

Erdahl, L. O., 1975, "Economic Aspects of Joint Ventures," *Mining Engineering,* September.

Frohling, E. S., and McGeorge, R. M., 1975 "How Stepwise Financing Can Turn Your Prospect into an Operating Mine," *Mining Engineering,* September.

Gentry, D.W., 1975, "The Buy-or-Lease Decision for Capital Equipment," *Mining Engineering,* September.

Gentry, D.W., 1971, "Two Decision Tools for Mining Investment—And How to Make the Most of Them," *Mining Engineering,* November.

Guccione, E., 1970, "The Financing of Mines in the 1970's," *Engineering and Mining Journal,* December.

Holland, F. A., and Watson, F. A., 1977, "Putting Inflation into Profitability Studies," *Chemical Engineering,* Feb. 14.

Joralemon, P., 1975, "The Ore Finders," *Mining Engineering,* December.

Lewis, C. K., 1969, "An 'Economic Life' for Property Evaluation," *Engineering and Mining Journal,* October.

Lillico, M., 1973, "How to Maximize Return on Capital When Planning Open Pit Mines," *World Mining,* June.

Mackenzie, B. W., 1969, "Economic Evaluation Techniques Applied to the Mine Development Decision," SME Preprint 69K326, SME Fall Meeting, Salt Lake City, September.

O'Neil, T. J., 1974, "The Minerals Depletion Allowance: Its Importance in Nonferrous Metal Mining," *Mining Engineering*, Pt. 1, October, Pt. 2, November.

Parks, R. D., 1949, *Examination and Valuation of Mineral Property,* Addison-Wesley Press, Cambridge, MA.

Pillar, C. L., and Lipkewich, M. P., 1969, "Economic Requirements for Placing Marginal Ore Bodies into Production," SME Preprint 69B309, SME Fall Meeting, Salt Lake City, September.

Sani, E., 1977, "The Role of Weighted Average Cost of Capital in Evaluating a Mining Venture," *Mining Engineering,* May.

Stermole, F. J., 1974, *Economic Evaluation and Investment Decision Methods,* Investment Evaluations Corp., Golden, CO.

Verner, W. J., and Shurtz, R. F., 1966, "For Mine Evaluation—A Fresh Model," *Mining Engineering,* November.

Vickers, E. L., 1961, "Marginal Analysis—Its Application in Determining Cutoff Grade," *Mining Engineering,* June.

Wild, N. H., 1977, "Program for Discounted Cash-Flow Return on Investment," *Chemical Engineering,* May.

8 Cost Records of Open Pit Mining

Robert F. Winkle
Pincock, Allen & Holt, Inc.

Robert F. Winkle is the vice-president for mining for Pincock, Allen and Holt, Tucson, AZ. He has managed an iron mine in the Dominican Republic and was involved in premining construction for the Aluminum Co. of America. He worked with the Atomic Energy Commission; he was a supervisor at the Chuquicamata copper mine in Chile; and was mine superintendent for the Ray Mines Div. of Kennecott Copper Corp.

He received his B.S. degree from the Missouri School of Mines and has lately earned an M.S. from the University of Arizona.

A detailed breakdown of mining costs, available to management on monthly and year-to-date bases, is mandatory for a controlled and efficient mining operation. A simple lump sum reporting of costs may indicate a profit or loss differential, but it does not allow an effective search for areas of possible cost reduction. Moreover, functional reporting is necessary on a timely basis so as to note trends of increasing costs in specific areas, which in turn allow possible preventive action by management. Definitive cost information helps to eliminate the tendency to overlook areas where cost improvements can be made. A good cost control system points out salient operational weaknesses by: (1) allowing comparison of functional costs with similar costs of other mines and (2) pinpointing sudden increases over unit empirical costs within the organization.

In general, the cost accounting structure is relatively uniform within large open pit mines, but surprisingly some systems will overlook the need for detailed breakout in comparatively high cost areas. In some other systems, unrelated costs will be lumped together, preventing careful analysis of cost accrual.

Format

Most mine cost reporting formats incorporate the following basic operating functions or cost centers: (1) drilling, (2) blasting, (3) loading, (4) truck haulage, and (5) roads and dumps.

In addition, most organizations include a breakout of air drilling and blasting costs, general mine costs, and general maintenance costs. In many cases, drainage costs are included in the basic functions. Utility services are normally included to indicate water, pumping, and electrical costs, which cannot be allocated to a specific function. Service equipment allocation charges are also properly included in the cost sheets.

To allow management a comprehensive summary of costs, related information should be included. Performance by the operating function plus a salary and wage summary and force report should be incorporated in the cost report.

A format (Exhibit 1, Tables 1-15) is appended to illustrate typical cost structure reporting in large open pit mines. It is intended only to serve as a guide and should be amended to meet local conditions and needs. It should be understood that "Production Cost and Performance Report" in the Appendix is a summary and that more detailed information from which this type of report is generated should be maintained in readily accessible form so explanation of the statistics shown can be further examined and analyzed as required.

These additional cost and performance reports should be prepared for the same time intervals. These are support documents and will furnish a more detailed exposure of cost accumulation.

Exhibit 2 in the Appendix is a sample of a support document designed to give more detailed information on truck haulage. This illustration is basic and can be expanded to include all fleet sizes and configuration. For example, it may be desirable to keep separate information on one truck of a similar fleet if it is powered by a different engine than the rest of the fleet.

Similar support reporting should be done for other major functional cost centers, particularly loading (shovels) and drilling (blasthole drills).

Areas of interest for shovels might consist of the following: (1) direct operating costs which include labor, labor (moving equipment), lubricants, electrical power, service equipment, depreciation, and other operating expenses; and (2) maintenance and repairs consisting of the following areas: hoist, propel, swing, boom, sticks, bucket, track, sprockets, tumblers, undercarriage and frame, electrical, cables, teeth, adapters, and accidents.

This list of 15 areas in maintenance and repair (M&R) may have to be combined if local data processing cannot handle 15 different entries, but if the information is needed, computer limitations should be secondary.

Rotary drill costs should also be reported in detail with a supplementary report in which the following M&R area costs would be delineated: mast, hoist, rotary, undercarriage, track, sprockets, pulleys, compressors, electrical system, diesel engine (if applicable), hydraulic system, drill steel, bits, and accidents.

The direct operating costs for the drills would be similar to shovel costs with diesel fuel costs added, if applicable.

A part of the shovel and drill supplementary report would also include production statistics similar to Table 17 following Table 16, "Truck Haulage Supplemental Statement" (Exhibit 2) or Tables 4 and 5 in "Production Cost and Performance Report" (Exhibit 1). The repetition is simply to eliminate the need for cross reference when examining the supplemental sheets.

Comparable cost breakdowns can be developed for all types of mining equipment including dozers, rubber-tired loaders, air drill equipment, cranes, and service trucks.

In many mines, these types of equipment, except large rubber-tired loaders used as major production units, are reported as service equipment, and their hourly costs are charged to whichever function(s) they are assigned. Normally, the charge of any specific type of equipment is based upon a running yearly average with the latest month's cost being included in the average cost while the 13th month cost is dropped out. Exhibit 3 shows a possible monthly service equipment charge sheet of the type of equipment normally used in the support of the major mining equipment. It will be noted that depreciation is not included in the hourly costs as service equipment is charged against a major function when used within that function, and functional costs are carried with depreciation as a special charge. If depreciation was charged as a part of the service equipment hourly rate, the depreciation could not be readily isolated. In the case of service equipment, depreciation is usually carried as a plant cost.

Attached to Exhibit 3 showing "Service Equipment Costs" are Tables 19 and 20, which are suggested methods of showing maintenance and repair costs of each individual piece of service equipment. Only a few of the various types of service equipment have been included, but in practice, this type of reporting can be expanded to include all types of ancillary units. This portion of the reporting provides management with an appraisal of the condition of the various individual service equipment units so as to better judge the trade-in or disposal time for the machines.

Cost Accumulation

The need for careful presentation to management of functional costs, which makes possible accurate analytical examination of operational procedures, is dependent upon accurate reporting of accumulated expenditures. Every expense whether for materials, labor, or services must be charged to the proper function, if the final unit cost is to be truly representative.

Forms for reporting must be designed to provide accurate information, and the timely completion of these forms by concerned personnel is vital, if the accounting department is to properly accumulate and allocate charges to the areas where the expenses have occurred.

No sample forms have been included in this chapter, but the following list indicates many of the areas where expense is generated, and record of these expenditures must be made.

General

Time Cards: All personnel must show proper allocation of their working time so as to provide the accounting department with a breakdown of labor time, which is in turn converted into dollar charges against the various mining activities. In the case of supervision, frequently a supervisor's time is arbitrarily split against two or more functions.

Personnel Actions: Forms showing vacation schedules, notice of disciplinary action, permission to be off work, absence follow-ups, and notice of promotions, demotions, transfers, etc., signal the accounting department of changes that can cause an adjustment in cost allocation.

Personal Chargeable Items: These include safety glasses replacement, clothing or lunch boxes, etc., replacement. (These items may be insignificant in some organizations but can prove costly in large operations).

Force Control Reporting: This is necessary to arrive at efficiency determinations [metric tons (short tons) per man shift, etc.].

Charge Cards: These are needed for all warehouse withdrawals so as to properly allocate materials used.

Operations

Shovel, Haulage Truck, Drill, etc.: Use (time) records. These records are vital for functional efficiency evaluations.

Equipment Performance Records for Shovels, Haulage, Trucks, Blasthole Drills, etc.: These would be reported in appropriate units [metric tons, truckloads, meter per shift (short tons, truckloads, feet per shift, etc.)]. The shovel and truck units should be reported in direct ore or waste tonnage or in units which are readily convertible to tonnage through use of empirical factors.

Availability and Utilization Records: These should be done for all equipment. This is not a charge but is very necessary for unit cost analysis.

Daily Account of Blasting Supplies: This item is covered in "Charge Cards" of the previous section, but is mentioned a second time because of federal government regulations relating to use of explosives and blasting agents.

Maintenance

Work Order Sheets: Because the work order system is used by all departments to request and report expenditures, these might be more properly listed under the General heading, but they are listed under maintenance because these sheets are mainly used in that department. The work order system is the heart of maintenance reporting which sums up the parts, labor, and service (if any) on every job performed by the maintenance department. When used to report equipment repair, it should note the equipment type and company number. Every work order should contain a charge number to allow proper allocation of expenditures. The explanation given here of the work order function is basic and is normally much more sophisticated with work order numbers being broken down into various series or groups to assist the accounting department in more rapid allocation of charges. No attempt has been made to explain a complete work order system but only to stress the need for a comprehensive, well-organized cost collection procedure.

Equipment Working Hours: Records of hours worked should be kept on individual mining equipment. This is particularly necessary for service and preventive maintenance schedules and is documentary evidence where warranty considerations are involved. Hours on components should be recorded as well as on entire units when warranty is to be protected.

Equipment Repair: This includes repair of shovels, trucks, drills, loaders, graders, dozers, etc. Many other reporting forms are required besides those covered by the work order system, and any company must analyze its needs to obtain the desired information.

Records of Grease Pit Lubrication: As an example of the numerous forms that are required for the various maintenance operations, the following list indicates the volume of paper work involved in just this one operation:

Lube 1, Haulage Truck Service Schedule—Service schedule for haulage trucks made out by grease pit foreman; filed at grease pit.

Lube 2, 8, 24, and 48-Hr Service Form—Used by grease pit foreman to calculate when haulage trucks need 8, 24, and 48-hr service. Information is taken from pit production's Availability and Utilization Record form; filed at grease pit.

Lube 3, Record of Heavy Truck Service—Record of 8, 24, and 48-hr service performed on each haulage truck; filed at grease pit.

Lube 4, Record of Equipment Services—Record of 8, 24, and 48-hr service performed on each loader, dozer, and rubber-tired piece of equipment; filed at grease pit.

Lube 5, Heavy Truck Service Consumption—Record of quantities of oil and fuel put in haulage trucks; three copies: accounting department, truck shop, and grease pit.

Lube 6, Pit Equipment Service Consumption—Record of amounts of oil and fuel put in dozers, loaders, and drills; three copies: accounting department, truck shop, and grease pit.

Lube 7, Haulage Truck Lubrication, Service & Component Change—Sent to accounting for computer printout of hours on oil, filters, engine, transmission, and differential.

Lube 8, Field Equipment Oil and Filter Change Record—Monthly record of oil and filter changes made by pit grease trucks on the drills, dozers, loaders, and fork lifts; filed at grease pit.

Lube 9, Equipment Lubrication, Service, & Component Change—Sent to accounting for computer printout of hours on oil, filters, engines, transmissions, and differentials on small equipment.

Lube 10, Haulage Truck Oil Analysis—Sample of engine oil in haulage trucks taken every 100 operating hr; sample analyzed by spectograph and results sent to truck shop.

Lube 11, Oil Change Record—Monthly record of oil and filter changes made in gas shop on small trucks; filed at grease pit.

All of the information to be recorded on these forms obviously is not directly used for cost accumulation, but a low cost operation is dependent upon efficient record keeping of all action. It will be noted that reference is made to computer printouts as today even comparatively small mining companies have or have access to data processing equipment.

Tire Shop Forms: Records of the various tire shop activities should be equally as extensive as those of the grease pit because tire costs are major items in shovel truck operations and tire care both in maintenance and in operations is mandatory.

Application

The best possible cost accumulation system will fall short of maximum accomplishment without the proper integration of procedures.

The various work or cost centers must be clearly defined and the appropriate supervisors should be held responsible for all costs within their area of responsibility. Operational and maintenance supervisors working within a functional cost area must work together if high maintenance costs are to be avoided. Obviously, maintenance costs tend to be reduced by careful handling of equipment.

Investigation of causes of high component cost can frequently lead to the discovery and the elimination of a problem. The following examples prove this point:

1) High tire costs indicate poor road, dump, or shovel pit conditions. If haulage surfaces are generally good, the problem area may be traced to one bad curve where rocks are falling in the path of returning trucks, etc.

2) Sudden increase in frame or torque tube costs may well be discovered to be caused by an isolated spot in a haul cycle where unnecessary flexing stress results from poor road conditions.

3) Increased shovel costs might be found to be due

to an overzealous blasting foreman interested primarily in reducing his blasting costs, or frequent hoist cable breakage, or underbelly boom damage may well indicate need for additional training of one or more operators.

4) Suddenly climbing bit costs in blasthole drilling suggest that harder rock is being encountered, and if regular roller bits are being used, it might be prudent to go to bits with carbide inserts (at least in certain areas). It might also suggest that the delivered air pressure to the bit on one or more drills has dropped, and compressor repairs are indicated.

The proper application of cost record information is vital to a truly successful business, and few single elements can improve operating costs as well as detailed cost information made available to the first level of front-line supervision. Even the morale is improved by effectively making the foreman level a more significant part of management. Needless to say, scheduled periodic review of costs between front-line and middle management is essential to obtain optimum results.

Cost awareness intensifies the daily search for better and cheaper methods and promotes improved scheduling. It suggests taking advantage of routine situations to minimize unproductive time of equipment and labor.

All mining companies should from time to time review their cost record system to insure that this major management tool is providing optimal informative data.

Appendix

Exhibit 1. Any Mine USA, Production Cost and Performance Report (Month and Year To Date), November

Table of Contents

Table 1. Pit Mining Summary

	Month & Year (i.e., November 1977)			Year to date (i.e., Jan/Nov. 1977)		
	Budget	Actual	Variance	Budget	Actual	Variance
Pit Mining						
Drilling						
Blasting						
Loading						
Truck haulage						
Haul roads	Enter total money either spent or budgeted and variance					
Waste dumps						
Drainage						
General mine						
General maintenance						
Total mine operation						

Table 1. Pit Mining Summary (Continued)

	Month & Year (i.e., November 1977)			Year to date (i.e., Jan./Nov. 1977)		
	Budget	Actual	Variance	Budget	Actual	Variance
Cost transferred to:						
Concentrate inventory						
Ore mining						
Waste stripping						
Depreciation						
Total cost transferred	(Should balance with total mine operation amounts)					
Cost/ton* ore mined						
Drilling						
Blasting						
Loading						
Truck haulage						
Haul roads						
Waste dumps						
Drainage						
General mine						
General maintenance						
Total cost/ton ore mined						
PRODUCTION STATISTICS						
Days operated						
Ore						
Waste						
Material moved (tons)*						
Ore						
Waste						
Total material mined, tons						
Stripping ratio						
Material moved/day, tons						
Ore mined						
Waste removed						
Total material						
Total material moved/man shift, tons						
Total material moved/man-hour, tons						
Cost/ton all material						
Cost/ton ore mined						

* Metric equivalent: 1 st × 0.907 184 7 = t.

Table 2a. Stripping and Mining Production Cost Summary, Month and Prior Month

	All material		Ore		Waste	
	Total cost	Cost/ton*	Total cost	Cost/ton	Total cost	Cost/ton
Applicable tons						
Production cost						
Drilling						
† Blasting (primary plus secondary drilling, blasting)						
Loading						
Truck haulage						
Haul roads						
Waste dumps						
Drainage						
General mine						
General maintenance						
Total mine operations						
Less: depreciation						
Mining cost less depreciation						
Net cost, stripping, mining						
Net cost, prior month						

* Metric equivalent: 1 st × 0.907 184 7 = t.
† Should be segregated if secondary work is of a significant amount.

Table 2b. Stripping and Mining Production Cost Summary, Year to Date (Net Cost Prior Year)

	All material		Ore		Waste	
	Total cost	Cost/ton*	Total cost	Cost/ton	Total cost	Cost/ton
Applicable tons						
Production cost						
Drilling						
† Blasting (primary plus secondary drilling, blasting)						
Loading						
Truck haulage						
Haul roads						
Waste dumps						
Drainage						
General mine						
General maintenance						
Total mine operations						
Less: depreciation						
Mining cost less depreciation						
Net cost, stripping, mining						
Net cost, prior year						

* Metric equivalent: 1 st × 0.907 184 7 = t.
† Should be segregated if secondary work is of a significant amount.

Table 3. Drilling

	Month			Year to date		
	Budget	Actual	Variance	Budget	Actual	Variance
Drilling (rotary)						
Labor						
Labor (moving equipment)						
Other operating supplies						
Drill bits						
Drill steel						
Lubricants						
Compressed air						
Electric power						
Service equipment						
Maintenance, repairs						
Depreciation						
Total						
Allocated to other services						
Total drilling						
Cost/ton* material mined						
Production statistics						
Feet* drilled						
Feet drilled/drill shift						
Cost/foot drilled						
Material mined/foot drilled						

* Metric equivalents: 1 st × 0.907 184 7 = t; 1 ft × 0.304 8 = m.

Table 4. Blasting

	Month			Year to date		
	Budget	Actual	Variance	Budget	Actual	Variance
Primary blasting						
Labor						
Other operating supplies						
Explosives						
Blasting agents						
Tires, tubes						
Fuel						
Service equipment						
Maintenance, repairs						
Depreciation						
Total primary blasting						
Cost/ton* material mined						
Secondary drilling, blasting						
Labor						
Other operating supplies						
Drill bits						
Drill steel						
Explosives						
Blasting agents						
Tires, tubes						
Fuel						
Service equipment						
Maintenance, repairs						
Depreciation						
Total secondary drilling, blasting						
Cost/ton material mined						
Total blasting						
Cost/ton material mined						
Production statistics						
Tons all material mined/lb* powder						

* Metric equivalents: 1 st × 0.907 184 7 = t; 1 lb × 0.453 592 4 = kg.

Table 5. Loading

	Month			Year to date		
	Budget	Actual	Variance	Budget	Actual	Variance
Loading						
Labor						
Labor (moving equipment)						
Other operating supplies						
Lubricants						
Fuel						
Electric power						
Service equipment						
Maintenance, repairs						
Depreciation						
Total						
Allocated to other services						
Total loading						
Cost/operating hour						
Production statistics						
Operating loading hours						
Ore						
Waste						
Other						
Total						
Total						
Hours available, but not used						
Hours available, but not used due to restrictions (road blockage, etc.)						
Hours not available						
Material loaded, ton*						
Ore						
Waste						
Stockpile						
Total						
Tons* per loading hours						
Ore						
Waste						
All material						
Cost/ton material loaded						

Table 5. Loading (Continued)

	Month			Year to date		
	Budget	**Actual**	**Variance**	**Budget**	**Actual**	**Variance**
Availability, %						
Utilization, %						
Used of available, %						

* Metric equivalent = 1 st × 0.907 184 7 = t.

Table 6. Truck Haulage

	Month			Year to date		
	Budget	Actual	Variance	Budget	Actual	Variance
Truck haulage						
Labor						
Other operating supplies						
Lubricants						
Tires, tubes						
Fuel						
Maintenance, repairs						
Equipment rentals						
Depreciation						
Total						
Allocated to other services						
Total truck haulage						
Cost/operating hour						
Cost/ton* material hauled						
Production statistics						
Operating truck hours						
Ore						
Waste						
Other						
Total operating truck hours						
Material hauled, dry tons*						
Waste						
Stockpile						
Other						
Total material hauled, dry tons*						
Tons/operating truck hours						
Ore						
Waste						
All material						
Availability, %						
Utilization, %						

* Metric equivalent: 1 st × 0.907 184 7 = t.

Table 7. Haul Roads

	Month			Year to date		
	Budget	Actual	Variance	Budget	Actual	Variance
Haulage roads						
Labor						
Other operating supplies						
Lubricants						
Tires, tubes						
Fuel						
Service equipment						
Maintenance, repairs						
Depreciation						
Total haulage roads						
Cost/ton* material hauled						
Production statistics						
Material hauled, tons*						

* Metric equivalent: 1 st × 0.907 184 7 = t.

Table 8. Waste Dumps

	Month			Year to date		
	Budget	Actual	Variance	Budget	Actual	Variance
Waste dumps (trucks)						
Operating supplies						
Electric power (dump lighting)						
Service equipment						
Maintenance, repairs						
Depreciation						
Total waste dumps (truck)						
Cost/ton* waste removed						
Production statistics						
Loading shifts (waste)						
Truck shifts (waste)						
Waste removal (total)						
Tons*						
Cubic yards*						
Waste removed/truck shift						
Tons*						
Cubic yards*						

* Metric equivalents: 1 st × 0.907 184 7 = t; 1 cu yd × 0.764 554 9 = m^3.

Table 9. Drainage

	Month			Year to date		
	Budget	Actual	Variance	Budget	Actual	Variance
Labor						
Electric power						
Service equipment						
Maintenance, repair						
Outside services						
Other expense						
Depreciation						
Total						
Total gal* pumped						

* Metric equivalents: 1 gal × 3 785 412 = L; 1 gal × 0.003 785 4 = m^3.

Table 10. General Mine

	Month			Year to date		
	Budget	Actual	Variance	Budget	Actual	Variance
General						
Salaries						
Labor (undistributed) (Vacation pay, holiday pay, holiday premium, overtime premium, sick leave, back wage bonus—any of these if applicable)						
Training						
Severance benefits (if applicable)						
Operating supplies (undistributed)						
Company automobile supplies						
Inventory adjustments						
Office supplies						
Water						
Power transmission lines						
Maintenance, repair (undistributed)						
Undistributed service equipment						
Scrap preparation, disposal						
Scrap credits						
Outside services						
Travel						
Telephone, telegraph						
Property damage						
Equipment rentals						
Others						
Depreciation (buildings, etc.)						
Total General						
Mine engineering and surveying						
Salaries						
Supplies						
Maintenance, repairs						
Outside services						
Other expense						
Depreciation						
Total engineering and surveying						
Development drilling						
Operating supplies						
Service equipment						
Maintenance, repairs						
Outside services						
Other expense						
Depreciation						
Total development drilling						

Table 10. General Mine (Continued)

	Month			Year to date		
	Budget	Actual	Variance	Budget	Actual	Variance
Geological engineering						
Salaries						
Supplies						
Maintenance, repairs						
Other expense						
Depreciation						
Total geological engineering						
Pumping						
Operating supplies						
Electric power						
Maintenance, repairs						
Depreciation						
Total Pumping						
Building, roads, walks, yards						
Supplies						
Electric power						
Water						
Fuel						
Service equipment						
Maintenance, repairs						
Outside services						
Depreciation						
Total building, roads, etc.						
Employee transportation						
Labor						
Supplies						
Lubricants						
Fuel						
Tires, tubes						
Maintenance, repairs						
Depreciation						
Total employee transportation						
Other costs, i.e., deferred expense amortization, assessment work, etc.						
Total General Mine						

Table 11. General Maintenance

	Month			Year to date		
	Budget	Actual	Variance	Budget	Actual	Variance
General						
Salaries						
Labor (undistributed) (vacation pay, holiday pay, holiday premium, overtime premium, sick leave, back wage bonus—any of these if applicable)						
Training						
Severance benefits (if applicable)						
Operating supplies (undistributed)						
Company automobile supplies						
Inventory adjustments						
Office supplies						
Water						
Power transmission lines						
Maintenance, repair (undistributed)						
Undistributed service equipment						
Scrap preparation, disposal						
Scrap credits						
Outside services						
Travel						
Telephone, telegraph						
Property damage						
Equipment rentals						
Others						
Depreciation (buildings, etc.)						
Total General						
Field repairs (shovels, drills)						
Salaries						
Labor						
Supplies						
Maintenance, repairs						
Depreciation						
Total field repair						
Truck shop (trucks, dozers)						
Salaries						
Labor						
Supplies						
Maintenance, repairs						
Depreciation						
Total truck shop						

Table 11. General Maintenance (Continued)

	Month			Year to date		
	Budget	Actual	Variance	Budget	Actual	Variance
Shop support						
Salaries						
Labor						
Supplies						
Maintenance, repairs						
Depreciation						
Total shop support						
Electric shop						
Salaries						
Labor						
Supplies						
Maintenance, repairs						
Depreciation						
Total electric shop						
Total maintenance shops						
Total General Maintenance						

Table 12. Utility Services

Utility	Month			Year to date		
	Budget	Actual	Variance	Budget	Actual	Variance
Water service						
Supplies						
Electric power						
Maintenance, repairs						
Outside service						
Other						
Depreciation						
Total water service						
Cost M/gal*						
M/gal used						
Electrical service						
Power cost						
Transmission lines						
Maintenance, repair						
Other						
Depreciation						
Total electrical service						
Cost/kw-hr*						
kw-hr used						

* Metric equivalents: 1 gal × 3 785 412 = L; 1 gal × 0.003 785 4 = m³; 1 kw-hr × 3 600 = J.

Table 13. Service Equipment

	Month			Year to date		
	Budget	Actual	Variance	Budget	Actual	Variance
Trucks (large)						
Labor						
Operating supplies						
Lubricants						
Tires						
Fuel						
Maintenance, repairs						
Depreciation						
Total truck (large)						
Trucks (small)						
same as large						
Total truck (small)						
Tractors (crawler, large)						
Labor						
Operating supplies						
Lubricants						
Fuel						
Maintenance, repairs						
Depreciation						
Total tractors (crawler, large)						
Tractors (small)						
Same as large						
Total tractors (small)						
Tractors (rubber-tired)						
Labor						
Other operating supplies						
Lubricants						
Tires, tubes						
Fuel						
Maintenance, repairs						
Depreciation						
Total tractors (rubber-tired)						

Note: List any other equipment used as service equipment to any mining function or outside service in same manner as those listed. Among others would be service trucks, front-end loaders, mobile cranes, graders, etc.

Table 14. Mine Salaries and Wages Paid

	Month			Year to date		
	Salaries	Day pay	Total	Salaries	Day pay	Total
Administration and general						
Administration						
Accounting department						
Accounting						
Warehouse stores (if applicable)						
Purchasing						
Safety						
Security						
Personnel						
Quality control						
Total administration & general						
Mining						
General mine						
Mine administration						
Mine surveying, planning						
Geological engineering						
Operations						
Pit administration						
Drilling, blasting						
Loading						
Truck haulage						
Other						
Maintenance						
Maintenance administration						
Field repairs						
Truck shop repairs						
Shop support						
Electrical repairs						
Total mine salaries and wages paid						
Shifts worked						
Average rate/shift worked						

Table 15. Mine Force Report

	Previous month date	Hired	Transferred or promoted	Separated	Present month date
Administration and general					
Administration					
Accounting					
Warehousing					
Purchasing					
Safety					
Security					
Personnel					
Quality control					
Total administration and general					
Mining					
General mine					
Mine surveying, planning					
Geological engineering					
Operations					
Pit administration					
Drilling, blasting					
Loading					
Truck haulage					
Other					
Maintenance					
Maintenance administration					
Field repairs					
Truck shop repairs					
Shop support					
Electrical repairs					
Total mining					
Total mine force					

Exhibit 2. Any Mine USA, Supplemental Statement: Mine Haulage Cost and Performance Report, November 1977

Table of Contents

* Truck reporting should be broken down into fleets of similar manufacture and configuration.

Table 16. Truck Haulage Summary*

	November 1977			Jan./Nov. 1977		
	Actual, $	Cost/ton†	Cost/hr	Actual, $	Cost/ton†	Cost/hr
Truck Haulage						
Truck type No. 1 (description)						
Truck type No. 2						
Truck type No. 3, etc						
Truck Haulage Combined						
Labor						
Tire hardware						
Tires, tubes						
Fuel						
Lubricants						
Other operating supplies						
Total operations						
Maintenance and repairs						
Running repairs						
Engines, turbos						
(power transmission)‡						
(differential and drive axle)‡						
(generator/electric wheel)§						
Bed						
Brakes						
Hoist						
Suspension, front axles						
Cab, chassis, other						
Accidents						
Total maintenance and repair						
Tire warranty credits						
Depreciation						
Total						

* Truck reporting should be broken down into fleets of similar manufacture and configuration.
† Metric equivalent: 1 st × 0.907 184 7 = t.
‡ For mechanical drive trucks.
§ For electrical drive trucks.

Table 17. Truck Type No. 1, Mechanical Drive, Haulage Records

	November 1977			Jan./Nov. 1977		
	Actual, $	Cost/ton*	Cost/hr	Actual, $	Cost/ton*	Cost/hr
Equipment Statistics						
Labor						
Tire hardware						
Tires, tubes						
Fuel						
Lubricants						
Other operating supplies						
Total operations						
Maintenance and repairs						
Running repairs						
Engines, turbos						
Power transmission						
Differential, drive axle						
Bed						
Brakes						
Hoist						
Suspension, front axles						
Cab, chassis, other						
Total maintenance, repair						
Tire warranty credits						
Depreciation						
Total						
Production statistics						
Operating truck hr						
Ore						
Waste						
Other						
Total						
Hr available, but not used						
Hr not available						
Hr available, but not used due to restriction						
Material hauled, dry tons						
Ore						
Waste						
Other						
Total						

Table 17. Truck Type No. 1, Mechanical Drive, Haulage Records (Continued)

	November 1977			Jan./Nov. 1977		
	Actual, $	**Cost/ton***	**Cost/hr**	**Actual, $**	**Cost/ton***	**Cost/hr**
Tons per operating hr						
Ore						
Waste						
All material						
Cost per ton material hauled						
Availability, %						
Utilization, %						
Used available, %						

* Metric equivalent: 1 st × 0.907 184 7 = t.

Table 18. Truck Type No. 2, Electrical Drive, Haulage Records*

	November 1977			Jan./Nov. 1977		
Truck type #2 (electrical)	Actual, $	Cost/ton†	Cost/hr	Actual, $	Cost/ton†	Cost/hr
Equipment Statistics						
Labor						
Tire hardware						
Tires, tubes						
Fuel						
Lubricants						
Other operating supplies						
Total operations						
Maintenance and repair						
Running repairs						
Engines, turbos						
Generator/electric wheel						
Bed						
Brakes						
Hoist						
Suspension, front axles						
Cab, chassis, other						
Accidents						
Total maintenance and repairs						
Tire warranty credits						
Depreciation						
Total						

* Should be similar to Table 17 except for an electrical-type truck.
† Metric equivalent: 1 st × 0.907 184 7 = t.

Exhibit 3. Any Mine USA, Service Equipment Cost, November 1977, Without Depreciation

Table 19. All Service Equipment

	Month			Year		
	All cost, $	Oper. hr	Cost per oper. hr, $	Total costs, $	Oper. hr	Cost per oper. hr, $
Trucks						
Trucks						
Tractors-crawler						
Tractors, rubber-tire						
Mobile crane						
Loaders, rubber-tire						
Small yd. crane						
Graders						
Scrapers						

Code		Standard hourly cost to be charged in (Date)
1	Trucks	$______
2	Trucks	$______
3	Tractors-crawler	$______
4	Tractors, rubber-tire	$______
5	Mobile cranes	$______
6	Loaders, small rubber-tire	$______
7	Small yd cranes	$______
8	Graders	$______
9	Scraper, rubber-tire	$______

Table 20. Maintenance and Repairs Only, Tractor

	Month			Year		
Code 3 Crawlers	**M&R* cost, $**	**Hr oper.**	**Hourly cost, $**	**M&R* cost, $**	**Hr oper.**	**Hourly cost, $**
D-8 Class						
101						
102						
103						
104						
105						
106						
D-9 Class						
110						
111						
112						
113						
114						
Total						

* Maintenance and repair.

Table 21. Maintenance and Repairs Only, Small Rubber-Tired Loaders

	Month			Year		
	M&R* Cost, $	**Hr oper.**	**Hourly cost M&R* only, $**	**M&R* Cost**	**Hr Oper.**	**Hourly cost M&R* only, $**
Code 6						
Small rubber tire						
200						
201						
202						
203						
204						
Total						

* Maintenance and repair.

Table 22. Maintenance and Repairs Only, Graders—Mine

	Month			Year		
Code 8 Graders	**M&R* cost, $**	**Hr oper.**	**Hourly cost M&R* only, $**	**M&R* cost, $**	**Hr oper.**	**Hourly cost M&R* only, $**
300						
301						
302						
304						
Total						

* Maintenance and repair.

9 Price Forecasting and Sensitivity Analysis for Economic Analysis of Final Pit Limit

Alan C. Noble
United Nuclear Corp.

Alan C. Noble has a B.S. in mining engineering from the Colorado School of Mines. He was technical systems analyst and senior planning engineer for Cities Service Co. from 1970-1977. He is currently the division planning engineer and manager of the Scientific and Engineering Systems Div. of United Nuclear Corp.

He has specialized in the development and application of scientific and engineering technology to problems in the mineral industry and has been responsible for the development and use of numerous computer systems for ore reserves, mine planning, and economic analysis. He is currently involved in business economic planning, development of uranium geostatistics, and research into the cause and effects of uranium disequilibrium.

INTRODUCTION

Forecasting of metal and mineral prices is a complex and hazardous undertaking. In order to accurately forecast prices, the complexities of both marketing and production must be modeled for each commodity. Even with the most detailed model, the economy may change unexpectedly, invalidating the forecast.

Because of these factors, this chapter will not attempt to present specific methods for price estimation. It will, however, review the fundamentals of price forecasting as applied to planning and design of the final pit shell. Recognizing that no price estimate can be exact, sensitivity analysis will be reviewed as an approach to coping with these unknown variations.

PRICE PROJECTION

One of the more difficult problems in pit planning is to estimate a price for final pit design. The solution to this problem is found through two interrelated processes. First, a time frame must be defined during which price will be estimated. Second, a method for price forecasting must be developed and used to estimate prices during that time period.

Time Frame of Price Projection

There is often a good deal of confusion associated with the time frame of price estimation as it relates to final pit design. This confusion may be eliminated by reviewing the concept of the final pit as it relates to the reality of mining.

The final pit limit is the pit shape reached when mining is no longer profitable and operations cease. The essential concept is that the ultimate pit is not known until the last ton of ore is mined and the pit abandoned. The planned final pit is only an estimate of the size and shape of the actual final pit, which is defined by economic conditions at the end of mining. Thus, the economic and operating parameters used for design of the final pit must be estimated during the time frame of the final pit mining.

The price to be used for final pit design is then the expected value of price during mining of the final pit increment. Estimation of this price usually requires a projection 10 to 20 years in the future (with the exception of small, single slice pits). The difficulty of this estimation is complicated by factors other than the length of forward projection.

First, the differential between price and cost is the variable more relevant to pit design than price alone. This implies that the assumptions used for the price forecast must also be used for the cost forecast. Secondly, a change in price can affect the time frame of price estimation by changing the reserve tonnage and mining life. This effect can be very important when evaluating a small pit of three to four years life where fluctuations in the price trend are more clearly defined.

Price Forecasting

Having defined the time frame of the final pit, it is necessary to estimate the price during that time. In the easiest case, the planner will consult his economic forecasting group, which will provide him with the magic number. If this option is available, it should be used but with some caution. First, the assumptions used for estimating price should be clearly stated for use in cost forecasts. Secondly, the error range and major sources of error should be specified for use in sensitivity analysis. If the error range is not specified, the engineer is reduced to making blind assumptions about errors in the forecasting model.

When the in-house forecast is not available, the planning engineer must develop his own model for forecasting cost. Many options are available from complex econometric models to simple trend analysis. Several of these are discussed from the point of view of pit planning.

Current Pricing: A common expedient is to use the current price or to escalate the previous year's price. The current price is usually one of the worst estimators for purposes of final pit design. Since the prices of metals and commodities tend to fluctuate cyclically, the current price will generally be higher or lower than the long-term trend.

It is noted that it is common practice to use current price when stating ore reserves. The statement against using the current price is not a contradiction with ore reserves practice but reflects the use of the ultimate pit. The design pit is an estimate of the final shape of the pit. This shape is used as a design target in developing the detailed schedule of mining. Only after examining all economic effects of ore scheduling can reserves be stated.

Trend Analysis: The concept of trend analysis is to fit a trend line to historical price data by least squares regression. This method can often provide a good first approximation of future prices. In addition, the calculations can be performed on most programmable calculators, eliminating the tedium of computing the regression line. For these reasons, an example is included, demonstrating the use of the regression line. Several precautions must be observed when using regression lines for price forecasting:

1) Metal and mineral prices usually follow cyclic patterns. The trend line should be computed using data

over a period of several cycles. Otherwise, the regression line will follow the short-range swing of market price rather than the long-range trend.

2) The regression line represents the long-term average trend with the cyclic fluctuations removed. Sensitivity analysis will be required since actual price will be above or below the trend.

3) Correlation between time and price does not necessarily imply a dependent relationship which will continue in the future. The trend line represents the cumulative effect of all market and production conditions over a period of time. If those conditions change, the trend will be broken (this happened at the beginning of the depression of the 1930's).

4) The trend may be computed using either the actual prices or prices converted to constant dollars. The advantage of using constant dollars is that the trend line is independent of inflation rates and both prices and costs may be expressed in constant dollars.

Ratio Methods: There has been some discussion of the use of price/cost ratios to forecast metal prices. The problem with these models is twofold. First, they apply to an industry not to the individual operation. For instance, if the price/cost ratio for a commodity is 1.25 for the industry, it could be 1.10 for an individual operation. Therefore, the position of the individual mine must be evaluated relative to the industry. Secondly, while the industry price/cost ratio may remain constant, costs will vary as grades get lower and production methods change. Thus, costs must be forecast independently for the operation and the industry in order to compute a valid price/cost relationship for the mine.

Econometric Methods: Econometric price forecasting methods involve modeling the interrelationships between the supply, consumption, and cost elements for one or more commodities. These models can be extremely complex and are usually developed and operated by service bureaus or specialized economic forecasting groups.

The primary advantage of these models is the ability to predict short-range market cycles and to evaluate the effect of changes in the market-production structure. The main disadvantage is that complexity places them beyond reach of the planning engineer.

SENSITIVITY ANALYSIS

It is obvious that price cannot be forecast with the precision necessary for defining the absolute location of the final pit limits. The best estimate of the final price will be the average of all possibilities. The final pit shape, however, will be determined by a future price which may be drastically different from the best estimate. Thus, the final pit may be much smaller or larger than planned, depending on whether price is in a high or low cycle at the end of mining. Because of this variability, it is essential to perform a well-designed sensitivity analysis of the final pit limits. The purpose of this chapter is to analyze the variability of the final pit limits with respect to price and other important parameters. This information may then be used to develop a mine plan with the maximum possible range of development.

Computing of Sensitivity Pits

There are two possibilities for generating the sensitivity pit shapes. The simplest method is to select one or more price increments on either side of the best estimate and to compute new pit limits with each price. A problem with this approach is that large amounts of computer time may be used.

Computer time may be minimized by computing the first pit with the lowest price, then saving the resulting topography. This topography is then used as a base topography for computing the pit with the next highest price. Repeating this procedure will eliminate large numbers of blocks from the computations of each run, holding computing time to a minimum.

Another alternative for computing sensitivity pits was suggested by Francois-Bongarcon and Marechal (1976). The authors describe a method for parameterizing the block matrix as a function of the pit economics. Thus, each block is labeled with a parameter representing profitability. These values are then mapped and contoured to generate the pit contours. This method is not fully developed at present but shows great promise since the full range of pits can be computed in a single run.

Relating Pit Variability to the Mine Plan

One of the least studied areas of pit design has been the integration of variability in the final pit with the design of the mining sequence. Currently, each ore body must be analyzed as a special case. The engineer must design his mining increments (pushbacks) so that expandability is available when prices are high, but he must maintain flexibility for mining during periods of low prices. This demands that access be maintained to areas of expansion and that the final pit increment also be the least profitable. Design requirements for low- and high-price pits may be contradictory. This in itself will provide valuable information for future management and reevaluation of the mining plan.

EXAMPLE OF PRICE FORECASTING BY TREND ANALYSIS

The purpose of this example is to illustrate the use of trend analysis in forecasting price for two metals: copper and lead.

Method

The data for this study were tabulated (Table 1), then graphed on semilog paper (Fig. 1). Inspection of these graphs indicated a linear correlation between the logarithm of price and year of occurrence. For this reason, it was decided to perform a least squares fit of an exponential function of the form:

$$y=ae^{bx} \tag{1}$$

where y is the price value; x is the relative year; and a, b are regression coefficients.

$$b=\frac{\Sigma x_i 1ny_i - \frac{1}{n}(\Sigma x_i)(\Sigma 1ny_i)}{\Sigma x_i^2 - \frac{1}{n}(\Sigma x_i)^2} \tag{2}$$

$$a=\exp\left[\frac{\Sigma 1ny_i}{n} - b\frac{\Sigma x_i}{n}\right] \tag{3}$$

The regression coefficients were computed for both metals using a programmable calculator and data from 1935 through 1976. The correlation coefficient r was also calculated to evaluate the quality of the regression. Results of these calculations were tabulated in Table 2 and the corresponding curves plotted in Fig. 1.

Discussion of Results

The level of significance of each regression line was evaluated by comparing the correlation coefficient, r, against tabulated values of significance for r. For 40 pairs of data, a correlation coefficient of 0.393 is significant at the 1% level. With correlation coefficients of 0.96 and 0.80, both regressions may be accepted as due to correlation between year and price rather than chance.

It is also possible to compute confidence limits for estimates made using the following formula:

$$CL(y)=ae^{bx\pm c},$$

$$c=t_{\alpha/2}\,(1-r^2)^{1/2}\left[Sy^2+\frac{(x-\bar{x})^2}{(n-2)Sx^2}\right]^{1/2}$$

where α is the probability of y being outside the confidence limits; $t\alpha/2$ is the student's t value for a cumulative probability of 1-α/2 and (n-2) degrees of freedom; Sy^2 is the population variance of $1ny$ (square of standard deviation); Sx^2 is the population variance of x (square of standard deviation); and $\bar{x}$ is mean x (arithmetic average).

Thus, price projection and confidence limits may be computed for a confidence band of 20% as follows:

Price Projection:

$$y=9.5186 \exp\ (0.04809x) \text{ for copper}$$
$$y=5.9469 \exp\ (0.03108x) \text{ for lead.}$$

The multiplier for confidence limits:

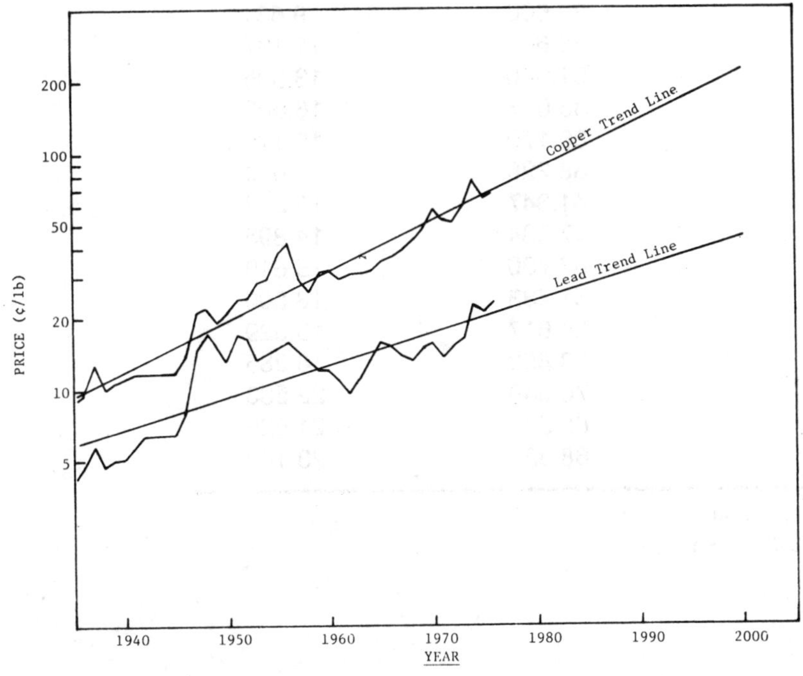

Fig. 1. Price forecasting. Metric equivalent: 1 lb×0.453 592 4=kg.

Table 1. Average Annual Prices 1935–76*

Calendar year	Relative year	Domestic copper Ref. (¢ per lb)†	Lead common NY (¢ per lb)†
1935	0	8.649	4.065
	1	9.474	4.710
	2	13.167	6.009
	3	10.000	4.739
	4	10.965	5.053
1940	5	11.296	5.179
	6	11.797	5.793
	7	11.775	6.481
	8	11.775	6.500
	9	11.775	6.500
1945	10	11.775	6.500
	11	13.820	8.109
	12	20.958	14.673
	13	22.038	18.043
	14	19.202	15.364
1950	15	21.235	13.296
	16	24.200	17.500
	17	24.200	16.467
	18	28.798	13.489
	19	29.694	14.054
1955	20	37.491	15.138
	21	41.818	16.013
	22	29.576	14.658
	23	25.764	12.109
	24	31.182	12.211
1960	25	32.053	11.948
	26	29.921	10.871
	27	30.600	9.631
	28	30.600	11.137
	29	31.960	13.596
1965	30	35.017	16.000
	31	36.170	15.115
	32	38.226	14.000
	33	41.847	13.212
	34	47.534	14.895
1970	35	57.700	15.619
	36	51.433	13.800
	37	50.617	15.029
	38	58.852	16.285
	39	76.649	22.533
1975	40	63.535	21.529
	41	68.824	23.102

*Source: Engineering and Mining Journal.

† Metric equivalent: 1 lb × 0.453 592 4 = kg.

Table 2. Results of Regression Analysis

		Copper	Lead
Regression coef. a	=	9.5186	5.9469
b	=	0.04809	0.03108
Correlation coef. r	=	0.9612	0.8010
S_{1ny} (std. dev. of $1ny$)	=	0.61372	0.4768
x (average of x)	=	20.50	20.5
S_x (std. dev. of x)	=	12.268	12.268

$$c = 1.303\ (0.2759)\left(0.3767 + \frac{(x-20.5)^2}{40(150.50)}\right)^{1/2}$$

$$= 0.3595\left(0.3767 + \frac{(x-20.5)^2}{6020.0}\right)^{1/2} \text{ for copper.}$$

$$c = 1.303\ (0.5987)\left(0.2273 + \frac{(x-20.5)^2}{40(150.5)}\right)^{1/2}$$

$$= 0.7801\left(0.2273 + \frac{(x-20.5)^2}{6020.0}\right)^{1/2} \text{ for lead.}$$

The estimates and 20% confidence limits were then computed (Table 3). The confidence belt for copper is approximately ±30% of the regression line for lead; it varies from ±44% to ±61% as the year increases. This confirms the visual impression of the quality of the regression lines; the copper trend is more stable than lead and will provide more accurate forecasts. Furthermore, the variability of the lead trend is enough to render it relatively useless for forecasting.

REFERENCES

Anon., 1977, "Annual Average Metal Prices, 1910-1976," *Engineering and Mining Journal,* Vol. 178, No. 3, March, p. 68.

Anon., 1975, "Exponential Curve Fit," *HP-25 Applications Programs,* Hewlett-Packard Co., pp. 92-94.

Francois-Bongarcon, D., and Merechal, A., 1976, "A New Method for Open Pit Design: Parametrization of the Final Pit Contour," *Proceedings,* 14th APCOM, AIME, New York, pp. 573-583.

O'Neil, T. J., 1973, "Estimating Minimum Copper Price Levels Through Production Cost Projections," *Proceedings,* 11th APCOM, University of Arizona, Tucson, pp. G1-G30.

Strongman, J. E., et al., 1976, "The Dynamics of the International Copper System," *Proceedings,* 14th APCOM, AIME, New York, pp. 688-700.

Tinsley, Richard C., 1976, "Computer Applications of Nonferrous Econometric Models from the Raw Materials Consumer Perspective," *Proceedings,* 14th APCOM, AIME, New York, pp. 701-713.

Table 3. Estimated Prices and Confidence Limits

Year	Relative year (x)	Estimated price (y)	20% Confidence Limits Lower	Upper
		Copper		
1977	42	71.7	56.3	91.4
1980	45	82.9	64.7	106.2
1990	55	134.0	102.0	176.0
2000	65	216.8	160.3	293.2
		Lead		
1977	42	21.9	14.3	33.7
1980	45	24.1	16.5	35.1
1990	55	32.9	19.8	54.6
2000	65	44.8	25.1	80.2

10 Economic Evaluation of Open Pit Mines

Franklin J. Stermole
Colorado School of Mines

Franklin J. Stermole has a B.S., M.S., and Ph.D. in chemical engineering from Iowa State University. He is a professor of mineral economics and chemical and petroleum refining engineering at Colorado School of Mines where he has taught since 1963. He has done economic evaluation consulting for numerous mineral and nonmineral companies and since 1970 has taught more than 100 "Economic Evaluation" short courses to over 3500 persons from mineral and nonmineral industry companies and government organizations in the US, Australia, Canada, Colombia, France, Indonesia, Saudi Arabia, Trinidad, and Venezuela.

He authored and published *Economic and Investment Decision Methods* in 1974 and revised it in 1977. He served in an administrative capacity as director of research development at Colorado School of Mines from 1972 to 1975.

INTRODUCTION

Economic evaluation of an open pit mine is similar to the economic analysis of any major investment project in any industry. It requires (1) taking into account all of the costs and revenues that are estimated to be incurred over the economic life of the project; (2) crediting the project with all the tax savings and tax credits that will be realized over the project life from tax deductions and credits related to various project costs; (3) charging the project with all taxes to be paid on taxable income over the project life; and (4) properly accounting for the time value of money for all project costs, revenues, taxes, tax savings, and tax credits when making the overall project economic evaluation. These considerations must be taken into account for any valid method of economic analysis such as discounted cash flow rate of return (DCFROR) or net present value (NPV) analyses.

All economic analyses fall into one of two categories: (1) comparison of income-producing alternatives, or (2) comparison of service-producing alternatives. The analysis of alternatives in either of these categories can be made by using correctly any of five basic economic evaluation approaches which are: (1) present value, annual value, (3) future value, (4) rate of return (ROR), and (5) break-even analysis. Of these evaluation techniques, DCFROR and NPV analyses are used most widely for evaluating income-producing alternatives while present worth cost and equivalent annual cost analyses are used most often for the evaluation of service-producing alternatives. In this chapter income-producing alternative analysis is emphasized, so DCFROR and NPV are utilized for evaluating the case studies presented.

It should be mentioned that the DCFROR method of analysis often is referred to by other names such as internal rate of return or after-tax rate of return. All of these names and probably others refer to the same after-tax compound interest rate of return. Defining annual cash flow as the money that is left from annual project revenues after paying all out-of-pocket costs including taxes, DCFROR is the compound interest rate that makes the present worth of all project cash flow equal to the present worth of all project net costs (costs netted against the effects of appropriate tax savings and tax credits not included in cash flow calculations). For economic decision purposes, the project DCFROR is compared with the minimum rate of return, or opportunity cost of capital, which is the after-tax rate of return that is thought to be attainable by investing available capital in other investment opportunities.

Net present value (NPV) is defined as the difference between present worth project cash flow and present worth project net costs with the present worth calculations made using the minimum rate of return. If the present worth cash flow is greater than the present worth net costs, then NPV is positive, which means there is more than enough revenue to pay off the project investment costs at a DCFROR that is greater than the minimum rate of return. Therefore, a positive NPV result indicates a satisfactory project from an economic viewpoint for the same reasons that a project DCFROR greater than the minimum rate of return indicates a satisfactory project from an economic viewpoint. If you think of project net costs as being analogous to mortgage loan money and project cash flow as being analogous to the mortgage payment dollars that are available to pay off the mortgage loan money plus mortgage interest due, then the project DCFROR is analogous to the mortgage loan interest rate. The use of DCFROR and NPV analyses reduces project economic evaluations to a comparison of project alternatives on a basis similar to comparing different mortgage loan investment opportunities.

Before proceeding to the development of additional economic analysis considerations, it should be emphasized that although this chapter is concerned primarily with the application of proper economic analysis techniques to evaluate alternative open pit mine plans, investment decision-making generally involves three considerations: (1) economic analysis, (2) financial analysis, and (3) intangible analysis.

Economic analysis as used here refers to analysis of the relative profitability potential of various investment alternatives, while financial analysis refers to analysis of where the investment funds for proposed projects will be obtained. Intangible analysis involves consideration of investment factors which are difficult to quantify in terms of dollars such as political considerations, public opinion and goodwill, ecological and environmental factors, and uncertain legal grounds to name a few. Many times a project that looks good economically may be rejected for financial or intangible reasons and vice versa. The fact that this chapter emphasizes economic analysis considerations should not be considered to imply that economic analysis considerations always are more important than financial and intangible factors. All three factors are important in most investment decisions, and any of the three considerations may be most important in a specific evaluation situation.

INFLATION AND ESCALATION

Inflation and escalation effects on project costs and revenues must be taken into account in economic evaluation calculations for the results to be meaningful (Stermole, 1977). The term inflation generally refers to a persistent rise in prices that is not accompanied by an offsetting rise in productivity, whereas the term escalation refers to a persistent rise in the price of specific commodities, goods, or services due to the combined effects of inflation, supply/demand, environmental and engineering changes, or other factors that cause prices to change. In economic evaluation work we must be concerned with the effects of escalation on all costs and revenues. Note that inflation is one of several components that contribute to escalation. The rate of inflation in any country is usually measured by using a basket of goods similar to the US Consumer Price Index; therefore, the inflation rate in a given country for a given period of time is the average percentage price change of all goods and services in the economy. Note that some specific goods and services may drop in price while others increase. In economic analysis work we are concerned with potential price changes of specific goods, services, and commodities for any reason, so we must be concerned with escalation effects rather than just inflation.

Now, recognizing that escalation of all costs and revenues must be accounted for in valid economic analyses, we must further recognize that different people use different techniques and assumptions to achieve this. There are three different kinds of dollars that people use in making various economic analyses: (1) today's dollars, (2) escalated dollars (also called current dollars, inflated dollars, or nominal dollars), and (3) constant dollars (also called real dollars and deflated dollars).

Today's dollars represent what project costs and revenues would be today if the project took place in an instant. Since major projects do not take place instantaneously, it is necessary to project how today's project costs and revenues will escalate between the today's dollar base time and when the costs or revenues will actually be realized. There are an unlimited number of different assumptions that can be used to convert today's dollars to the actual escalated dollar values that are estimated to be incurred. In general, it may be projected that all costs and revenues will escalate at different rates each year and that these rates may all be different than the assumed inflation rate. There is no reason to think that the inflation rate represents the rate at which capital costs, operating costs, and revenues will change from year to year on a specific project. You should make that assumption only if you feel it is better than all other possible assumptions for the specific physical situation you are evaluating. Some people make analyses using today's dollar estimates of all costs and revenues. This involves the assumption that the effect on economic analysis results of any escalation of capital costs and operating costs over the project life will be washed out (offset) by escalation of revenues. Once again you should be satisfied with that assumption only if you feel that it realistically represents a physical evaluation situation better than any other escalation assumption. In recent years, many industries such as the general mining industry, the potential shale oil from oil shale industry, and the farming industry have found capital costs and operating costs rising much more rapidly than revenues. In this situation, or when revenues rise more rapidly than costs, today's dollar analyses will not give meaningful and valid economic results.

Constant dollars are escalated dollars that have had the effects of inflation washed out of them by determining the present worth of escalated dollars at the rate inflation to some time zero base. This gives all project costs and revenues in terms of what people often call constant purchasing power dollars at the time zero base time. It has been shown that valid economic analyses using constant dollars always give the same economic conclusions reached with valid escalated dollar analyses (Stermole, 1977). Since more calculations are required for constant dollar analysis than for escalated dollar analysis, and since after-tax cash flow calculations and borrowed money calculations must be done in escalated dollars for either constant or escalated dollar analysis, it is felt that escalated dollar analysis generally is desirable compared to constant dollar analysis. Escalated dollar analysis is used by a large majority of companies in practice and that is the approach utilized for the economic analysis case study presented in this chapter.

EVALUATION INCOME TAX CONSIDERATIONS

Valid after-tax economic analyses must be based on using the appropriate tax laws for the country in which the project is to be done. However, regardless of the country in which a project is to be carried out, there are two basic financial situations that the company (or individual) may be in from an economic evaluation viewpoint. These two situations are (1) a company or individual has sufficient taxable income and tax obligations from other revenue sources to enable it to utilize all tax deductions and tax credits from new projects in the year incurred, and (2) a company or individual does not have other income or tax obligations against which to use tax deductions and credits from new projects in the year incurred, so tax deductions and credits in early project years must be carried forward and used against project income and taxes as soon as available. Both of these approaches are illustrated in the case study evaluation presented later in this chapter.

BORROWED MONEY CONSIDERATIONS

Most major investment projects done today in mining or other industries involve some amount of borrowed money. However, often at the time a major project is being evaluated from an economic viewpoint, the financing details are not known or firmly known. Also, the risks and uncertainties associated with achieving a leveraged (borrowed money analysis) DCFROR are very different from the risks and uncertainties associated with achieving a cash investment DCFROR (Stermole, 1977). Primarily for these reasons, many people feel that it makes sense for economic decision-making purposes to evaluate projects on a cash equity investment basis to determine if the project would be a desirable investment if we had the money to pay for it. If a project looks satisfactory on a cash investment basis, and if we can borrow money for an after-tax cost that is less than the project cash investment DCFROR, we know that borrowed money will work for us and that the leveraged DCFROR on equity investment dollars will be greater than the cash investment DCFROR. Because leveraged DCFROR results relate to smaller and smaller equity investment dollars as you go to larger amounts of borrowed money, leveraged DCFROR results generally become very large with large borrowed money percentages. However, these large leveraged DCFROR results are very sensitive to changes in a given parameter such as revenue, and it can be deceiving and financially dangerous to base economic investment decisions on leveraged results alone without also looking at cash equity investment analysis results. The case study economic analysis presented in this chapter is based on a cash equity investment.

MINE PLAN ECONOMIC ANALYSIS

Economic analysis of alternative mining plans requires estimations of project capital costs, operating costs, and replacement costs over the mine life along with estimations of annual revenues and salvage values. The project must be credited with all the tax savings and tax credits that will be realized over the mine life from income tax deductions and tax credits related to various project costs. And, of course, the project must be charged with all taxes to be paid on taxable income over the mine life. Then, the time value of money must be taken into account using an analysis method such as DCFROR or NPV to evaluate the relative economic merit of various alternatives to be considered.

In the United States, different mining costs may or must be treated in different ways for tax deduction purposes. Following is a concise summary of the US tax deduction choices available for different mining capital costs. The term expense means treat as an operating cost and deduct in the year incurred, while the term capitalize means deduct over a period of time greater than a year, such as by depreciation (Table 1).

Exploration refers to any operation the purpose of which is to delineate an ore body; to determine the location, grade, extent, amount or quality of a mineral deposit. Exploration may include costs for core drilling, assaying, engineering, and geological fees, exploratory shafts or stripping costs, pits, drifts, etc. At the point in time when a determination is made that com-

Table 1. US Tax Deduction Choices for Different Mining Capital Costs

Mining capital costs	US Tax deduction choices
Exploration	Expense or capitalize into depletion basis
Minerals rights acquisition or lease bonus or equivalent	Capitalize into depletion basis
Development	Expense or capitalize for units of production depreciation
Plant and equipment for mine, mill, buildings, etc.	Capitalize and depreciate
Land and working capital	Nondeductible except against terminal sale value, i.e., capitalize and deduct against liquidation value

mercially marketable quantities of a mineral exist, and the decision is made to develop a property, costs from that time on are termed development. The difference between exploration and development can be very important because if exploration costs are expensed for tax deduction purposes and the project results in a producing mine, the expensed exploration costs must be recaptured by deducting them from depletion allowable in the first years of mine production until such exploration costs are recaptured for tax deduction purposes.

When all project costs have been estimated and the tax deduction choice made for all costs, then tax savings, tax credits, and tax on taxable income generally are accounted for in a cash flow calculation analogous to the format in Table 2.

The net cash flow represents what is left from sales revenue each year after paying all out-of-pocket costs including taxes and capital cost expenditures. The project costs and cash flows are handled in DCFROR and NPV calculations as illustrated in the following case study.

AN OPEN PIT COPPER MINE EVALUATION CASE STUDY

Economic evaluation case studies all fall into one of two categories as mentioned earlier: (1) analysis of alternatives that provide a service, and (2) analysis of income-producing alternatives. Evaluation work is split about equally between these two types of problems. Analysis of the overall economics of a proposed revenue-producing open pit copper mine has been chosen for presentation here. The reader should recognize that a series of service-producing alternative analyses would be needed to determine the economically optimum equipment and service facilities needed to develop the operation described in the following case study statement.

A proposed open pit copper mine will produce 30 000 t/day of copper ore, 350 working days/year for 20 years. The stripping ratio will be 2 t of overburden rock per metric ton of copper ore with an average 0.8% ore grade. An overall metallurgical recovery of 83.5% is assumed based on 87% mill recovery and 96% smelter recovery. Copper price is assumed to be $0.80 per lb in production year 1, and custom smelter costs will be $0.28 per lb of pay copper less an $0.08 per lb rebate for precious metal value, giving net smelting cost of $0.20 per lb. Assume that any escalation of smelting costs will be offset by a like-dollar increase in precious metal rebate value, keeping the net smelting cost constant at $0.20 per lb. This gives year 1 net smelter return of $0.60 per lb. Assume the net smelter return value will escalate 5% per year for production years 2 through 5. In production years 6 through 20, assume net smelter returns escalate sufficiently to wash out any escalation of operating costs from the year 5 operating cost and net smelter return base. Operating costs are assumed to be $3.80/t of copper ore produced in year 1, escalating 8% per year in years 2 through 5,

Table 2. Annual Cash Flow Calculation Procedure

1. Project tons to be produced x grade x recovery
2. Net smelter return (value of minerals produced)
3. —Operating costs for mining, concentration, transportation, marketing, administration, royalties, severance taxes, etc.
4. —Expensable exploration/development costs
5. —Noncash cost tax deductions for depreciation, amortization, depletion, and loss carry-forward deductions
6. Taxable income
7. —Taxes (both federal and state)
8. +Tax credits
9. Net profit (or net income after tax)
10. +Noncash cost deductions (depreciation, depletion, etc.)
11. Cash flow
12. —Capital costs
13. Net cash flow

with a washout of escalation of operating costs and sales revenue assumed in years 6 through 20. The year 1 operating costs of \$3.80/t of ore is based on mining costs of \$0.60/t of material or \$1.80/t of ore, milling costs of \$1.60/t of ore, environmental and tailings handling operating costs of \$0.10/t of ore, and general and administrative operating costs of \$0.30/t of ore for costs such as incremental main office costs, property taxes, road and environment upkeep costs, townsite costs, and so forth.

Capital costs for the mine total \$211 million and will be spread over two years, referred to as evaluation years −1 and 0, with time 0 representing the start of production. All capital costs are given in the actual escalated dollars expected to be incurred (Table 3).

There are three major year −1 capital costs. These include mine equipment costs of \$25 million for shovels, dozers, graders, haulage trucks, flatbed trucks, pickup trucks, and so forth. Double declining balance depreciation switching to straight-line for a life of five years will be used for the mine equipment investment. The second year −1 cost is \$30 million for mill shell and buildings (real property) to be depreciated straight-line over the 20 years of producing mine life. In practice, 150% declining balance depreciation might be used, but straight-line depreciation simplifies the analysis. The third year −1 cost is \$20 million for development costs for preproduction stripping to be expensed as an operating cost for tax purposes in the year incurred. There are four major year 0 capital costs. These include (1) an \$80 million mill equipment cost to be depreciated double declining balance and switching to straight-line for a ten-year depreciation life; (2) a \$16 million tailings line and dam cost to be depreciated straight-line over the mine life; (3) a \$20 million development cost again for preproduction stripping; and (4) a \$20 million working capital cost based on three months operating costs of \$10 million plus \$10 million for inventories of spare parts, equipment, raw materials, and product. Working capital cost is not tax deductible except when a project is liquidated. Replacement costs for mining equipment will be incurred

Table 3. Project capital costs, Years −1 and 0, All Values in Millions of Dollars

Mine equipment = 25	Working capital = 20
Mill shell and buildings = 30	Mill equipment = 80
Development = 20	Tailings line and dam = 16
	Development = 20
−1	0

in years 5, 10, and 15 in escalated dollar amounts of \$20 million, \$30 million, and \$40 million, respectively. Assume that these costs will be depreciated straight-line over the five years following when the cost is incurred. In fact, assume that all depreciable costs will be depreciated starting in the first year following when the cost is incurred, except that the mill shell and buildings depreciation starts in year 1.

Assume that a 50% effective income tax rate covers federal and state income tax. Assume that there are no significant mineral rights acquisition or mineral lease bonus costs so cost depletion is not applicable. Also assume that no mineral exploration costs were expensed in earlier years from exploration work in the region of this proposed mine, so there will be no effect on percentage depletion deductions due to recapture of previously expensed exploration costs. Percentage depletion for minerals is based on net smelter return and after year 5, neglect the effect that escalating net smelter return would have on percentage depletion. Neglect all salvage values and working capital return at the end of the project.

Calculate project DCFROR and net present value for a minimum DCFROR of 15% for the following company or individual tax situations: (1) other taxable income and income tax obligations exist against which to use project tax deductions and tax credits in the year incurred; (2) other taxable income and income tax obligations do not exist against which project tax deductions and tax credits can be used in early project years, so negative taxable income and tax credits must be carried forward and used against project income and taxes.

OPTIMUM MINE PLAN AND CUTOFF GRADE ANALYSIS

Economic analysis of alternative ways of developing a new mine involves evaluation of many different mutually exclusive alternatives from which we must pick the one best choice. We probably want to consider the analysis of a proposed new mining operation for different production rates such as 10 000 t/day of ore vs. 15 000 t/day or 20 000 t/day. Production rate affects most mine costs including mine equipment capital cost, mill size and capital cost, development costs and working capital, as well as annual operating costs and revenues. Cutoff grade analysis in turn affects many of the mine costs since changing the cutoff grade changes stripping ratios and therefore affects the mine equipment cost to produce a given number of tons of ore per day. Changing the cutoff grade also may affect metallurgical recovery and the desired design and cost of mill facilities. Cutoff grade also affects tailings and

Table 4. Cash Flow Calculations for a Company with Other Taxable Income and Income Tax Obligations Against Which to Use Project Tax Deductions and Tax Credits in the Year Incurred. All Values in Millions of Dollars

Year	^{c}equip.=25 ^{c}mill=30 shell & bldgs., ^{c}dev.=20 −1	^{c}w.cap=20, ^{c}mill equip.=80, ^{c}tail.=16, ^{c}dev.=20 0	1	2	3	4	^{c}equip. =20 5	6-9	^{c}equip. =30 10	11-14	^{c}equip. =40 15	Final value =0 16-20
Net smelter return	–	–	92.6	97.2	102.1	107.2	112.5	112.5	112.5	112.5	112.5	112.5
−Op. costs	–	–	−39.9	−43.1	−46.5	−50.3	−54.3	−54.3	−54.3	−54.3	−54.3	−54.3
−Development	−20.0	−20.0	–	–	–	–	–	–	–	–	–	–
−Deprec. 25 & replacement	–	−10.0	−6.0	−3.6	−2.7	−2.7	–	−4.0	−4.0	−6.0	−6.0	−8.0
−Deprec. 30	–	–	−1.5	−1.5	−1.5	−1.5	−1.5	−1.5	−1.5	−1.5	−1.5	−1.5
−Deprec. 80	–	–	−16.0	−12.8	−10.2	−8.2	−6.6	−5.2	−5.2	–	–	–
−Deprec. 16	–	–	−0.8	−0.8	−0.8	−0.8	−0.8	−0.8	−0.8	−0.8	−0.8	−0.8
Taxable before depletion	−20.0	−30.0	+28.4	+35.4	+40.4	+43.7	+49.3	+46.7	+46.7	+49.9	+49.9	+47.9
−15% depletion	–	–	−13.9	−14.6	−15.3	−16.1	−16.9	−16.9	−16.9	−16.9	−16.9	−16.9
or 50% limit	–	–	−14.2	−17.7	−20.2	−21.8	−24.6	−23.3	−23.3	−25.0	−25.0	−23.9
Taxable income	−20.0	−30.0	+14.5	+20.8	+25.1	+27.6	+32.4	+29.8	+29.8	+33.0	+33.0	+31.0
−Income tax @50%	+10.0	+15.0	−7.2	10.4	−12.6	−13.8	−16.2	−14.9	−14.9	−16.5	−16.5	−15.5
+Invest. tax cr.	+1.7	+8.0	–	–	–	–	+1.3	–	+2.0	–	+2.7	–
Net profit	−8.3	−7.0	7.2	10.4	+12.5	+13.8	+17.5	+14.9	+16.9	+16.5	+19.2	+15.5
+Deprec/depl.	–	+10.0	+38.2	+33.3	+30.5	+29.3	+25.8	+28.4	+28.4	+25.2	+25.2	+27.2
Cash flow	−8.3	+3.0	+45.4	+43.7	+43.0	+43.1	+43.3	+43.3	+45.3	+41.7	+44.4	+42.7
−Capital costs	−55.0	−116.0	–	–	–	–	−20.0	–	−30.0	–	−40.0	–
Net cash flow	−63.3	−113.0	+45.4	+43.7	+43.0	+43.1	+23.3	+43.3	+15.3	+41.7	+4.4	+42.7

Net present value at year −1 = present worth cash flow @ minimum DCFROR of 15%
−present worth costs @ minimum DCFROR of 15%

$$= 45.4\,(P/F_{15,2}) + 43.7\,(P/F_{15,3}) + --- + 42.7\,(P/F_{15,20}) - 113.0\,(P/F_{15,1}) - 63.3$$

$$= 54.3$$

where $P/F_{15,n}$ = 15% single payment present worth factor = $\left(\frac{1}{10+15}\right)^n$.

Project DCFROR = 20.8% which is the rate of return that makes the NPV equation equal to zero.

Payback period = years of production required for cash flow to recover costs at zero interest = 4.1 years.

mine dump costs. In addition to looking at cutoff grade changes and different production rates, it usually is desirable to look at the economic effects of taking high-grade ore first, maybe at the expense of higher stripping ratios, vs. taking lower-grade ore with lower stripping ratios first.

All these analyses are mutually exclusive alternatives from which we want to select the one best alternative. NPV analysis is the easiest way to evaluate the relative economic potential of mutually exclusive alternatives, because you always want the alternative with the biggest NPV on total investment (Stermole, 1977). With DCFROR analysis of mutually exclusive alternatives, you must make incremental DCFROR analysis for the various alternatives to obtain a valid economic decision, and incremental DCFROR analysis gets messy and creates special economic analysis difficulties when alternatives have different lives (Stermole, 1977). Changing cutoff grade or production rate generally changes mine life, which creates the situation where NPV is much easier and more effective to use than DCFROR. The scope and space for this chapter do not permit reproducing the analysis reasons that support the desirability and even necessity of using NPV over DCFROR analysis in this evaluation situation.

To illustrate the general technique for applying NPV

Table 5. Cash Flow Calculations for a Company That Does Not Have Other Taxable Income and Income Tax Obligations Against Which Project Tax Deductions and Tax Credits Can Be Used, So Negative Taxable Income and Tax Credits Must Be Carried Forward and Used Against Project Income and Taxes.

Year	−1	0	1	2	3	4	5	6-9	10	11-14	15	16-20
Net smelter return	—	—	92.6	97.2	102.1	107.2	112.5	Year 6-20 net cash flow calculations are same as in Table 1.				
−Op. costs	—	—	−39.9	−43.1	−46.5	−50.3	−54.3					
−Development	−20.0	−20.0	—	—	—	—	—					
−Deprec. 25 & replacement	—	−10.0	−6.0	−3.6	−2.7	−2.7	—					
−Deprec. 30	—	—	−1.5	−1.5	−1.5	−1.5	−1.5					
−Deprec. 80	—	—	−16.0	−12.8	−10.2	−8.2	−6.6					
−Deprec. 16	—	—	−0.8	−0.8	−0.8	−0.8	−0.8					
Taxable before depletion	−20.0	−30.0	+15.6	+35.4	+40.4	+43.7	+49.3					
−15% depletion	—	—	−13.9	−14.6	−15.3	−16.1	−16.9					
or 50% limit	—	—	−7.8	−17.7	−20.2	−21.8	−24.6					
Loss forward	—	−20.0	−50.0	−42.2	−21.4	—	—					
Taxable income	−20.0	−50.0	−42.2	−21.4	+3.7	+27.6	+32.4					
−Income tax @50%	—	—	—	—	−1.8	−13.8	−16.2					
+Invest. tax Cr.	—	—	—	—	—	+9.7	+1.3					
Net profit	−20.0	−50.0	−42.2	−21.4	+1.8	23.5	+17.5					
+Deprec./depl.	—	+10.0	+32.1	+33.3	+30.5	+29.3	+25.8					
+Loss forward	—	+20.0	+50.0	+42.2	+21.4	—	—					
Cash flow	−20.0	−20.0	+39.9	+54.1	+53.7	+52.8	+43.3					
−Capital costs	−55.0	−116.0	—	—	—	—	−20.0					
Net cash flow	−75.0	−136.0	+39.9	+54.1	+53.7	+52.8	+23.3	+43.3	+15.3	+41.7	+4.4	+42.7

NPV @ minimum DCFROR of 15% at year −1 = +36.9; DCFROR = 18.5%; payback period = 4.4 years.

analysis to mine planning and cutoff grade economic decisions, we will now look at three specific variations from the case study described earlier in this chapter and analyzed in Tables 4 and 5 for the situation where other income and taxes exist against which to use tax deductions and tax credits in the year incurred. Consider case 1 to be the evaluation as described earlier and evaluated in Table 4 with cases 2, 3, and 4 having variations from case 1 as follows:

Case 1: Base Case

30 000 t/day of ore, 350 days/year

0.8% average ore grade based on 0.4% cutoff grade

Reserves = 210 million tons of ore, 20-year mine life

Metallurgical recovery = 83.5%

Total rock mined/day = 90,000 t based on 2:1 strip ratio

Mill cost = \$30 m shell + \$80 m equip. = 110 m total, m = million

Mine equipment cost = \$25 m

Tailings line and dam cost = \$16 m

Operating cost = \$3.80/t ore.

Case 2: Increase Production Rate and Reduce Cutoff Grade

40 000 t/day of ore, 350 days/year

0.7% average ore grade based on 0.3% cutoff grade

Reserves = 270 million t of ore, 19.3-year mine life

Metallurgical recovery = 81.0%

Total rock mined/day = 100 000 t based on 1.5:1 strip ratio

Mill cost = \$135 m total = \$35 m shell + \$100 m equip.

Depreciate mill shell over 20 years

Mine equipment cost = \$27 m

Tailings line and dam cost = \$18 m

Operating cost = \$3.40/t of ore based on \$1.50/t mining, \$1.55/t milling, \$0.10/t tailings, and \$0.25/t admin.

Economic analysis results: NPV $_{@\ i=15\%}$ = +\$55 m, DCFROR = 21%.

Case 3: Reduce Cutoff Grade

30,000 t/day of ore, 350 days/year
0.7% average ore grade based on 0.3% cutoff grade
Reserves = 270 million t, 25.7-year mine life
Metallurgical recovery = 81.0%
Total rock mined/day = 75 000 t based on 1.5:1 strip ratio
Mill cost = \$110 m total = \$30 m shell + \$80 m equip.
Mine equipment cost = \$23 m
Tailings line and dam cost = \$16 m
Operating cost = \$3.50/t ore
Economic analysis results: NPV $_{@\ i=15\%}$ = +\$18 m, DCFROR = 18%.

Case 4: Increase Production Rate

40 000 t/day of ore, 350 days/year
0.8% average ore grade based on 0.4% cutoff grade
Reserves = 210 million t, 15-year mine life
Metallurgical recovery = 83.5%
Total rock mined/day = 120 000 t based on 2:1 strip ratio
Mill cost = \$135 m total = \$35 m shell + \$100 m equip.
Depreciate mill shell over 15 years
Mine equipment cost = \$30 m
Tailings line and dam cost = \$18 m
Operating cost = \$3.70/t
Economic analysis results: NPV $_{@\ i=15\%}$ = +\$93 m, DCFROR = 24%.

CONCLUSIONS

Both the NPV and DCFROR results for the different cases analyzed for this project are presented in the last section and indicate that all of the alternatives evaluated are acceptable compared to investing the money elsewhere at the 15% minimum DCFROR. However, to select the best of these alternatives, select case 4 with the largest NPV of +\$93 million. This project with the largest NPV has incremental capital investment dollars, compared to any smaller investment project, earning at a DCFROR greater than the minimum DCFROR, which is a necessary requirement of a satisfactory mutually exclusive investment alternative. It turns out that for the four cases evaluated in the last section, the economic choice by NPV analysis, case 4, also has the largest DCFROR on total investment. This is not always the case. When project investments differ significantly, it often turns out that a bigger investment project with a DCFROR that is smaller than the DCFROR of another smaller investment project is the best economic choice because the incremental investment dollars are earning more than the minimum DCFROR. Selecting case 4 gives us the mine plan with the higher production rate and higher cutoff ore grade. For the conditions evaluated in these four cases, going to a lower cutoff grade is not economically desirable compared to the higher cutoff grade results. With respect to the DCFROR and NPV analyses results, keep in mind that these analyses have been made for the most expected capital costs, operating costs, and revenues; and we should look at the sensitivity of the DCFROR and NPV results to changes in the project parameters that are considered possible, to obtain a feel for the range over which the DCFROR or NPV results might vary if things go from very good to very bad. DCFROR and NPV analyses results are tools that enable us to systematically and quantitatively evaluate the economic potential of various investment alternatives. Economic analysis results are only as good as the input costs, revenues, and project lives used. If there is significant uncertainty associated with project costs, revenues, and project life, for the analysis presented in the last section, it would be foolhardy to talk about case 1 project DCFROR being 20.8%. It might be much more reasonable to speak in terms of the possibility the DCFROR could vary over some range of values, say from a high of 26% to a low of 12% for a given range of project parameters.

In summary, DCFROR and NPV analyses techniques to be effective and useful must be applied with the same common sense and good judgment required in all good management decision-making situations. Applied properly, DCFROR and NPV analyses will assist you in doing a better job of economic decision-making than you can do without using these techniques.

REFERENCES

Stermole, F. J., 1974, *Economic Evaluation and Investment Decision Methods,* 1st ed., Investment Evaluations Corp., Golden, CO.

Stermole, F. J., 1977, *Economic Evaluation and Investment Decision Methods,* 2nd ed., Investment Evaluations Corp., Golden, CO.

2D Pit Limit Slope Design

B. L. Seegmiller, editor

Contents

INTRODUCTION TO SUBSECTION 2D

The final or ultimate slopes of an open pit mine are a very important factor in pit design. The steeper the final slope can be mined, the greater the savings in stripping. However, as the steepness increases, the probability of its failing increases. Therefore, an optimum mine plan should have the steepest final pit limit slopes which will not fail until mining has been completed in that area of the pit. To achieve such properly designed slopes pertinent data relative to geologic discontinuities, rock mass strength, and ground water must be collected. The data may then be analyzed and the final pit limit slopes analytically designed using factor of safety and/or probability of failure criteria. The chapters contained in this subsection describe methods that may be used to collect stability data, techniques used to analytically design the slope angle, and remedial measures that may be taken to lessen the effects of slope failure and/or prevent it entirely.

11 General Comments, Data Collection, Remedial Stability Measures

Ben L. Seegmiller
Seegmiller Associates

Ben L. Seegmiller's mining and rock mechanics experience spans more than 20 years and has ranged from hard rock mining to university teaching. He is a graduate of the University of Utah. In 1969, Dr. Seegmiller received his doctorate in mining engineering and began employment with the Anaconda Co. as a rock mechanics specialist. For the next five years, he traveled extensively throughout the Anaconda operations, working on open pit and underground rock mechanics problems. In 1974, he formed his own consulting organization. Dr. Seegmiller is a registered professional engineer and a member of SME-AIME, ASCE, CIM, and ISRM. He has presented papers on rock mechanics at a number of professional meetings and has had feature articles published in various mining journals.

Introduction

The profitability of an open pit operation depends to a large extent on the use of the steepest pit slopes possible, provided they do not fail during the life of the mine. Optimum pit slopes can be designed by using rock mechanics technology. Such technology includes gathering and analyzing pertinent data prior to pit design, rigorous application of rock mechanics principles during the design process, remedial actions to improve slope stability, and monitoring slopes during the life of the mine. Ignoring rock mechanics technology in pit design may place the entire mining enterprise in jeopardy.

Disruption of mining operations and safety hazards caused by slope instability can be severe. A classic case in point is the major slope failure (Seegmiller, 1972) that occurred at a southwestern US copper mining operation in 1970-1971. Minor slope stability problems initially amounted to no more than a nuisance, but in time, the unstable zones became larger, causing greater disruption of production and threatening safety. Mining at or near the slide toe had continued throughout the period of slope failure, aggravating the already unstable condition. The original plan for this pit called for mining the overall pit slope to 37.5°. Mine operators were forced to lessen the backslope angle in the hope that the failing rock mass would stabilize. However, stabilization did not take place, and when mining in the area of failure was completed, the overall slope had been reduced to approximately 27°. Mining had been successful in removing only a small portion of the ore body at a high stripping ratio.

The case of this southwestern mine is not unique. Mining operations are disrupted, and safety problems are caused by slope instability in numerous other mines throughout the world. In South America, a major slope failure halted all pit operations when sections of the main-line rail haulage system were destroyed. For more than three years, a giant iron mine in Australia has been hampered by safety hazards and mining interruptions caused by slope failure extending over 2 km. In 1975, an operation in the western US suffered one of the few fatalities ever recorded in an open pit copper mine as the direct result of slope failure. Many mining companies have experienced damage to fixed structures located near the pit rim or inside the pit; many crushing facilities in such locations have not been able to function properly because of slope instability. Examples of slope failures and instability problems which mine operators must typically cope with are presented in Figs. 1-5.

Fortunately, the study of failing rock masses in open pits has received much attention from experts over the past 12 years. Remedial action, including dewatering (Seegmiller, 1973), slope modification (Stewart and Seegmiller, 1972), and artificial stabilization (Seegmiller, 1974; Coates and Sage, 1973; and Seegmiller, 1975), is available to the mine operator. Safety hazards and disruptive effects of failure can be minimized. The safety aspects of slope instability have been greatly improved through the use of displacement monitoring techniques. Not only has monitoring and recording equipment been improved, but better techniques (Seegmiller, 1974) have been developed to interpret data on instability.

Prevention of slope instability may be achieved if optimum slope angles (Seegmiller, 1976) are determined when the mine is still in the planning stage. Using rock mechanics technology, planners can determine a sound engineering choice for optimum bench profiles and overall slope angles by using the following procedures: (1) begin an initial assessment of the

Fig. 1. Slope failure in weak sedimentary rocks.

Fig. 2. Wedge failure on a 15-m bench.

Fig. 3. Catastrophic slope failure.

important stability parameters as soon as it is apparent that a future open pit will be a reality; (2) start a detailed stability study as soon as possible after completion of the initial assessment; and (3) collect as much stability data as possible during exploration and/or development drilling.

Stability data include the spacial locations of geologic discontinuities and their shear strengths, as well as the ground water and seismic forces that may affect the slopes. Potential modes and probabilities of failure and safety factors against slope failure are determined. Recommendations for optimum slope angles and bench profiles are then developed. Methods of stability improvement and potential remedial action measures are reviewed. A displacement monitoring system may then be developed for the life of the pit. The payoffs of such stability studies have previously been discussed by engineers (Seegmiller, 1972, 1973). The emphasis is now on achieving practical stability results as defined

Fig. 4. Planar-toppling failure mode combination.

Fig. 5. Tension crack displacement in a main haulage ramp.

by safety and economic considerations as opposed to simply achieving stability.

An increase or decrease in slope angle in a medium to large open pit will significantly alter stripping requirements. Consider the open pit plan and section shown in Fig. 6. The hypothetical pit is 1900 m in length from crest to crest, the width of the bottom floor 75 m from toe to toe, the vertical depth 150 m, the overall slope angle 40°, and the specific gravity of material 2.65. If the slope angle could be increased safely to 45°, stripping would be reduced by approximately 23 million t. Conversely, if the slope had originally been laid out at 45° and slope instability factors subsequently forced a reduction in slope angle to 40°, the increased stripping of 23 million t required to maintain stability would significantly alter operations. Significant changes in stripping required by small changes in slope angles are graphically pointed out in Fig. 7. Such changes strongly affect the long-term profitability of a mining operation, emphasizing the need for deter-

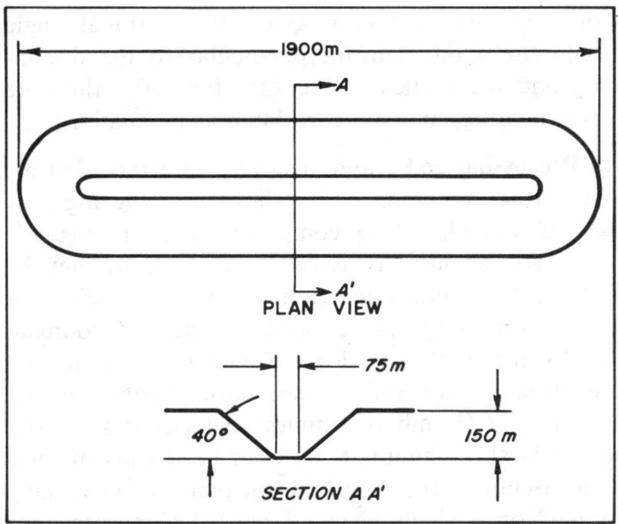

Fig. 6. Hypothetical open pit mine.

mining optimum slope angles during the planning stage. In essence, the desired mining plans must be closely studied, data affecting stability must be collected and analyzed, and a sound engineering approach implemented for operations. Recent studies (Kim and Hall, 1977) of the economic analysis of pit slope design have resulted in the development of benefit-cost and risk-analysis models. These models require broad stability and economic data bases from which a slope failure risk to economic benefit ratio may be computed.

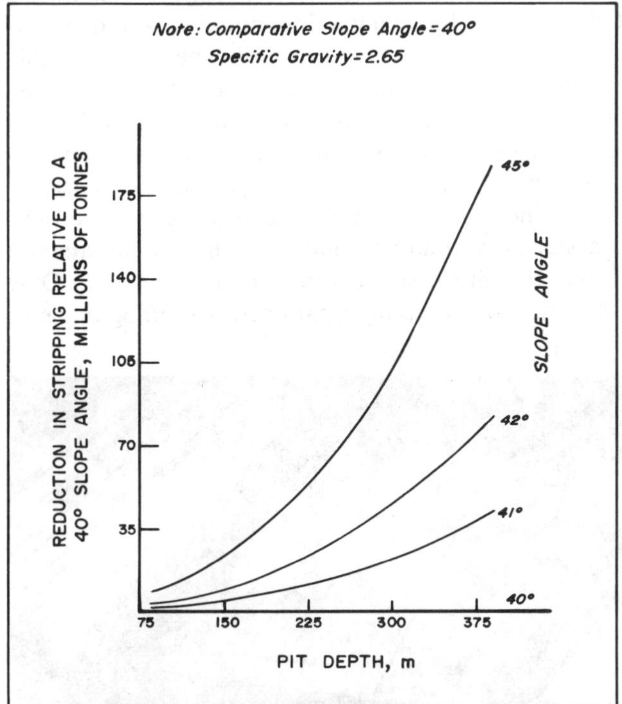

Fig. 7. Stripping reduction with changes in slope angle.

To achieve optimum pit slopes at the lowest possible cost, rock mechanics studies should begin as soon as practicable, preferably during exploration and/or development drilling. Combining the collection of stability data with the collection of data for ore delineation is the surest way to minimize costs of rock mechanics studies. The core can be oriented to obtain discontinuity data; selected pieces of core can be used for direct shear tests; and the borehole can be used for emplacement of a piezometer. Obtaining such stability data during ore delineation drilling in no way interferes with the appraisal of the economic geology. It does, in fact, broaden the knowledge about the prospective pit.

Stability Data Collection

Geologic Discontinuities

Collection of geologic discontinuity data on faults, joints, and bedding planes is for the purpose of determining the possible modes, sizes, and locations of potential slope failures. These discontinuities or planes of weakness are the most important factors governing the stability of a pit slope. To design optimum slope angles, the location and orientation of planes of weakness must be known. Methods of obtaining discontinuity data include on-site surface mapping and oriented borehole core logging. Where rock outcrops are available, surface mapping may provide important data on the occurrence and orientation of faults, dikes, bedding planes, and joint sets. However, such data may not be truly representative of the important structures at depth. Additional discontinuity data should be obtained from borehole core by orienting the core during drilling, using one of the several methods of orienting core available (Seegmiller, 1976). Illustrated in Figs. 8 and 9 is one core orienting method. Owing to the fact that borehole core is commonly collected to investigate, sample, and delineate ore bodies, it is simple and in-

Fig. 8. Craelius core orientator in use.

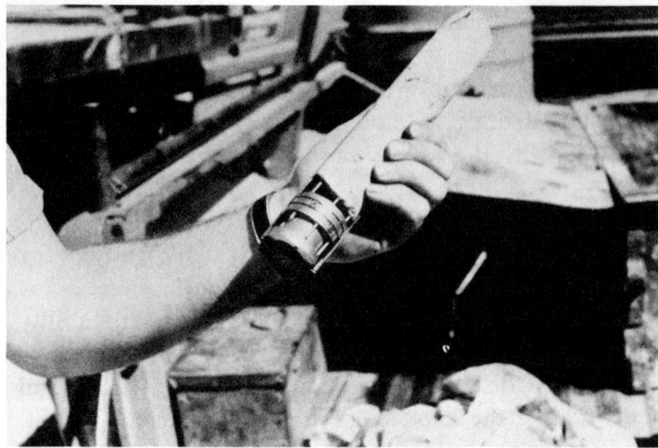

Fig. 9. Matching orientator and core.

expensive to extend the use of the core to rock mechanics studies. The core gives a reasonably good indication of the unweathered conditions that will prevail in a future slope at depth. The discontinuities may be obtained virtually intact with infilling materials if drilling systems utilize split-tube core barrels. Where loose decomposed overburden or dense plant growth covers a deposit, oriented core may provide the only means for determining discontinuity orientations. Where both surface outcrop mapping and core logging can be performed, each method can be used to substantiate the findings provided by the other.

Surface Mapping: A detailed program of surface mapping should normally be conducted if suitable exposures of bedrock occur. A typical mapping program may involve collecting the following data for each discontinuity: (1) discontinuity type (fault, joint, bedding, etc.); (2) strike and dip (the spacial orientation); (3) location (the coordinates and elevation of occurrence); (4) continuity (the exposed length in both strike and dip directions); (5) roughness and planarity (the relative roughness and planarity); (6) spacing (the distance between approximately parallel discontinuities); (7) filling (the type, thickness, and hardness of various infilling materials); (8) rock (the type and hardness of the surrounding rock type); and (9) miscellaneous data (the occurrence of moisture, recent movement, or other data).

Borehole Core Logging: Oriented borehole core may be logged for similar data including discontinuity type, roughness, spacing, filling, and rock type. In addition, the following should be recorded: (1) depth (the distance from the hole collar); (2) tightness (broken or unbroken nature); and (3) reference angles [the strike and dip of the discontinuity may be determined by recording four angles: the minimum angle between the discontinuity and the core axis, the rotational angle between the axial plane perpendicular to the discontinuity and the vertical plane (see Fig. 10), the core interval bearing, and the core interval inclination].

Data Processing and Handling: The data recorded for either surface mapping or borehole core logging may be coded and placed on computer cards or magnetic tape. The discontinuity reference angle data may be combined to produce individual strikes and dips for each discontinuity using stereographic (Goodman, 1976) or mechanical analog techniques. A useful tool for determining the various discontinuity sets is to use an equal-area (Lambert-Schmidt) net (Coates, 1970). These nets allow hundreds or even thousands of individual discontinuity strike and dip pairs to be visually examined on a single sheet of paper. Concentrations of strike and dip pairs, which are called poles, may then be contoured to determine specific discontinuity sets. Statistical calculations performed on the discontinuity data in each set show the relative quality characteristics, which may be assigned to each specific set. An example of a contoured equal-area net is presented in Fig. 11. The quality characteristics for each of the discontinuity sets on the equal-area net are exemplified in Table 1. Such quantity and quality data may then be used in an analytical slope design.

Shear Strength

While it may be determined that displacement in a pit slope is possible along a discontinuity, soil, or weakness zone, this does not necessarily mean it is highly probable. The mechanical characteristics of rock and soil materials determine how those materials will react to stress concentrations. The stress concentrations become greater with respect to higher and steeper pit slopes. Therefore, if a future open pit is to have optimum slopes, a complete understanding of the amount of stress the pit slope will withstand is necessary. Only by knowing shear strength parameters (sliding angle of

Fig. 10. Oriented core logging.

Table 1. Quality Characteristics of Major Joint Discontinuity Sets

Location	Set	Strike	Dip	Quantity	% Having Specified Characteristic																								
					Filling											Planarity			Spacing					Continuity					
					Type						Thickness																		
					None	Chlorite	Serpentine	Graphite	Clays	Iron Oxide	0 mm	<1 mm	1-3 mm	3-5 mm	>5 mm	Planar	Wavey	Irregular	<0.25 m	0.25-0.50 m	0.50-0.75 m	0.75-1.00 m	>1.00 m	<1 m	1-3 m	3-5 m	5-10 m	10-20 m	>20 m
Zone 1	A	N42±07°E	50±04°W	242	52	5	27	1	12	3	52	5	22	19	2	42	36	22	5	46	43	6	0	0	0	0	2	40	58
	B	N05±10°W	33±10°E	161	40	22	29	5	4	0	40	34	23	2	1	68	26	6	20	48	22	8	2	3	7	15	26	42	7
	C	N83±03°W	69±05°W	83	32	31	3	14	14	6	32	47	8	8	5	24	71	5	10	22	48	20	0	3	6	10	44	30	7
Zone 5	A	N42±05°E	45±06°W	146	63	0	13	0	24	0	63	11	10	13	3	51	31	18	10	52	31	7	0	0	0	0	1	47	52
	B	N16±07°E	38±11°E	126	45	18	34	2	1	0	45	44	9	2	0	54	41	5	37	41	16	5	1	1	10	20	13	49	7
	C	N88±04°W	71±06°W	181	26	39	1	18	12	4	26	53	13	3	5	33	63	4	15	31	32	21	1	5	2	14	51	27	1
	D	N52±03°W	11±02°W	72	79	0	6	2	5	8	79	15	3	3	0	22	71	7	39	52	9	0	0	10	15	30	40	4	1
Zone 8	A	N58±06°E	43±05°W	53	81	1	4	2	10	2	81	7	0	12	0	32	55	13	7	58	30	4	1	0	0	0	3	46	51
	B	N01±05°W	28±03°E	112	38	27	22	5	5	3	38	13	42	4	3	71	25	4	41	46	9	3	1	5	5	35	40	15	5

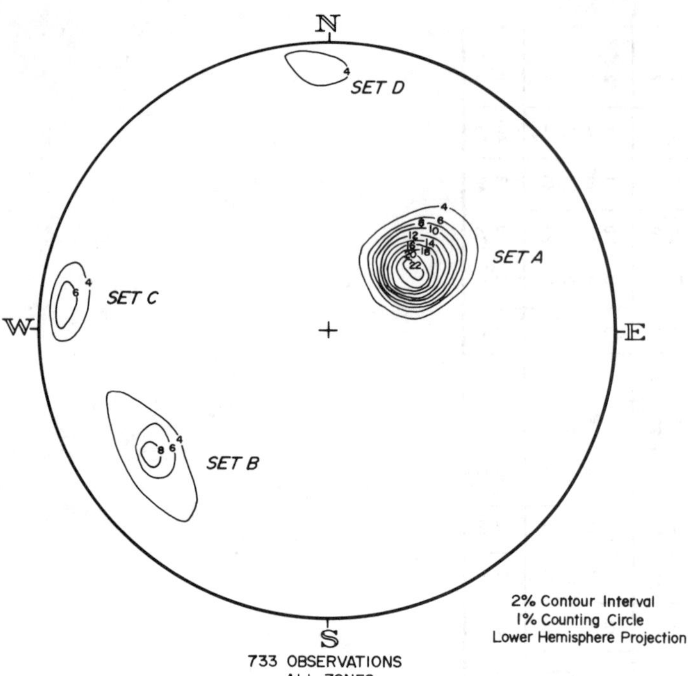

Fig. 11. Contoured equal-area net showing various sets of discontinuities.

friction and cohesion) can one reasonably decide what the pit slope's reaction to various stresses will be. Once that reaction is understood, it may be ascertained whether or not displacement will take place, or even if it is very likely to occur.

Discontinuities: A method whereby engineering shear strength parameters for discontinuities may be readily determined is the direct shear test. The test allows an estimate of the field shear strength parameters to be made using laboratory techniques. In essence, the method consists of (1) selecting a piece of rock containing a discontinuity, (2) casting plaster molds onto each end of the rock, (3) placing the molded rock into a laboratory direct shear machine (see Fig. 12), and (4) shearing one end of the rock relative to the other end along the discontinuity. The larger the normal force, the greater the shearing force required for displacement along the discontinuity. The same rock sample may be tested two, three, or more times under differing normal loads. The displacement is recorded as a function of shearing force during each test. A plot of shearing stress vs. strain may then be made for each normal load applied to the sample as shown in Fig. 13. The first test conducted on a discontinuity will generally yield a peak shearing value, which is the maximum shearing force which may be applied to the discontinuity. Once the peak value is reached, the discontinuity shears with less applied force. What has happened, in

Fig. 12. Direct shear machine.

effect, is that interlocking portions of the discontinuity have become sheared, and there is less resistance to displacement. A relatively smooth surface condition then exists, and a lower, but constant, shearing force will cause the strain to increase. The discontinuity is then said to be at residual strength. Should a second test at a lower normal force be made on the same discontinuity, a curve which gradually increases with stress will usually develop. Such a curve has no true peak because the interlocking mechanisms along the discontinuity surface were sheared during the first test. The maximum shearing force that can be applied will usually occur at the residual strength condition. Because a lower normal force was applied during the second test, a lower residual shearing force will result. The residual shear stresses and their corresponding normal stresses may be plotted to develop a shear-normal line such as that shown in Fig. 14. This line represents a failure envelope for various combinations of shearing stress and normal stress for a particular discontinuity. In other words, if the ratio of shearing stress to normal

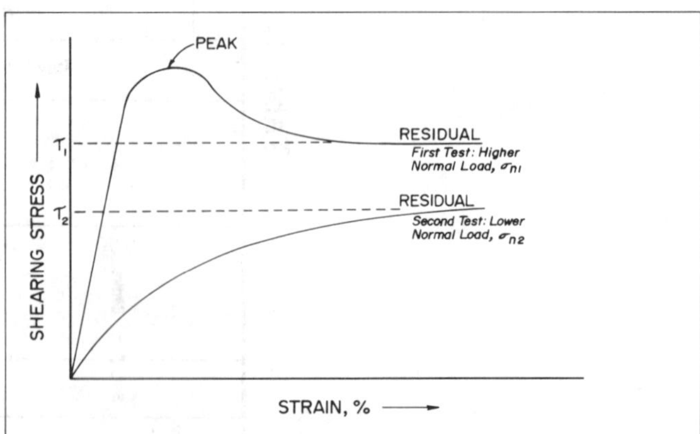

Fig. 13. Idealized stress-strain curve for a rock discontinuity.

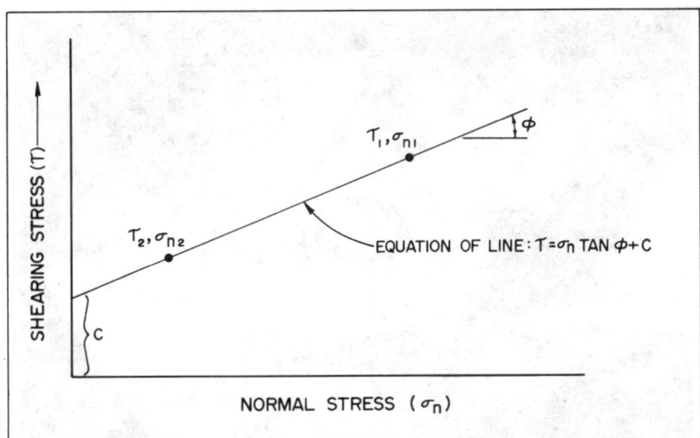

Fig. 14. Idealized shear-normal line for a rock discontinuity.

stress is above the line, displacement or failure will take place on the discontinuity. If the stress ratio is below the line, no displacement or failure will occur. The slope of the envelope is known as the tangent of the angle of sliding friction (ϕ), and the shearing stress intercept is known as cohesion (C). By combining the results of a number of tests on different samples with similar discontinuities (all clean joints, all chlorite joints, etc.), a failure envelope may then be developed for a particular set of joints, faults, or bedding planes. Owing to the fact that the failure envelope data have come from several different tests, the shear strength parameters may be treated in a statistical manner, i.e., a mean and standard deviation may be computed for both the angle of sliding friction and the cohesion. By specifying the shear strength, as well as other stability data in a statistical sense, it is possible to use analytical methods for slope design which yield a probabilistic estimate of a slope's stability.

Soils: Shear strength parameters for engineering design in soil slopes may be determined by laboratory testing. Direct shear or triaxial tests may be performed to determine the soil's shear strength. Triaxial testing is more expensive, but a greater insight into the soil's reaction to stress is obtained, particularly under differing pore pressure conditions.

Weakness Zones: Depending upon the type of mineral deposit, the pit slopes may contain zones of rock materials, which have distinctly contrasting shear strength. Such may be the case in a copper mine where hydrothermal alteration has caused weakening of an irregularly shaped portion of the pit slope. In such cases, definition of the weakness zones may be augmented with the use of surface seismic refraction and downhole seismic logging. In general, zones of lower velocity may be associated with lower shear strengths. Once these lower velocity zones have been adequately defined, relative shear strength parameters may be assigned for use in analytical slope design.

Back Analysis: Another method which may be used to estimate the shear strength parameters is to perform a back analysis of a slope failure. This method obviously may only be used where pit slopes already exist which have failures in them. The dimensions of the existing failure should be measured and the failure plane(s) located. Ground-water conditions prevailing at the time of failure will need to be known or closely estimated. Using analytical methods, the relationship of friction and cohesion may be calculated for the slope at the time of failure. Because the angle of sliding friction may usually be more closely approximated by laboratory testing than cohesion, the cohesion existing at the time of failure may be estimated from the back-analysis relationship using an appropriate laboratory friction value. Thus, the back-analysis method may be used to augment shear strength parameters obtained by other means.

Ground Water

Detrimental Effects: While geologic discontinuities are the single most important factor in slope stability, the presence of ground water is certainly a close second. The major adverse effect of ground water is to reduce the shear strength along potential failure planes. In analytical terms, this may be exemplified by rewriting the equation shown in Fig. 14 to allow for the presence of ground water:

$$\tau = (\sigma_n - u) \text{ Tan } \phi + C$$

where u is water uplift pressure.

The ground water causes an uplifting or unweighting of the normal stress across a discontinuity, thereby reducing the discontinuity's shear strength. The uplifting stress u is a function of the density and head of the water at the point in question on the discontinuity.

A further effect of ground water may be horizontal thrust. If an open tension crack exists in the slope, and surface runoff or ground-water flow fills the crack, a horizontal force will be exerted. The horizontal thrust, which also is a function of water density and head, will act in such a manner as to induce slope failure. In both uplift and horizontal thrust, it is ground-water head or pressure, not ground-water flow that causes a reduction in the shear strength.

Determination: Methods of determining the magnitude of ground-water pressure include analysis of the overall ground-water environment and direct measurement. Analysis of the overall ground-water environment may include studies of the sources of ground water and the permeability of the rock mass. Such studies may become quite complex, but reasonable estimates of the significant factors can usually provide data sufficient for stability analysis. Because pit slopes vary widely in their hydrologic environments (see Fig. 15), detailed hydrology studies may be more important for the design of one open pit mine than for another.

Direct measurement of ground-water pressures is essential during the process of collecting data on slope stability. The most convenient method of directly measuring ground-water pressures is placement of piezometers in boreholes located in the zones of interest. A piezometer can be an open hole or observation well, or it may be a sophisticated instrument placed at a specific location in a borehole. Open-hole piezometers are sufficient for relatively high permeabilities but may be entirely inadequate for situations of low permeability.

When a stability study for a future open pit mine is undertaken, there may be reasonable doubt as to the relative permeability of the rock mass. For this reason, the use of more sophisticated piezometers, such as the air-actuated variety (see Fig. 16), is recommended. They should be installed no later than the time when final ore-body delineation holes are drilled for two reasons. First, when a costly borehole is completed, piezometer installation is simple, and such multiple use of exploratory boreholes yields major cost savings. Second, early installation of a piezometer permits a history of ground-water pressure changes to be recorded for study. Because most ground-water systems exhibit a seasonal variation, a record over a period of at least one year can prove beneficial in predicting

Fig. 16. Piezometer readout unit.

Fig. 15. Mined-out pit in a semiarid environment.

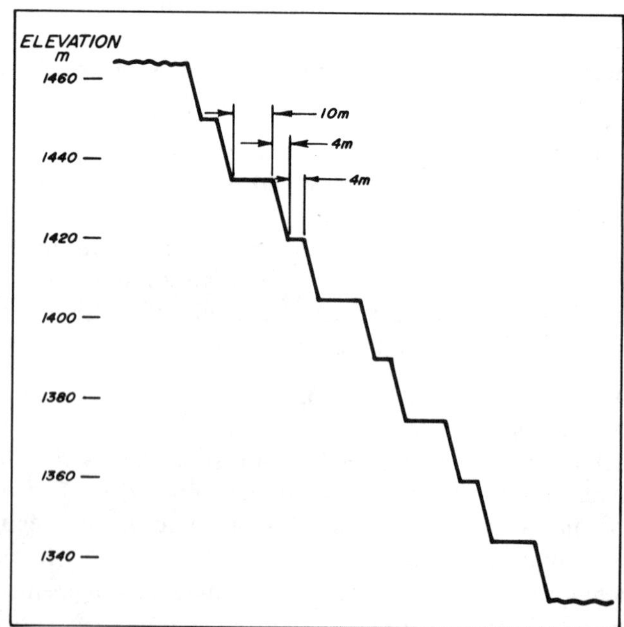

Fig. 17. Slope profile using alternating bench widths.

Fig. 18. Emplacement of horizontal drains.

potential pressures during the wettest months of the year. Direct measurement of the rainfall and its correlation to changes in ground-water pressure will not only help deduce maximum potential pressure, but will also help determine seasonal periods when remedial action must be taken to prevent surface runoff from flowing into tension cracks.

Seismic Forces: The effects of seismic forces are important for mines located in zones of high seismic activity. Collection of historic records on earthquake occurrence, intensity, and duration may provide adequate input data. The design of optimum pit slopes should take into account seismic factors, if such factors have a moderate to high probability of occurrence.

Remedial Stability Measures

The specific purpose of any remedial action is to improve slope stability and minimize potential safety hazards. Remedial measures include rubble cleanup, acknowledgment of critical zones, dewatering, blast control, artificial stabilization, and displacement monitoring.

Fig. 19. Horizontal drains.

Fig. 20. Presplit on final slope.

While loose rubble generally amounts to no more than a nuisance, it can present critical safety hazards that can be minimized by cleaning the rubble from loose raveling benches. However, access to such benches with cleanup equipment is often a problem during mining operations. Use of bench profiles with alternating bench widths, as shown in Fig. 17, may offer a solution. Such alternating bench widths allow a much wider bench every 30 m, to which long-term access could more likely be maintained. Periodic cleanup along benches, together with a 1-m berm along the outer edge of the bench, may prevent most rock falls.

Monitoring zones of potential slope failure is the ongoing response to previous work. Stability and sensitivity analyses allow in-depth investigation of potentially critical slope zones, pinpointing where special

Fig. 21. Artificial stabilization with cable anchors.

Fig. 22. Surface extensometer.

attention should be directed while mining is in progress. These zones should be inspected periodically for tension cracks and other signs of instability.

Ground water is one stability factor over which some control can usually be exerted during mining. Dewatering will almost always pay for itself as a low-cost remedy to improve slope stability specifically when sensitivity analyses have indicated that either a particular slope must be dewatered or its slope angle must be reduced to 5° or more. Horizontal drains often used in dewatering are illustrated in Figs. 18 and 19.

Blast-control techniques such as presplitting (see Fig. 20) and smooth wall blasting may preserve the natural strength of near-surface discontinuities.

Fig. 24. Surface extensometer with continuous recorder.

Stronger discontinuities would result in fewer rock falls particularly on individual benches. As most blast-control techniques are expensive, cost vs. anticipated results should receive close consideration before any large blast-control program is put into effect. Usually such control programs should be limited to final pit slopes.

Artificial stabilization techniques have proved to be effective in improving slope stability under the right

Fig. 23. 3D monitoring device.

Fig. 25. Borehole extensometer.

Fig. 26. Light ranging instrument.

conditions (Seegmiller, 1975). Such techniques, including the use of cable anchors (see Fig. 21), can represent a sizable investment, and a thorough study should be conducted before committing a mining operation to large-scale artificial stabilization projects. Small-scale projects, those which might be directed at improving slope stability in the vicinity of an in-pit crusher, may prove very favorable economically. Specific recommendations for the use of artificial support systems can be part of the total rock mechanics approach to maintaining slope stability.

A displacement monitoring plan, with implementation procedures spelled out in detail, should be part of the overall approach to achieving slope stability. Optimum slope angles may necessarily include some relatively steep slopes, which should be monitored along with the more sensitive zones to anticipate and prevent safety hazards. Displacement monitoring can serve as a tool for determining, in advance, the best time for undertaking appropriate adjustments or actions. The monitoring plan itself may require various degrees of sophistication and various types of hardware. Illustrated in Figs. 22 through 26 are some typical monitoring devices.

References

Coates, D. F., 1970, *Rock Mechanics Principals,* Mines Branch Monograph 874, CANMET, p. G-1.

Coates, D. F., and Sage, R., 1973, "Rock Anchors in Mining," Mines Branch TB181, CANMET, Nov., p. 50.

Goodman, R. E., 1976, *Methods of Geological Engineering,* San Francisco, p. 472.

Kim, Y. C., and Hall, T. E., 1977, "Economic Analysis of Pit Slope Design," *CIM Bulletin,* Vol. 70, No. 783, July, pp. 100-108.

Seegmiller, B. L., 1972, "Rock Stability Analysis at Twin Buttes," *Proceedings,* 13th Symposium on Rock Mechanics, American Society of Civil Engineers, New York, pp. 511-536.

Seegmiller, B. L., 1973, "Slope Stability Research: Its Payoff in Mining," *Mining Congress Journal,* Vol. 59, No. 7, July, pp. 32-39.

Seegmiller, B. L., 1974, "Artificial Stabilization of a Pit Slope at Twin Buttes, Arizona," *Mining Engineering,* Vol. 26, No. 12, Dec., pp. 29-34.

Seegmiller, B. L., 1974a, "Time-Dependent Output from In Situ Measurements: Its Meaning with Respect to Stability," 9th Canadian Rock Mechanics Symposium, Montreal, Dec.

Seegmiller, B. L., 1975, "Cable Bolts Stabilize Pit Slopes, Steepen Walls to Strip Less Waste," *World Mining,* Vol. 28, No. 7, July, pp. 36-41.

Seegmiller, B. L., 1976, "How to Cut Risk of Slope Failure in Designing Optimum Pit Slopes," *Engineering and Mining Journal,* Vol. 177, No. 12, Dec., pp. 53-59.

Seegmiller, B. L., 1977, "Site Characterization Using Oriented Borehole Core," *Monograph on Rock Mechanics Applications in Mining,* S. J. Green, W. L. Brown, W. A. Hustrulid, eds., AIME, New York, pp. 74-78.

Stewart, R. M., and Seegmiller, B. L., 1972, "Requirements for Stability in Open Pit Mining," *Geotechnical Practice for Stability in Open Pit Mining,* AIME, New York, pp. 1-7.

12 Analytical Design

Dermot M. Ross-Brown
Science Applications, Inc.

Dermot Ross-Brown obtained his B.S. and M.S. degrees in mining engineering and foundation engineering from Birmingham University in England. Between 1964 and 1968 he was employed by Ashanti Goldfields Corp. in Ghana. In 1973, he completed his Ph.D. in rock mechanics at Royal School of Mines, Imperial College, London. He then spent 4½ years with Dames & Moore in various capacities. He is currently with Science Applications, Inc. in Fort Collins, CO.

Dr. Ross-Brown has specialized in the stability of soil and rock slopes, tailings dam design, terrestrial photogrammetry, rock mechanics testing, the collection of geotechnical data, and the design of open pit and underground mines. He is a registered professional engineer in California and Colorado.

INTRODUCTION

Slope angles are one of the major factors affecting the shape of the final pit and the location of the walls. Due to differences in geology, the optimum slope angles vary from pit to pit and within different parts of the same pit. Indeed, it is not unusual to have different slope angles for the top and bottom portions of the same cross section.

The optimization of these slope angles is a major study in itself and this chapter is intended as a general introduction to the subject. It is not intended to be a comprehensive text for the absolute beginner. The objective is to put the various aspects of slope stability analysis into relative perspective by describing the various analytical techniques available and indicating when each may be most appropriately used. References are made to the many excellent texts and papers on slope stability for the reader who does not have a formal training in soil or rock mechanics.

In writing this chapter, an assumption has been made that a tentative mining plan has been drawn up, including a preliminary estimate of the slope angles. It has also been assumed that all the necessary field and laboratory data has been collected and reduced to a form suitable for input into the stability analyses.

Objectives in Designing Final Pit Limit Slopes

The main objectives of designing final pit slopes is to make them as steep as possible without incurring any major economic penalties due to slope instability and without endangering the safety of personnel.

Steeper slopes will generally minimize the quantity of waste rock that has to be mined for the same amount of ore recovered. Much will depend upon the physical dimensions of the ore body, its depth below surface, the types of ore encountered, and their distribution and grade. The type of cutoff is also important, whether it is a geological boundary or an assay cutoff.

The major economic penalties due to slope instability may be listed as follows: (1) loss of ore, (2) extra stripping costs resulting from a new pushback to recover the ore that would otherwise be lost, (3) cost of cleaning up failures, (4) costs associated with rerouting the haul road, (5) production delays, and (6) inefficient production due to inaccessibility of some working areas or cramped working conditions.

Although safety is a very important criterion, injuries caused by major slope failures are rare. However, the rockfall hazard may sometimes be of great concern. Such a hazard is normally found to be more dependent on how the benches are initially excavated and scaled rather than on the ultimate slope angle. In such cases, the concern should be for the frequency of such falls and the necessity of having unprotected men working beneath such slopes. Consequently, slope design is largely a problem of economic optimization in which the anticipated economic penalties due to slope instability are offset against the savings due to reduced stripping.

Geotechnical Factors Influencing Slope Stability

The geotechnical factors affecting slope stability depend primarily on the geology, in particular, the lithology; the presence of weak zones; and the frequency, location, and orientation of faults, joints, and other discontinuities. The mechanical properties of the slope materials, such as the intact strength of the materials and the shear strength mobilized along discontinuities, are also important. The influence of water acting in various ways and seismic accelerations also have to be taken into account.

Slopes encountered in open pit mines are composed of a variety of materials ranging from soils to competent rocks. Although generalizations are difficult to make, the competence of the slope materials often improves with depth so that a typical mine slope may be comprised of up to several hundred meters (feet) of overburden underlain by relatively competent rocks. Even then, layers or zones of weak soil-like material may occur anywhere in the slope and these often control the stability of the overall slope.

While soil masses are often homogeneous and relatively isotropic, rock masses rarely are. The main reason for this is the ubiquitous presence of discontinuities in rock masses as compared to soils. The behavior and stability of rock masses are mainly controlled by the nature and orientation of these discontinuities, while the stability of soil masses is mainly controlled by the strength of the intact material. While these statements are generally true, it should be noted that no two slopes are exactly the same, that there is no definite demarkation between soils and rocks and their predicted behavior but rather a gradual change, and that there are other factors which affect the stability of slopes. Nevertheless, it is generally true to say that the intact strength and the presence of discontinuities are the most important factors with intact strength being the more important in the case of soil masses and discontinuities in the case of rock masses.

Recognition of Different Failure Modes

As a result of observing a large number of slope failures in open pit mines and elsewhere, a number of

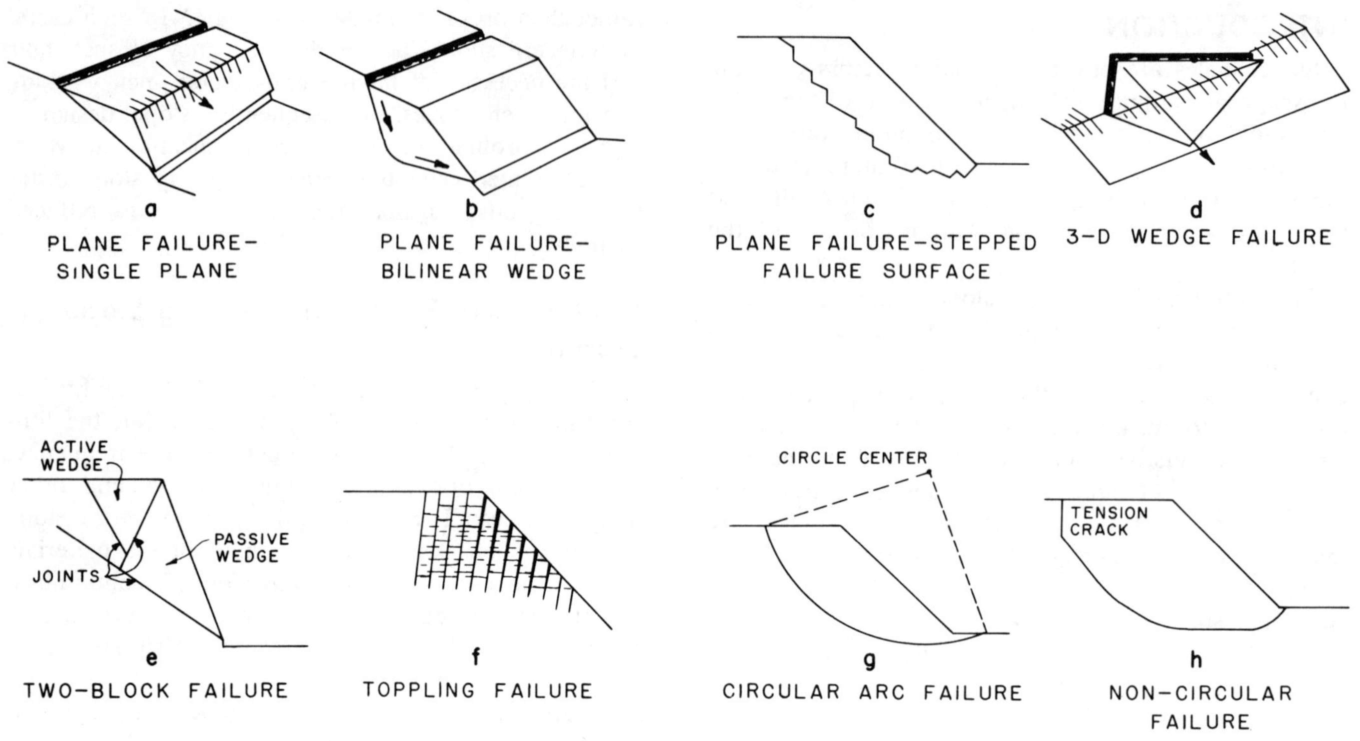

Fig. 1. Failure modes commonly seen in open pit mines.

recognizable modes of failure have been identified. Some of these are shown in Fig. 1.

Plane failure occurs when a discontinuity striking approximately parallel to the slope face and dipping at a lower angle intersects (or daylights) the slope face enabling the material above the discontinuity to slide. Variations on this simple failure mode can occur, as for instance when a tension crack forms or when the mean sliding plane is a combination of two joint sets which form a stepped path (Figs. 1a, 1b, 1c).

The 3D wedge failure occurs when two discontinuities intersect in such a way that the wedge of material formed above the discontinuities can slide out in a direction parallel to the line of intersection of the two discontinuities (Fig. 1d). Occasionally, these discontinuities can consist of stepped failure paths from one or more sets of joints.

Two-block failure occurs when an unstable wedge of rock produces an active force on a lower second wedge of rock which would otherwise be stable, resulting in the failure of both wedges. It occurs in slopes where major joints with various orientations can form active and passive wedges. Note that the lower passive wedge must rest on a low angle discontinuity in order to be stable by itself (Fig. 1e).

Toppling failures may occur in slopes having a near-vertical joint set. Very often the stability of the system will be dependent on the stability of one or two key blocks; once they are disturbed, the system may collapse (Fig. 1f).

All the failure modes mentioned are typical of rock slopes in which the mechanisms are dependent upon the orientations of the discontinuities and their relationship to the orientation of the cut slope. Sometimes in very high rock slopes, the stresses at the toe may be sufficiently high to cause crushing of a critical block, allowing an otherwise stable rock mass to fail by one of the modes previously described.

There are two other failure modes which can occur in soil slopes or in high rock slopes. These are the circular and noncircular failure surfaces common in soil mechanics, sometimes known as general slip surfaces. The circular method occurs when the joint sets are not very well defined and when the average block size is small compared to the height of the slope. In such cases, the rock slope is more akin to a coarse soil at the scale of a soil slope and a circular failure surface may then be generated. A noncircular failure surface occurs under similar circumstances to the circular arc. In this case, however, the failure surface tends to run parallel to a set of weakness planes such as bedding planes and consists of a mixture of curved and linear segments (Figs. 1g and 1h).

PRINCIPLES OF SLOPE STABILITY ANALYSIS

Limiting Equilibrium Methods

In limiting equilibrium methods of analysis, statics are employed to analyze the stability of the mass of rock and/or soil above the failure surface. If failure has already occurred, the geometry of the failure surface can be determined and analysis of the failure is known as back analysis. If it is a design situation, however, the failure surface is potential rather than actual. In this case, many potential failure surfaces may have to be analyzed to find the most critical geometry before a slope design can be considered acceptable.

In the case of the plane failure, the 3D wedge failure, the circular and the noncircular failure modes, the material above the failure surface will be on the point of slipping when the disturbing forces due to gravity are just counterbalanced by the forces tending to restore equilibrium (such as the forces due to friction and cohesion). The ratio of these two forces defines the factor of safety (F), which equals unity at the point of limiting equilibrium, i.e.,

$$\frac{\text{forces tending to restore equilibrium}}{\text{forces tending to disturb equilibrium}} = F$$

where $F=1$ at the point of limiting equilibrium.

Limiting equilibrium methods of stability have been used successfully in soil mechanics for over 100 years. Through back analysis of failed slopes, these methods have been improved and considerable confidence developed in their use. For instance, it is not uncommon for earth dams (in which the material properties and water conditions are carefully controlled) to be designed with factors of safety as low as 1.2. Limiting equilibrium methods of analysis have also been used successfully in the design of rock slopes, but due to uncertainties in the input parameters, much higher safety factors are usually required.

Stress Analysis Methods

Failures do not necessarily always occur along well-defined failure surfaces. An example of this is the block-flow failure mode in which the structural conditions do not favor sliding on discontinuities and where the material is not ductile enough to fail along a circular or noncircular surface. In this case, crushing of the rock material occurs at the points of highest stress. The high stresses are then transferred to other parts of the rock mass which in turn fail. Progressive failure of the rock mass can subsequently develop in such a manner as to cause considerable deformation of the slope and/or complete failure of the rock mass. Alternatively, a failure that starts off as a progressive failure of the rock substance can develop into a structurally controlled failure such as a 3D wedge.

The objective of stress analysis methods is to represent the rock mass by a series of structural elements (finite element method) or cells of constant material mass (one finite difference method) and perform an analysis to determine the stresses at points within the slope. The stress distribution can then be examined to determine where rock failure is likely to occur; rock failure occurs when the stresses to which the rock is subjected exceed the strength of the rock. Due to the interdependence of the stresses between the different rock elements, a computer solution is usually employed to determine the stress distribution within the rock mass. The computer analysis can then be used to re-examine the situation in the next instant of time. In this manner, the progressive failure of rock slopes may be studied.

Stereographic Techniques

Since most rock masses are composed of several different rock types and numerous discontinuities at various orientations, the number of potential failure geometries is also numerous. The task of deciding which failure geometries are likely to be the most critical is aided by the use of stereographic projections. Using these techniques, the orientations of the slope face and the joint sets within the rock mass may be represented on a plane diagram and the mechanisms of failure which are kinematically possible may be rapidly determined.

Basically, spherical projection is a method of representing 3D orientation data on a plane. It involves the use of a reference sphere to represent the orientations of lines and planes in space. The intersections of lines and planes with the reference hemisphere are projected onto a flat surface, such that lines are represented as points, and planes are represented as lines. It is only necessary to use one-half of the reference sphere in an analysis and, by convention, the lower hemisphere is usually used in geotechnical work.

Although several methods of projection are possible, the equal area projection in which a unit area on the projection plane represents the same fraction of the total area on the reference hemisphere is particularly useful (Fig. 2). For instance, the polar equal area net (Fig. 3) is often used for obtaining the density of plotted points, using a counter of fixed area and determining the centers of joint clusters. The Schmidt equal area net (Fig. 4) is usually used for determining which joint orientations might produce failures and for analyzing these situations.

A simple example of stereographic projection is

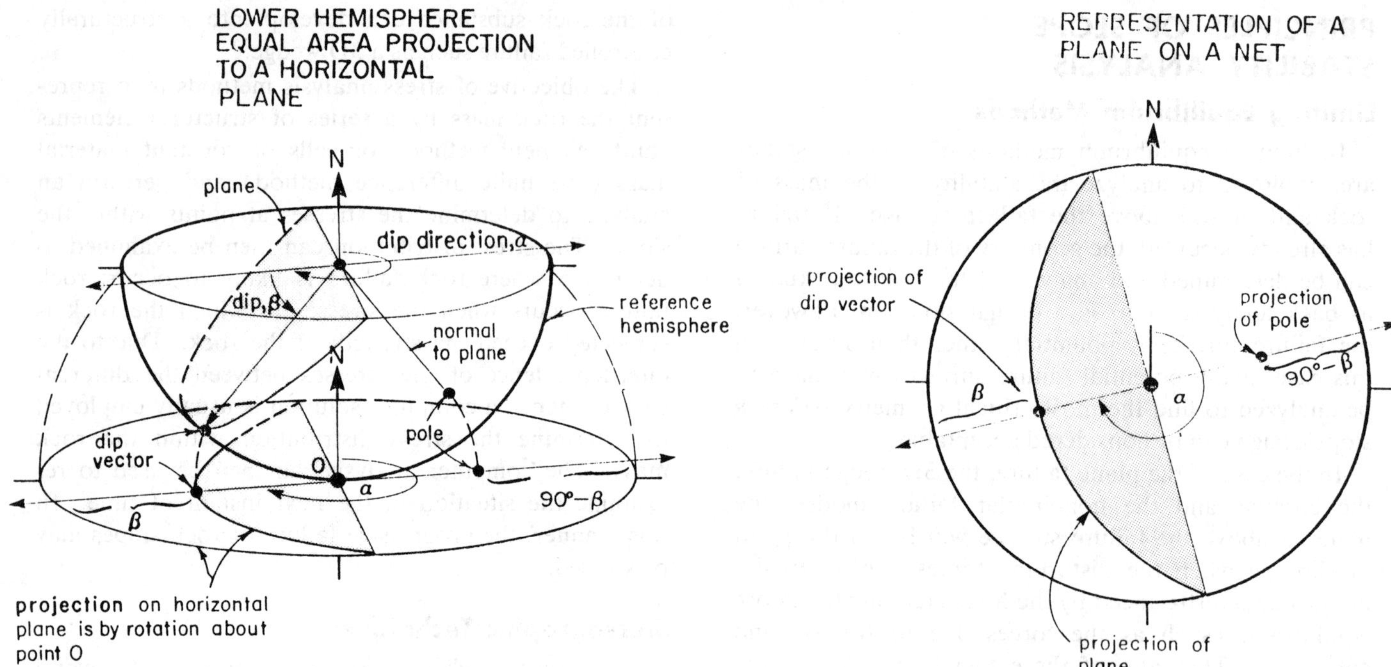

Fig. 2. Principles of equal area projection of a plane onto a net.

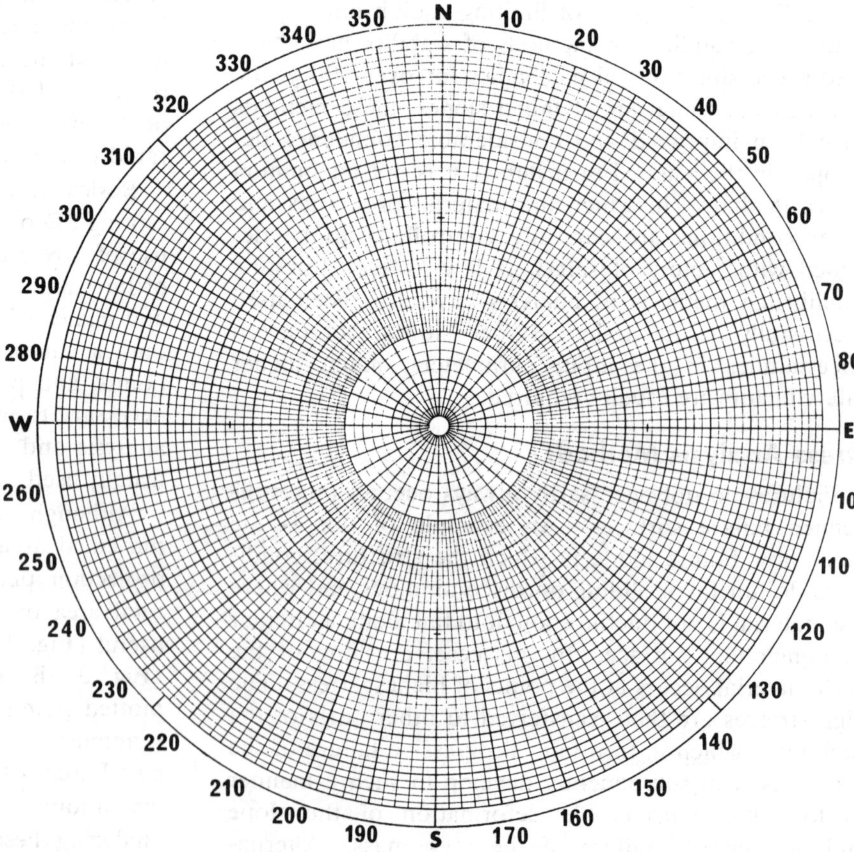

Fig. 3. Polar net (equal area).

POLAR NET (EQUAL AREA)

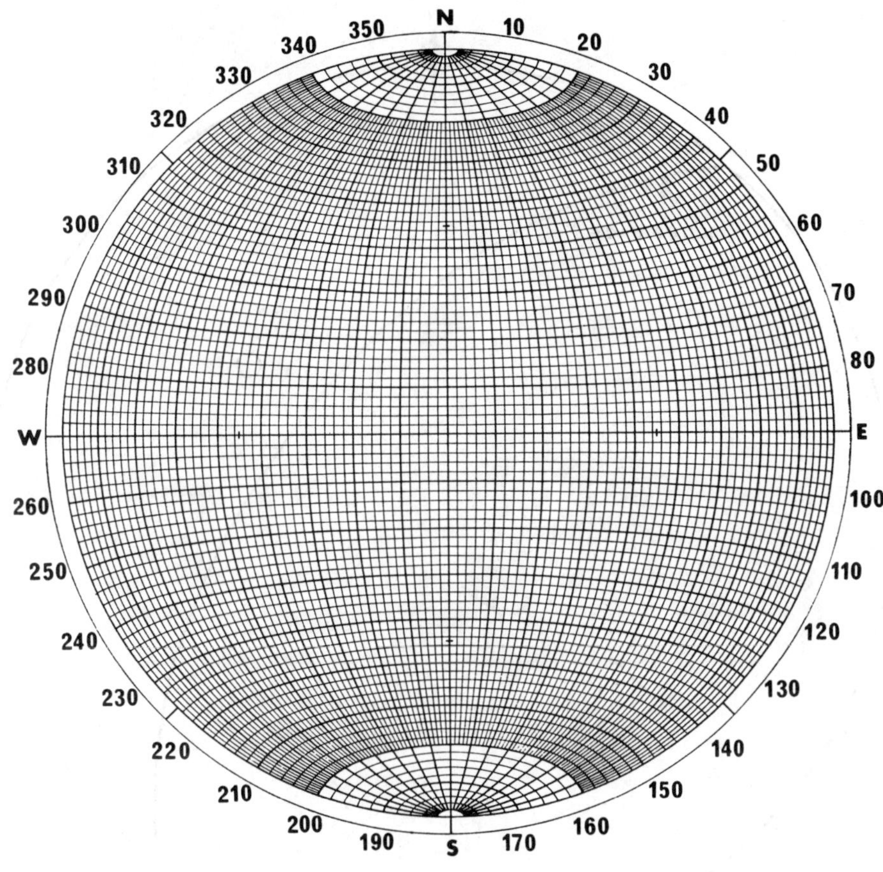

Fig. 4. Schmidt net (equal area).

shown in Fig. 5 where the slope face is represented by a great circle, and the dip vector of a joint set is represented by a point. The interpretation of the diagram is that plane failure is kinematically possible since the dip vector lies on the convex side of the line representing the slope face. In a similar manner, the intersection of the joint sets 1 and 2 on the convex side of the line representing the slope face in Fig. 6 indicates the possibility of a 3D wedge type of failure.

Excellent instructions on the use of stereographic projection are included in publications by Phillips (1971), John (1968), Hoek and Bray (1974), and Goodman (1975) among others. A thorough familiarity with these techniques is essential to anyone who wishes to be proficient in the design of rock slopes in open pit mines. As a general rule, no computer analysis of the stability of a rock slope should be made without a preliminary analysis on a stereonet. Even then the quality of the input data may dictate that the stereographic analysis is the most appropriate technique and that computer analysis is neither necessary nor more accurate.

Definitions of Stability

Traditionally, the stability of slopes has been analyzed using limiting equilibrium techniques and the results expressed in terms of a factor of safety. Calculations are usually based on mean values of the measured input parameters. Due to the scatter of test results and the uncertainty of these input parameters, a factor of safety greater than 1.0 is necessary to ensure an acceptably low chance of failure.

The sensitivity of the factor of safety to changes in the input variables may be determined by varying the input data and calculating the factor of safety for each combination of input variables. Although this involves much more work, it is much better than assuming that the first answer is correct and basing investment decisions on this single calculation. Even so, a sensitivity analysis should only be used to supplement engineering judgment in deciding on the likely variations of the various parameters and in determining the most likely value of the factor of safety.

In any case, there are several disadvantages to quot-

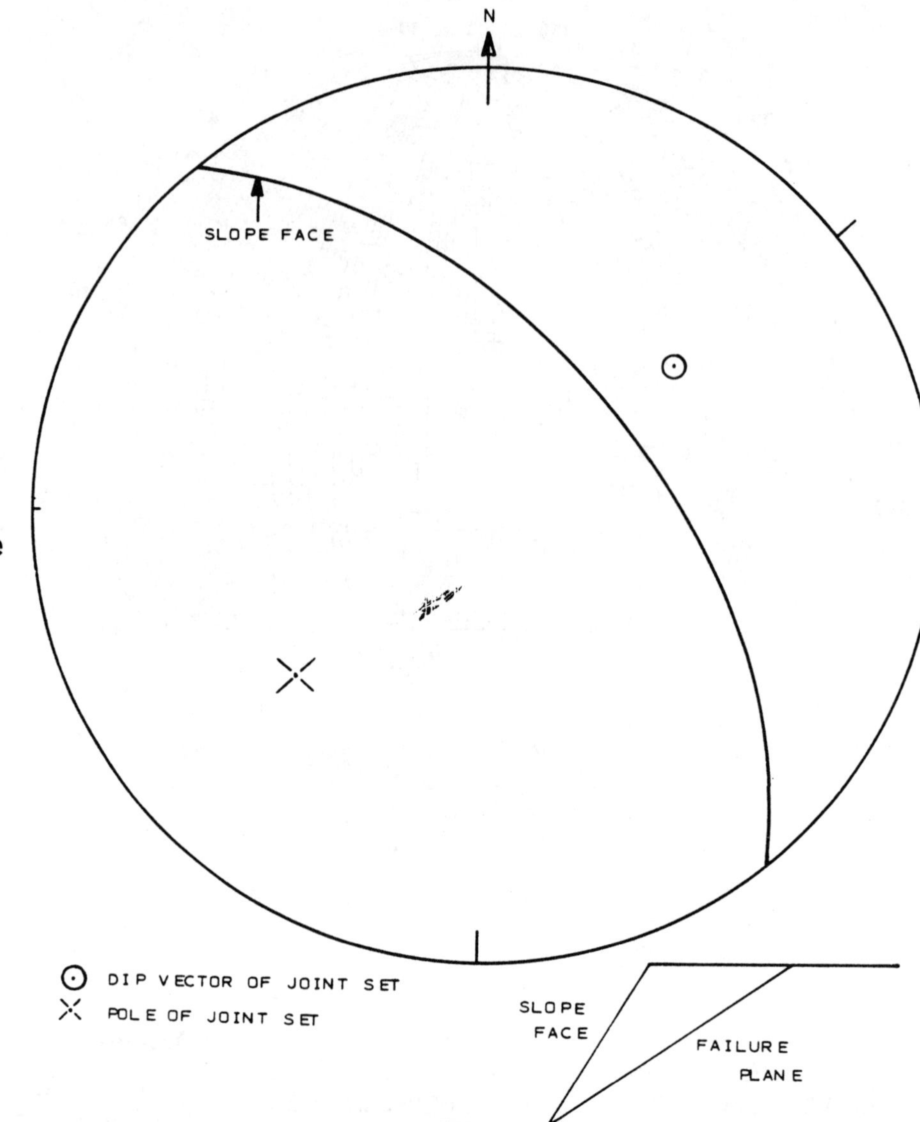

Fig. 5. Representation of a plane failure on a Schmidt net.

ing a single value for the factor of safety. For instance, in a long slope, failure may be kinematically impossible using mean values of the slope orientation and the joint orientations, and yet be kinematically possible for a certain percentage of the slope due to natural variations in the orientations about the mean. In addition, a given factor of safety will not always correspond to the same chance of failure.

Consequently, a more recent approach to the problem has been to express the stability of open pit slopes in terms of the chance or probability that the slope will fail. The probability is usually determined for various slope angles at a fixed slope height or for various slope heights at a fixed slope angle. In order to perform these analyses, it is necessary to quantify the input parameters by means of a careful program of testing and measurement. The results are more useful, however, since mine planners can choose a slope angle with a probability of failure which is acceptable to them. In addition, the results may be directly incorporated into a full risk analysis.

Although no method of analysis can fully replace engineering judgment, a probabilistic analysis can minimize the judgment required by correctly accounting for quantifiable uncertainties. A probabilistic analysis can be incorporated into most methods of slope stability analysis using a Monte Carlo simulation package which was recently developed as part of the *Pit Slope Manual* (Sage, 1977; Kim, Major, and Ross-Brown, 1978).

FAILURE ALONG A SINGLE PLANE OR A STEPPED PATH

Simple plane failure is the easiest form of rock slope failure to analyze. Although the various possible com-

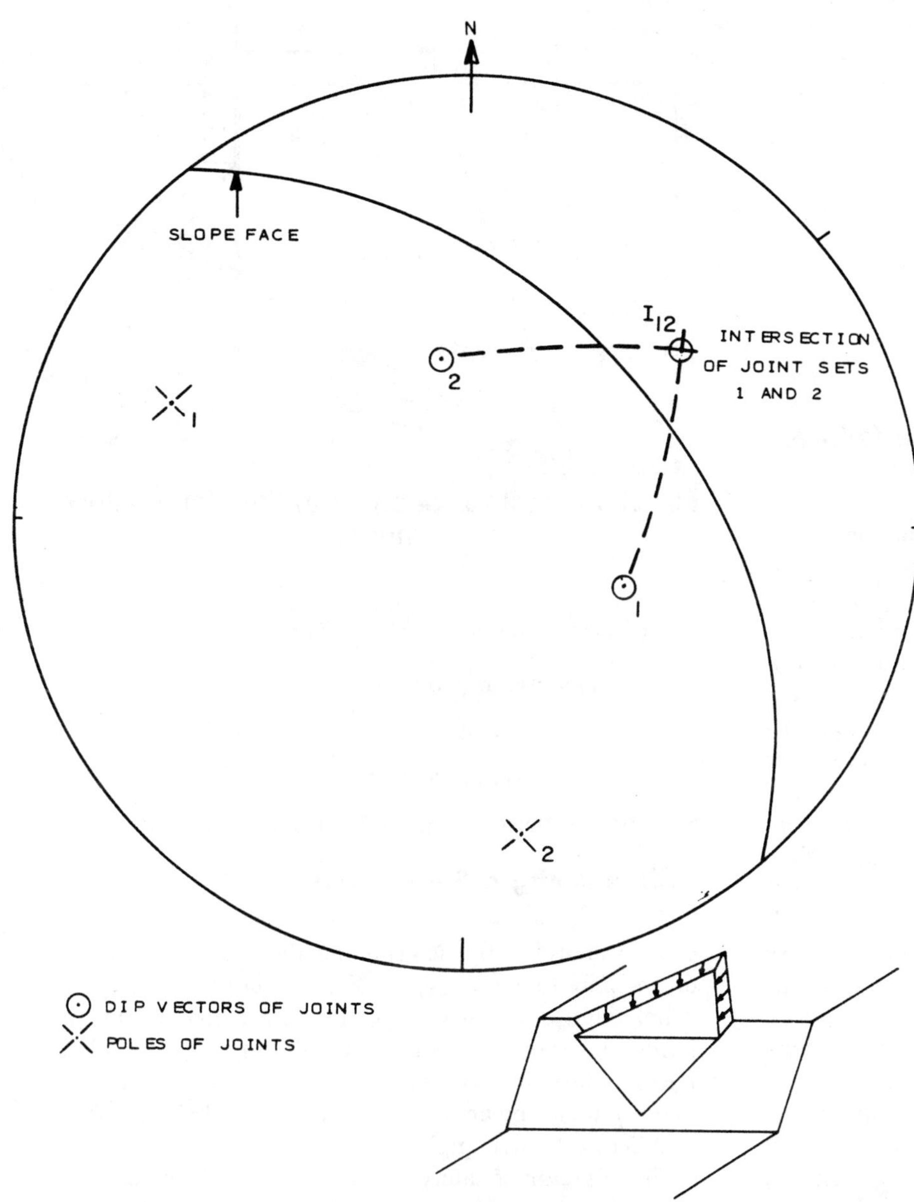

Fig. 6. Representation of a 3D wedge failure on a Schmidt net.

binations of tension crack, water pressure, and other external forces can make hand calculations lengthy, the problem can usually be adequately represented in two dimensions. This means that its solution is never anything more than the analysis of the equilibrium of a single block (the failure mass) resting on a plane and acted upon by a number of external forces (water pressure, reinforcement, earthquake loading, etc.). Deterministic (factor of safety) and probabilistic solutions in which some parameters are considered as being precisely known may be readily obtained by hand calculation if the effect of moments is neglected.

Factor of Safety Analysis of Plane Failure

The factor of safety for the general case of plane failure (Fig. 7) is the ratio of the forces acting to keep the failure mass in place (the cohesion times the area of the failure surface plus the frictional shear strength determined using the effective normal stress on the failure plane) to the forces attempting to drive the failure mass down the failure surface (the sum of the components of weight, water forces, and all other external forces acting along the failure surface). This is determined by resolving all the forces acting on the potential failure mass into directions parallel and normal to the potential failure surface. The general factor of safety expression which results is:

$$F=\frac{cA+(W\cdot\cos\Psi p+T\cdot\cos\theta-U-V\cdot\sin\Psi p)\tan\phi}{W\cdot\sin\Psi p-T\cdot\sin\theta+V\cdot\cos\Psi p} \tag{1}$$

where W is the weight of the sliding block, A is the

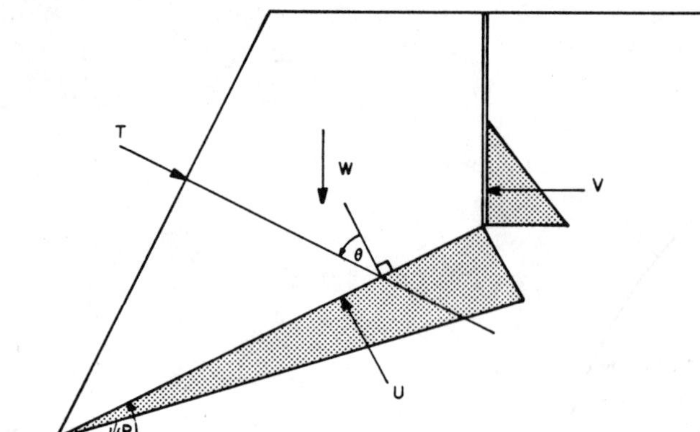

Fig. 7. The general case of plane failure.

area of the sliding surface, Ψp is the inclination of the sliding surface, U is the uplift force due to water pressure on the sliding surface, V is the force due to water pressure in the tension crack at the rear of the block, ϕ is the friction angle acting on the sliding surface, c is the cohesion acting on the sliding surface, T is an external force acting on the block, and θ is the angle between the external force and the normal to the sliding surface.

Note that any number of external forces can be included in this form of the factor of safety expression. In special cases of plane failure, e.g., if the slope is dry or there are no external forces, the expression is simplified because the associated terms become zero. All weights and forces can be determined either graphically or analytically.

A limitation of this type of analysis arises because the linear approximation to the shear strength envelope, which is used in defining the numerator, is only strictly valid at one normal stress. Although the stress across the failure plane is commonly calculated for the case of a dry slope with no external loading, the effective stress acting across the failure plane for the conditions of water and the external loading present in the slope should always be used. (Note that effective stress has a special meaning in geotechnical work and is equal to the total stress minus the pore water pressure.)

Example 1: Calculate the factor of safety of the slope shown in Fig. 8. Consider a unit width of the slope.

$$A=\frac{(100+200\cdot\tan\ 20^\circ)}{\cos\ 30^\circ}$$
$$=199.5\text{ sq ft }(21.5\text{ m}^2)$$
$$W=\left[\frac{(100+199.5\cos\ 30^\circ)}{2}\times 200 - \frac{172.9\times 172.9\ \tan\ 30}{2}\right]165$$
$$=3{,}077{,}000\text{ lb }(13\ 700\text{ kN})$$

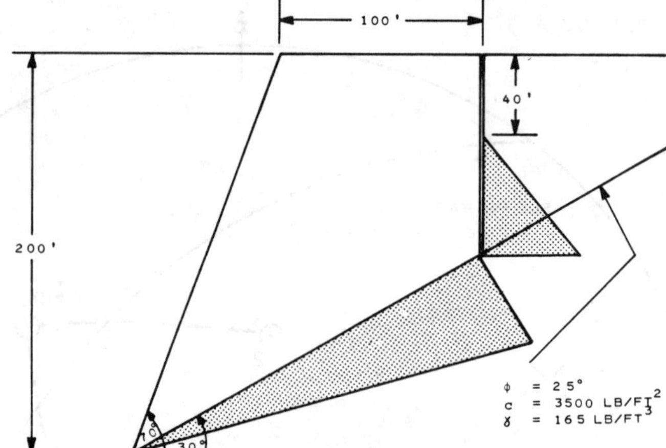

Fig. 8. Example 1, analysis of the plane failure mode.

$$V=\frac{\gamma H^2}{2}=62.4\times\frac{(100.2-40)^2}{2}$$
$$=113{,}000\text{ lb }(502\text{ kN})$$
$$U=\frac{\gamma HL}{2}=62.4\times\frac{(59.8\times 199.5)}{2}$$
$$=372{,}000\text{ lb }(1655\text{ kN}).$$

Substituting these values into Eq. 1 gives $F=1.06$.

Failure Along a Stepped Path

The case of rock slope failure along a stepped path can be analyzed by determining the mean orientation of the failure path (Ψp). Fig. 9 shows the way in which failure will occur by shearing along the major joints and simple separation on the cross joints. The mean failure surface, Ψp, dips more steeply than the major joint set by an amount δ, referred to by McMahon (1973) as the step angle.

For a factor of safety analysis, Ψp should be used to determine the weight of the failure mass and the effect

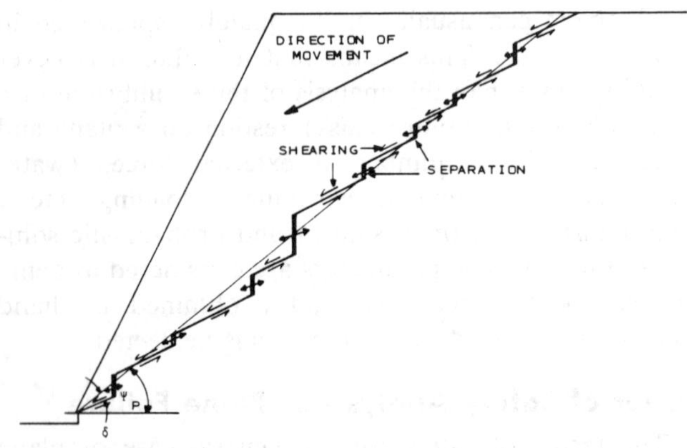

Fig. 9. Failure along a stepped path.

of water pressure. All forces should be resolved in the direction of movement of the failure mass, so the dip of the major joint set should be used in place of ψp in Eq. 1 (Jennings, 1970).

Defining a Stepped Failure Path: The step angle, δ, can be calculated in terms of the mean length ($\bar{L}$) and mean spacing ($\bar{s}$) of the major joint set as

$$\tan \delta = \frac{\bar{s}}{\bar{L}}. \qquad (2)$$

If spacing and length have log-normal distributions, the geometric means should be used (McMahon, 1973). This approach assumes that a cross joint will always exist running from the end of one major joint to the start of the next and so results in a single value for the step angle.

A different approach (Call, 1974) is to calculate δ using the distributions of joint lengths, dips, and spacings determined from a field investigation. Using a realistic failure criterion for failure through intact rock, Monte Carlo simulation can be used to model failure paths through the rock mass. Repeating this calculation results in a distribution of possible failure paths and an estimate of the probability that no failure path will occur. These results can then be incorporated into a complete probability analysis.

Allowances for Noncontinuous Joints

Several methods have been developed to define the path along which a plane failure will occur in a structurally controlled failure. The need for this arises from recognizing that the joints or other structures which define the failure surface(s) are frequently not continuous over the entire length of the failure surface. In these cases, the failure mechanism must be more complex than simple sliding along a continuous joint.

3D WEDGE FAILURE

The 3D wedge failure mode, as illustrated in Fig. 10, is a common mechanism of rock slope failure. It is particularly common on the individual bench scale but can also provide the failure mechanism for a large slope where structures are very continuous and extensive.

The 3D nature of the problem complicates the analysis. Certain assumptions are frequently used to simplify the analysis. It is, therefore, necessary to have an appreciation of the wedge problem and the methods of solution available in order to determine what assumptions are justifiable and what method of solution is the most appropriate.

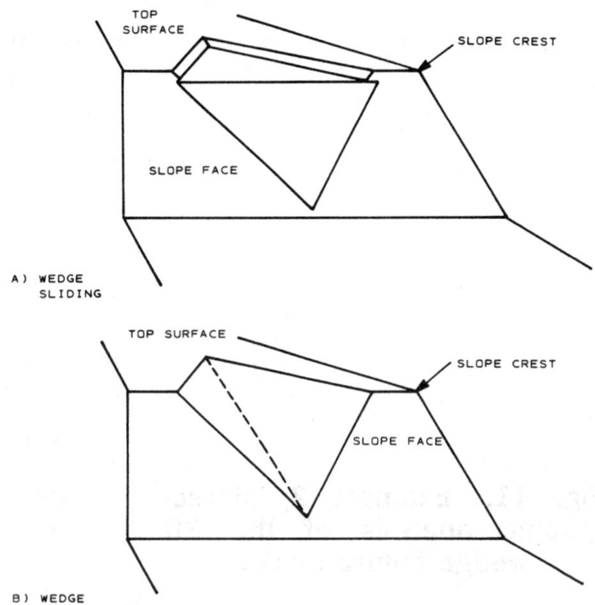

Fig. 10. The 3D wedge failure mode.

Spherical Projection Solution Using Factor of Safety

The 3D wedge problem can be analyzed using spherical projection techniques as described in detail by Hendron, Cording, and Aiyer (1971) and others.

When the shear strength of the shear surfaces is entirely frictional and there are no external forces, the problem is dimensionless and can be analyzed very simply by means of a stereonet analysis alone. The introduction of water pressure or other external forces requires the use of side calculations to determine the orientation of the resultant force acting on the wedge.

Use of a spherical projection rapidly establishes a zone of orientations for the resultant force for which the wedge will remain stable. The construction is illustrated by means of the following example.

Example 2: Determine the factor of safety against sliding down the line of intersection of the wedge formed by joints at the following orientations.

	Plane A	*Plane B*
Dip Direction	40	140
Dip	50	60

Assume the effective friction angle to be 25° on plane A and 32° on plane B. The normals to the shear planes, N_A and N_B, are plotted on a Wulff equal angle projection in Fig. 11 and friction circles of radius ϕ (25° and 32° as appropriate) are plotted about these normals.

The orientation of the line of intersection of the wedge is defined by the intersection of the great circles

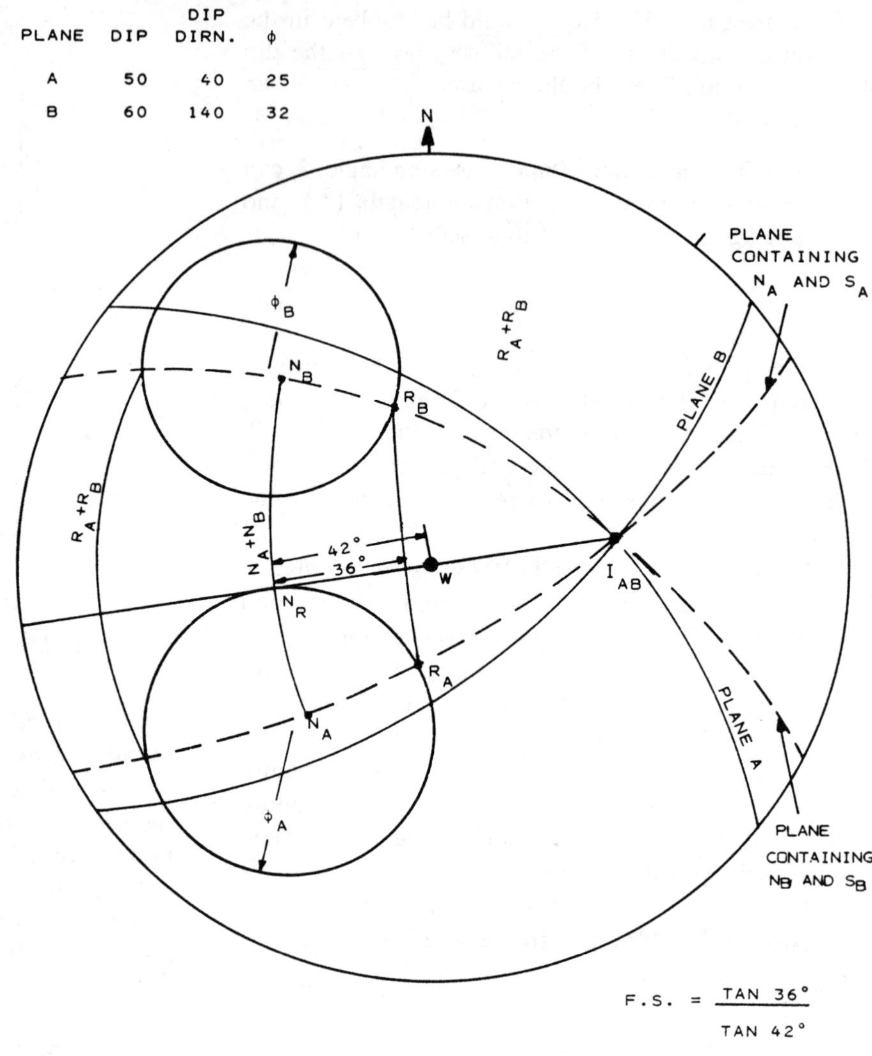

Fig. 11. Example 2, stereographic analysis of the 3D wedge failure mode.

which define the joints, I_{AB}. As 3D wedge failure is characterized by sliding in the direction of the line of intersection, this location is also the orientation of the shear forces on each plane, S_A and S_B.

The direction of the reaction on each plane at the limiting equilibrium condition, R_A and R_B, lies in the plane defined by the normal and shear forces and at an angle of ϕ degrees from the normal. This orientation is defined by the intersection of the great circle through both the normal and the line of intersection orientations with the friction circle about the normal.

The resultant of the reactions on both shear surfaces lies in the plane of these reactions, and this plane is defined by the great circle which passes through the orientations of these reactions. Note that the two great circles which fulfill these conditions define the limiting equilibrium conditions for sliding down and up the line of intersection. Along with the limits of the friction circles, these great circles define what is generally termed the stable zone. The wedge will remain stable as long as the resultant force acting on the wedge falls within this zone.

The resultant of the normal forces acting on each shear plane lies in the plane defined by these two forces. This is represented by the great circle containing both these points.

The direction of the resultant driving force acting on the wedge can be plotted directly on the net. If the wedge is acted on only by gravity, this direction is vertical and plots in the center of the net. Otherwise, the direction can be determined by a side construction of the force polygon. If this resultant force lies within the stable zone, the wedge will be stable.

To determine the factor of safety against sliding, the great circle containing both the resultant force acting on the wedge and the resultant shear force is drawn. The intersection of this great circle with the great circles through both normals and through both reactions on the shear planes defines the positions of the resultants of these normals and reactions, N_R and R_R. Measuring along this great circle gives the angles between the resultant normal, the resultant reaction, and the resultant of the forces acting on the wedge. The factor of safety can then be defined as the ratio of the resultant shear force acting along the line of intersection of the wedge to the resultant shear strength available to resist sliding in the same direction. This is given as

$$F = \frac{\tan 36^\circ}{\tan 42^\circ} = 0.81.$$

Chart Solutions

Hoek and Bray (1974) have produced a series of charts which can be used to rapidly assess the stability of rock wedges for which there is no cohesion or external forces. Under these conditions and for a given friction angle, the factor of safety is a function only of the dips and dip directions of the shear planes. These charts are convenient to use for simple wedge problems but do not give the insight or feel for the problem which is achieved by using a stereographic solution.

Spherical Projection Solution Using Probability

It is possible to account for the uncertainty in the orientations of the shear planes in order to determine with a specified confidence whether a given wedge will fail using a Monte Carlo analysis (Major, Kim, and Ross-Brown, 1977, and Major, Ross-Brown, and Kim, 1978).

TWO-BLOCK FAILURE

Two-block failures are a much less common mode of rock slope failure than single block failures such as the plane and the 3D wedge and, consequently, are only briefly considered here.

Several methods of solution exist and each may be appropriate at some level of investigation. The forces acting in the general two-block problem are shown in Fig. 12.

Stereographic Solution

A stereographic analysis is a convenient way of determining whether or not a two-block configuration will be stable (Goodman, 1975 and Kuykendall and Goodman, 1976). Any form of shear strength envelope can be accounted for by use of the secant angle of friction.

TOPPLING FAILURE

Toppling or overturning has been recognized by several investigators as being a mechanism of rock slope failure and has been postulated as the cause of several failures ranging from small to large in size.

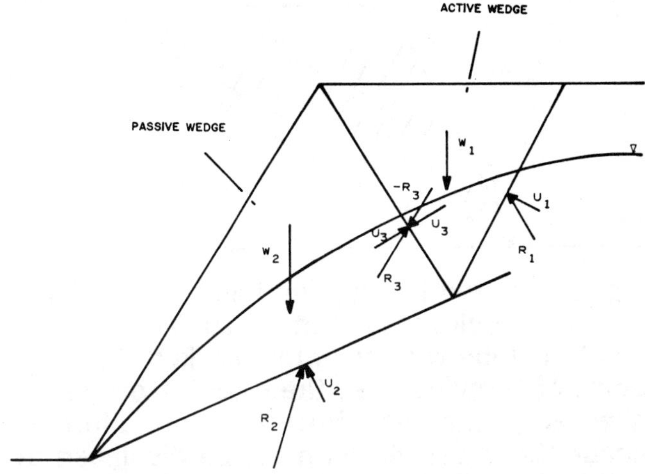

Fig. 12. The two-block failure mode showing forces acting at limiting equilibrium.

Simple analytical and model studies have defined the conditions under which toppling failure will occur, as shown in Fig. 13. This mode of instability is sensitive to the length:width ratio of the blocks involved (Fig. 13) and to the restraining conditions at the toe of the system of blocks (Fig. 14). Base friction models can provide a useful insight into the mechanism of failure.

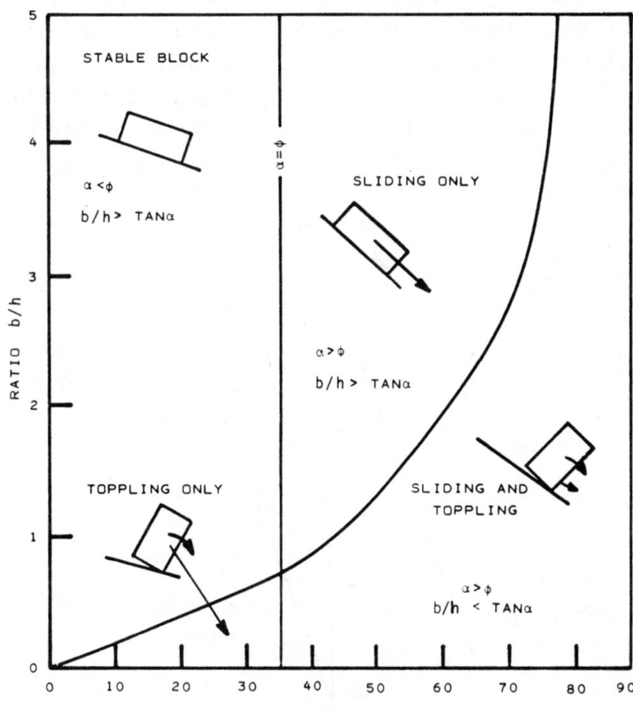

Fig. 13. Conditions for sliding and toppling of a block on an inclined plane (Hoek and Bray, 1974).

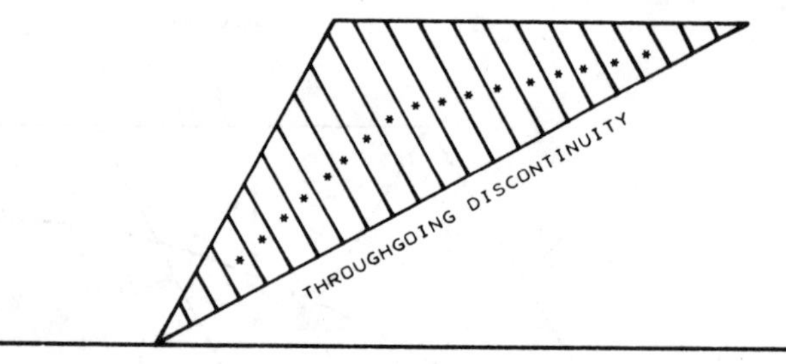

Fig. 14. Potential toppling failure in a jointed rock mass (Ashby, 1971). Jointed slope with columns marked thus* on point of toppling, i.e., their center of gravity overhangs the toe. However, for failure to occur the lower blocks must be displaced by sliding on the throughgoing discontinuity.

They can also be used to provide a qualitative assessment of the effect of possible slope stabilization procedures, such as by reducing the slope angle or installing horizontal reinforcement.

A factor of safety or stability factor can be defined for the case of a single block by considering the location of the resultant force in relation to the base of the block (Fig. 15). Although slopes will generally contain blocks of varying length:width ratios, the guidelines provided by Ashby (1971) will frequently provide an adequate quantitative basis for assessing the susceptibility of a slope to this mode of instability.

The availability of detailed geological data and/or the serious consequences of instability may justify the use of more sophisticated analysis (John, 1970) or numerical modeling (Cundall, 1974; and Bukovansky and Piercy, 1975). In most cases, it is probably adequate to be aware of the possibility of toppling-type failures and to be able to recognize the conditions under which toppling can occur. Then, although specific design criteria for the stabilization of such slopes are not available, stability may usually be achieved by employing the following guidelines: (1) ensure that any blocks in the toe which serve to key the system together are retained in place, and (2) apply horizontal or near horizontal reinforcement. Since this last measure can effectively increase the width:length ratio, large horizontal forces will generally not be necessary.

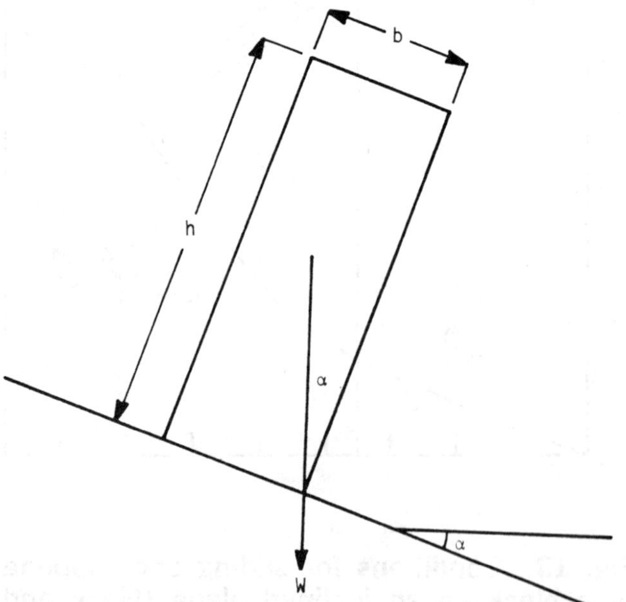

Fig. 15. Block at the point of toppling on an inclined plane.

FAILURE ALONG A GENERAL SLIP SURFACE

Historically, the commonest form of slope failure amenable to analysis has been the circular slip surface, which occurs in cohesive soils. These failures tend to be deep-seated such that the whole failure mass rotates about the center of the circle defining the slope surface.

For a stable slope, the shear strength resisting movement along a potential failure surface exceeds the shear stress tending to induce movement. Instability occurs when the shear stress on the failure surface equals the shear strength. This can happen in several ways. For instance, the shear stress might increase by increasing the height of the slope or the shear strength might decrease with time due to swelling, chemical weathering, creep strength loss, or increased pore water pressures.

None of the techniques described in this section allow for time effects in the method of analysis. The assumption of limiting equilibrium is made where failure is implied to be instantaneous. If the effects of time are important, they must be allowed for in a different way, such as: (1) empirically, when the length of time a newly cut slope will stand is known from experience of observing similar slopes in the same area, (2) by using a strength loss factor, such as the Brittleness Index (Bishop, 1971) which can be determined for a particular soil by laboratory testing, and (3) by means of a stress analysis (finite element or finite difference solution).

Many solutions to the circular arc type of failure have been devised over the years. Most of them involve a technique of dividing the soil mass above the failure plane into a number of vertical slices. Whatever method is used, it is usual to calculate the factor of safety for several trial slip surfaces in order to determine the slip surface having the lowest factor of safety. This is a tedious procedure by hand calculation, so that computer solutions are preferred for design purposes.

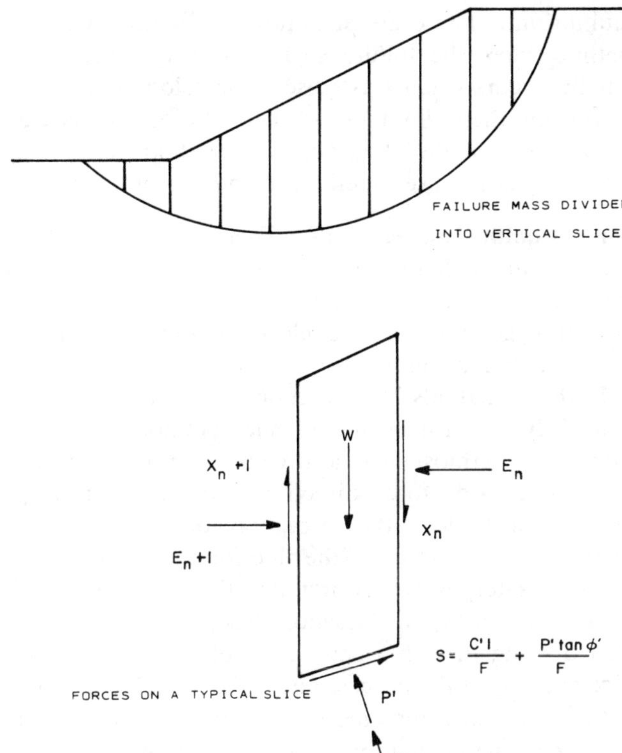

Fig. 16. General considerations for methods of slices.

The general problem of the circular slip failure arc is summarized in Fig. 16, in which: E_n and E_{n+1} denote the resultant horizontal forces on the sections n and $n+1$ respectively; X_n and X_{n+1} denote the resultant vertical shear forces on the sections n and $n+1$ respectively; W denotes the total weight of the slice above the sliding arc between n and $n+1$; P' denotes the total normal force acting on the base of the slice; S denotes the shear force acting along the base of the slice; ℓ denotes the length of the arc defining the base of the slice; c' denotes the effective cohesion of the soil; ϕ' denotes the effective friction angle of the soil; and F denotes the factor of safety of the slice.

If the soil mass is divided up into N slices and both force and moment equilibrium are considered, there are: N equations from considering horizontal equilibrium, N equations from considering vertical equilibrium, and N equations from considering moment equilibrium. This results in $3N$ equations in total.

On the other hand, the unknowns comprise: N-1 horizontal side forces, N-1 vertical side forces, N-1 locations of horizontal side forces, N normal forces acting on bases, N locations of normal forces acting on bases, and 1 factor of safety. Hence, there are a total of $5N$-2 unknowns.

Since there are more unknowns than equations, the problem is indeterminate and assumptions are required to make the equations balance the unknowns. The various methods of analysis differ mainly in the assumptions that are used to make the problem determinate.

In most methods of analysis, assumptions are made about the stress distribution acting on the failure surface. These assumptions are usually indirect in that assumptions are made about the side forces. For instance, in the Ordinary Method of Slices, the assumption is made that the resultant of the side forces is parallel to the base of the slice.

Methods of Analysis

Bishop's Simplified Method: An important step forward in the analysis of circular slip surfaces was made by Bishop (1955) and his method has virtually replaced most of the earlier methods such as the Swedish Circle Method for $\phi=0$, the Log Spiral Method, and the Ordinary Method of Slices.

In Bishop's method, which is summarized in Fig. 17, the assumption is made that the side forces are horizontal. Then, the forces are resolved vertically to determine the stress acting on the base of each slice.

In Bishop's Simplified Method (as opposed to his Rigorous Method), the sum of the terms containing the E and X forces is ignored, leading to the expression given in Fig. 17. A trial and error process is required to determine the factor of safety.

Bishop's method does not satisfy horizontal force

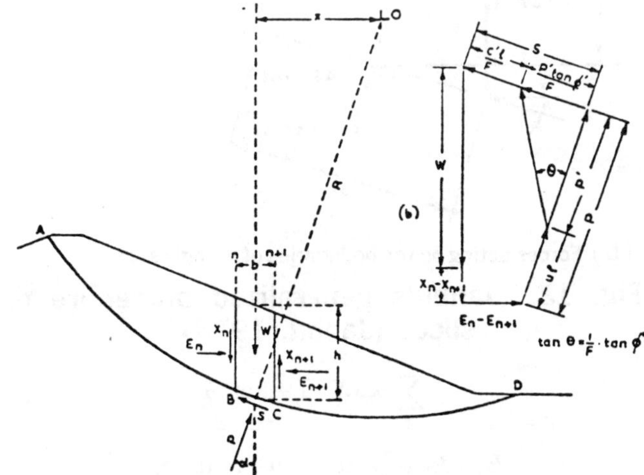

Fig. 17. The Simplified Bishop Method (Bishop, 1955).

$$F = \frac{1}{\Sigma W \sin \alpha} \Sigma \{c'b - W(1-\bar{B}) \tan \phi'\} \frac{1}{\dfrac{\cos \alpha - \tan \phi' \sin \alpha}{F}}$$

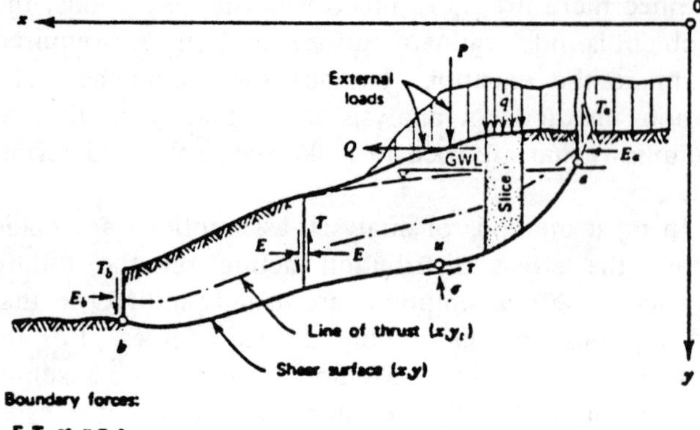

(a) Definitions and notations used for the generalized procedure of slices.

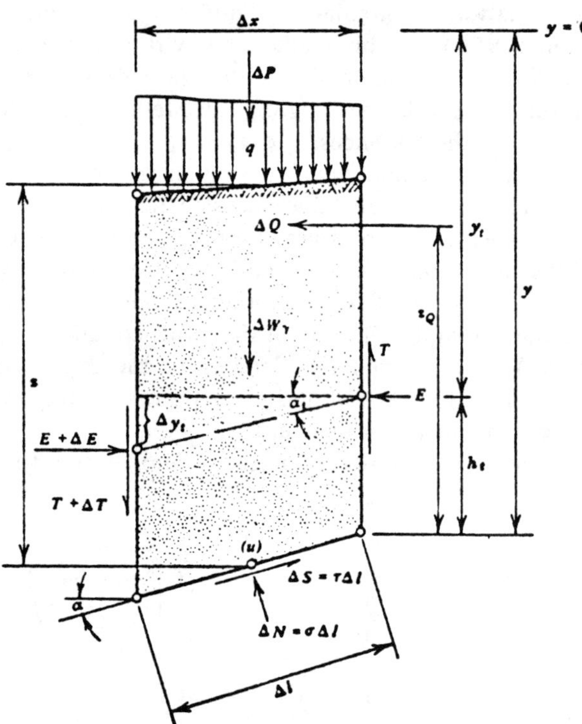

(b) Forces acting on the boundaries of a single slice.

Fig. 18. Janbu's generalized procedure of slices (Janbu, 1973).

$$F = \frac{\sum_{a}^{b} \tau_f \, \Delta X \, (1 + \tan^2 \alpha)}{E_a - E_b + \sum_{a}^{b} \Delta Q + (p + t) \tan \alpha \, \Delta X}$$

where τ_f is shear strength along the shear surface, ΔX is width of an individual slice, α is slope of the shear surface, ΔQ is magnitude of horizontal line load on a slice, $t = \frac{\Delta T}{\Delta X}$ with T = vertical interslice force, p is total overburden pressure, and E is total horizontal interslice force.

equilibrium. There are sometimes difficulties with the method when the failure surface is very steep. This usually occurs when F is close to or below unity.

Despite these limitations, the solution is accurate (within 5% of the most accurate method) and is the most commonly used method at the present time.

Force Equilibrium Methods: There are several force equilibrium methods, for example, Lowe (1967) and the US Corps of Engineers (1968). These employ force polygons which are closed either manually or numerically in a computer program.

In these methods, it is necessary to assume the direction of the side forces at each slice boundary. Starting with the uppermost slice, a force polygon is drawn with an assumed side force direction and factor of safety. In successive slices, the force polygons are closed by varying the magnitude of the side forces. If the correct factor of safety is chosen initially, the last polygon will close; if the polygon does not close, it is necessary to start over again with a new factor of safety for the top slice and repeat the process until the last polygon closes.

Lowe assumes that the side force inclination is the average of the inclinations of the top and bottom of the slice, while the Corps of Engineers method assumes that the side force inclination is parallel to the top of the slice.

The drawing of the force polygons is an iterative process. The method does not satisfy moment equilibrium and the value of F is dependent to some extent on the side force assumption that is used.

The advantages of these techniques are that they can be solved by hand and any shape of failure surface can be used. When computerized, convergence using the Newton-Raphson method is very rapid.

Janbu's Generalized Procedure of Slices: The Janbu method is intended to be used with noncircular failure surfaces, which correspond to failures along an irregular profile such as that shown in Fig. 18. As before, a cross section of the slope is divided into a number of vertical slices. These slices need not be of uniform width. Fig. 18a shows a typical slope cross section with general loading conditions and an irregular soil profile and shear surface. An enlarged version of an individual slice is shown in Fig. 18b, with all the appropriate forces acting on the slice.

Several equilibrium conditions must be met for both individual slices and the whole body; these are described in detail by Janbu (1973). Using these equilibrium equations, the overall factor of safety along the selected shear surface is defined as in Fig. 18.

The method satisfies all the conditions of equilibrium but assumes the location of the side forces between slices. These side force positions can be reasonably estimated, however.

Due to convergence problems, the method is more amenable to hand calculation than to computer solution. It is also relatively simple to use.

Spencer-Wright Method: An alternative to Bishop's Rigorous Method, in which both force and moment equilibrium conditions are satisfied, was developed by Spencer (1967). The method also takes the interslice forces into account.

Later, Wright (1969) extended Spencer's procedure to noncircular surfaces and made allowances for external forces. The method assumes that the side forces are parallel and satisfies all the conditions of equilibrium. It has been programmed for a computer. The program is very versatile and can cope with both circular and noncircular surfaces. It is also more convenient to use than the alternative solutions developed by Morgenstern and Price (1965) and Sarma (1973).

The program has recently been considerably modified and now includes options for an automatic sensitivity analysis and a probabilistic analysis (Major, Kim, and Ross-Brown, 1977). The input parameters, including the location of the soil profile lines, may be described by one of five different statistical distributions.

Use of Chart Solutions: In some cases a detailed analysis can be avoided, especially when the slopes being analyzed are relatively uniform and homogeneous. This is because there are a number of easy-to-use chart solutions available, which have been derived from the procedures mentioned before. For instance, circular failures for simple slopes can be analyzed from charts published by Bishop and Morgenstern (1960), Spencer (1967), and Hoek and Bray (1974). No chart solutions are available, however, for noncircular or general failure surfaces.

Application of the Various Methods

If the shape of the failure surface can be reasonably represented by the arc of a circle, a preliminary analysis should be carried out using one of the available chart solutions, such as the one described by Hoek and Bray (1974). Since it is easy to vary the slope height, slope angle, water conditions, and shear strength of the material, the engineer can soon develop a good feel for the problem and the sensitivity of the answer to changes in the various parameters.

Since the chart solution assumes a simple, uniform slope, the engineer must then decide whether the chart solution gives an adequate answer or whether he should then proceed to a more sophisticated method of analysis which will model the soil properties and the external and internal geometry of the particular slope more accurately.

If it is necessary to proceed further, it is recommended that either the computerized version of Bishop's Simplified Method or the Spencer-Wright method be used to solve problems involving circular failure surfaces. If it is necessary to do a hand solution as a check or because a computer is not available, it is recommended that either Janbu's generalized procedure of slices or Bishop's Simplified Method be used.

In the case of noncircular surfaces, either Janbu's generalized procedure of slices (hand solution) or the Spencer-Wright program (computer solution) should be used.

Since most mines have access to a computer (or alternatively the data can be mailed to a bureau service or consulting organization who can provide such a service), it is recommended that a verified computer solution be used. This is because it may be necessary to try hundreds of trial slip surfaces before the surface having the minimum factor of safety is located. Unless the engineer is very skilled in hand solution methods, it is possible that insufficient surfaces will be sampled to locate the minimum. On the other hand, computer programs often contain an automatic search option for locating the critical surface with the minimum factor of safety.

OTHER METHODS OF ANALYSIS

The methods of analysis that have already been considered are sufficient to deal with about 75% of the slope design problems encountered in open pits. Other methods may be required to deal with special situations; some of them are briefly considered in this section.

Finite element methods of stress analysis have been applied to the design of open pits for about ten years. An example from the *Pit Slope Manual* is illustrated in Fig. 19. It shows the stress concentrations as a result of excavating an open pit in a sequence of predominantly schistose rocks with a strong layer of quartzite in the footwall. Since the quartzite is much stiffer, the analysis shows that the stress concentrations in the quartzite are approximately five times greater than those occurring in the surrounding rocks.

Over the last few years, several workers (Hofmann, 1974-1977, Hammett, 1974, and Cundall, 1976), have been developing finite difference programs in which time is treated as an independent variable, i.e., explicitly. The large-displacement Lagrange codes that are often used are dynamic programs that have been damped to model quasistatic situations.

In many cases, these explicit programs show an ad-

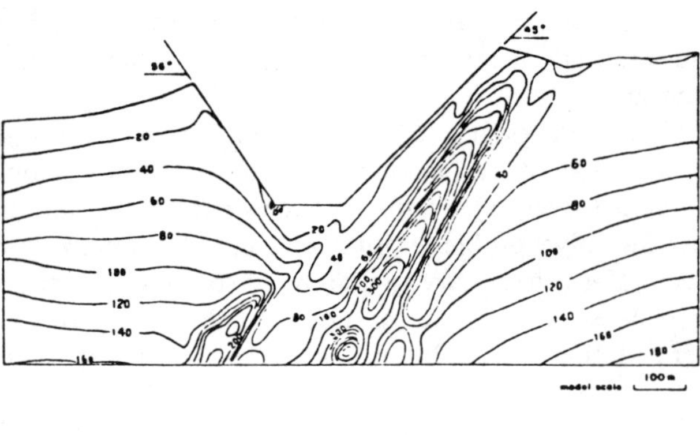

Fig. 19. Stress analysis of an open pit using the finite element method (Sage, et al., 1977).

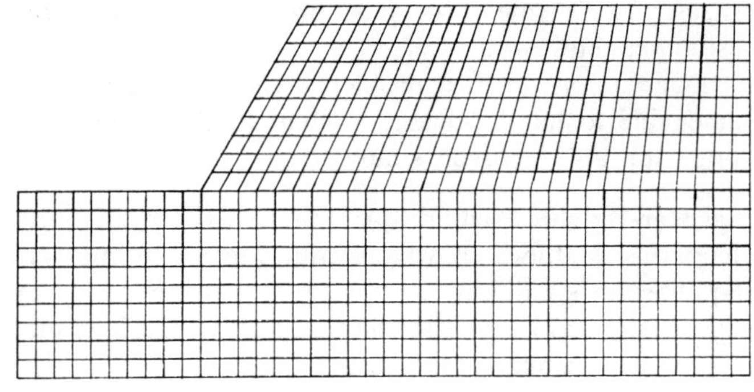

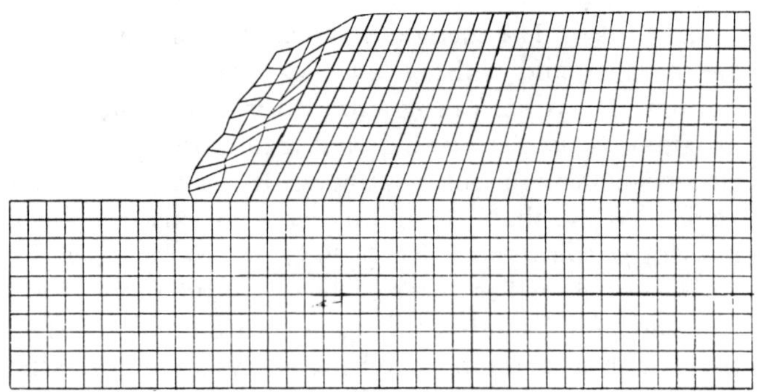

Fig. 20. Stress analysis of a soil slope using the explicit finite difference method (Hammett, 1974).

vance over many finite element formulations for studying open pit stability problems. These advantages may be summarized as follows: (1) they handle large deformations (unlike most finite element methods); (2) they handle complicated geometries; (3) nonlinear stress-strain laws can easily be incorporated; (4) they are very efficient in three-dimensions; (5) they can handle dynamic effects easily, e.g., effects of large blasts or earthquakes; (6) they are relatively inexpensive, especially when run on a minicomputer setup; and (7) they are highly user-oriented, and plots of stresses and displacements in any orientation can easily be obtained.

Both two and three-dimensional versions of these codes have been developed. An example of a two-dimensional slope stability study by Hammett (1974) is shown in Fig. 20. The soil slope was modeled using a Mohr-Coulomb failure envelope with an associated flow rule. The initial grid is shown in Fig. 20a, and the deformed grid after considerable movement has taken place is shown in Fig. 20b. The deformed region corresponds well with the slip surface given by Bishop's method. In particular, this example demonstrates the large deformation capability of the explicit Lagrange codes.

Another development of the explicit finite difference method is the distinct element method developed by Cundall (1974) which can be used to model discontinuities. The computer program models the behavior of assemblages of rock blocks and displays the behavior on the viewing screen of a cathode ray tube. A unique feature of this program is that the computer calculates the displacements, rotations, and interactions of the blocks as a function of time and generates failure surfaces where instability exists. The program is also unique in that the blocks are free to undergo large displacements and rotations. The blocks may even separate and fall freely as the rock mass is truly modeled as an assemblage of discrete blocks. An example of the progressive failure of a rock slope is shown in Fig. 21 in which a joint friction angle of 30° was used.

DESIGN PROCEDURE

The procedure for designing the final pit slopes is summarized as follows: (1) data collection consisting of lithology and structural geology, mapping of discontinuities, mechanical properties of intact rock and discontinuities, ground-water effects, and movement data; (2) defining design sectors; (3) analyzing all potential failure modes in each sector and drawing up probability of failure curves (or reliability schedules); (4) selecting slope angles corresponding to the acceptable degree

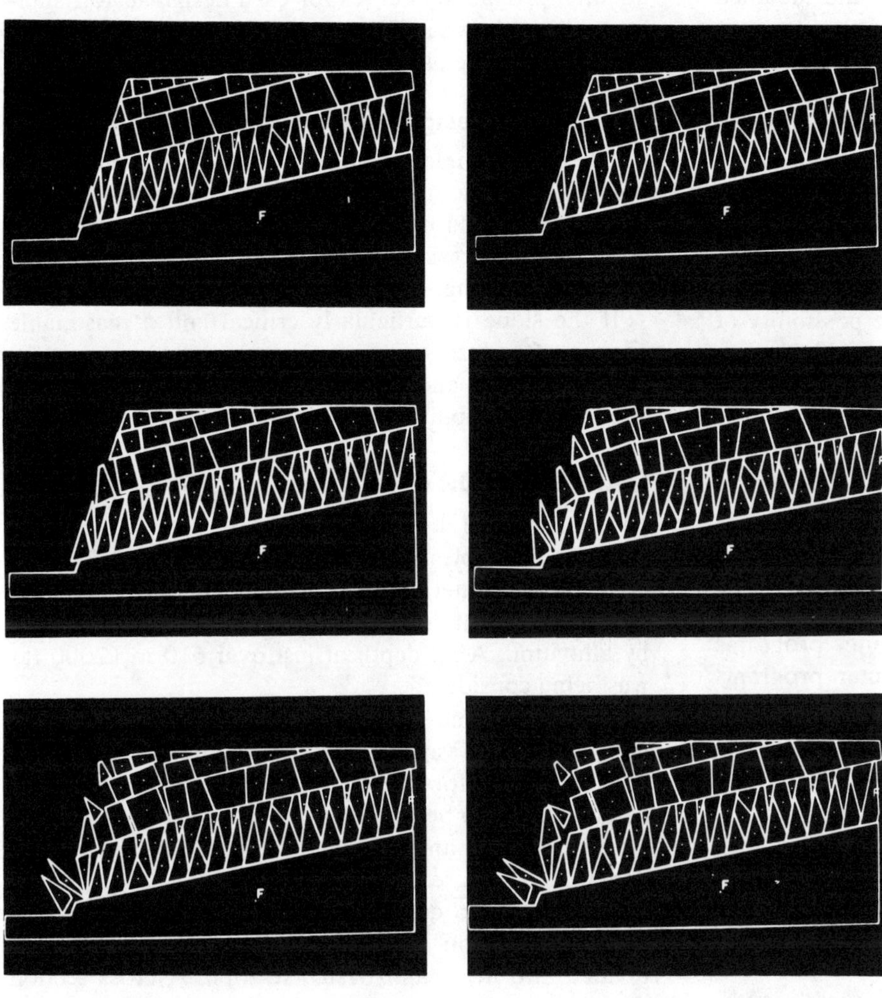

Fig. 21. Failure analysis of a slope comprised of rock blocks using the distinct element method (Cundall, 1974).

of risk; (5) performing associated designs as required, e.g., artificial support systems, ground-water control and perimeter blasting; and (6) designing slope monitoring systems, if these are required.

Data Collection

The data collection phase is the most time-consuming and expensive part of the operation. In particular, considerable care must be devoted to geological mapping and interpretation, in order to define the problems correctly. There is little point in performing an analysis to obtain the correct solution to the wrong problem, but it is frequently done.

Since the collection and interpretation of this data are covered in the previous chapter, it will not be considered further here. The importance of good data can hardly be overemphasized, however.

Design Sectors

After collection of the data and before any analysis, it is necessary to look at the preliminary mining plan and divide up the pit into design sectors. Each design sector is a portion of the pit which can be regarded as being homogeneous for design purposes, so that a single slope angle can be assigned to that area.

The main factors entering into the determination of design sectors are: rock type, type of discontinuities encountered and their orientations, orientation of the pit slope, and the type of failures anticipated. In a fairly uniform rock mass there will be at least four design sectors corresponding to the different slope orientations, e.g., footwall, hanging wall, and two end zones, and the different potential failure modes generated by the discontinuity sets intersecting the cut slopes. Since open pit geology is rarely very uniform, ten or more design sectors may sometimes be necessary.

Analysis of Potential Failure Modes

Each design sector is then considered in turn. First, the effects of major discontinuities such as faults, shear zones, and major joints are evaluated. This may involve consideration of various structurally controlled failures using stereographic techniques and detailed sensitivity analyses or probabilistic analyses using more sophisticated analytical methods. As a result of this work, a slope angle which is safe against failures caused by major discontinuities will be obtained. If a major failure is anticipated at this stage, remedial measures may have to be considered before proceeding further; otherwise, clean-up costs and delays, or a lower slope angle will have to be accepted.

The next stage is to design for the effects of minor discontinuities, such as joint sets, and the possibility of a general failure, such as a circular arc or a block flow failure mode. Various combinations of these methods, such as crushing at the toe leading to a structurally controlled failure may also have to be considered.

The analysis of each sector may be relatively simple or may be quite complex depending on the geological conditions. In general, it is recommended that the identification of structurally controlled failures is performed on a stereonet and that detailed analysis is performed using one of the many computer programs that are available. For instance, computer programs for the analysis of all of the commonly encountered failure modes may be purchased through the Canada Centre for Mineral and Energy Technology (555 Booth St., Ottawa, Ont., Canada, K1A 0G1).

With any computer program, however, great care must be exercised to ensure that the answers obtained are correct. The input data must be representative of the field conditions, the failure mechanisms to be analyzed must be realistic, and great care is required in reviewing the computer output. Under these conditions, computer programs are very useful and may be used to produce accurate answers relatively quickly.

Selection of Slope Angles

The slope angle chosen for a particular sector will depend upon the relative importance of the slope and the degree of risk that can be taken with it. A probabilistic analysis provides a convenient method of determining the variation in probability of failure with changing slope height or slope angle. Charts produced using these data can then form the basis for the design of rock slopes whereby the required slope geometry for a specified probability of failure can be determined. An example of such a chart is given in Fig. 24.

The curves shown in Fig. 24 are particularly useful to mine planners since it enables them to choose a slope angle for a given chance of failure for a particular slope. Since the relative importance of the slopes varies around the pit, depending on the pit geometry and the location of the haul roads and crushers, the acceptable risk of failure will vary from slope to slope. Consequently, probability curves for each design sector, similar to those shown in Fig. 24, will usually be an essential input to the mine planning process.

Associated Designs and Monitoring Systems

After slope angles have been assigned to the different design sectors, some associated design work will almost certainly be required. These may include the design of artificial support systems, ground-water control, and perimeter blasting.

If the slope is particularly critical and a reasonable chance of failure exists, a slope monitoring program may be devised and installed in order to provide early warning of instability.

CASE HISTORY

The results of a slope design for an open pit sector is given as Example 3 and is shown in Figs. 22-24.

The slope materials consist mainly of quartz monzonite porphyry intersected by dikes and sills and overlain by alluvium. A pit depth of just over 610 m (2000 ft) was being considered.

After a detailed study of the geology and a preliminary analysis of various potential failure modes, it was concluded that for overall slope design, jointing could be considered to be randomly oriented and that a circular arc mode of analysis could be used. Stability along noncircular failure surfaces and wedge-shaped failure blocks were also considered. Following a program of field investigation and laboratory testing, the pit was divided into five major design sectors. A cross section through one of these sectors showing the different geological zones is shown in Fig. 22; the density and shear strengths of the various materials are also shown on the figure.

First, the overburden slope was assumed to be completely drained, and this was designed as an independent exercise. The bedrock slopes were then analyzed assuming uniform slope angles from the top of bedrock to the toe of the slope. Four basic water conditions were considered: (1) undrained (assumption A), using an optimistic projection of the water table when the final slopes are being cut; (2) undrained (assumption B), using a pessimistic projection of the water table; (3) complete drainage in the material in the vicinity of the slope (e.g., from peripheral wells); and (4) partial drainage with the pessimistic projection of the water table (assumption B) by using horizontal drainage holes which have the effect of drawing back the water table 61 m (200 ft) behind the face, a less expensive alternative to complete drainage.

Five slopes were analyzed in which the overburden slopes were kept the same but the bedrock slope was

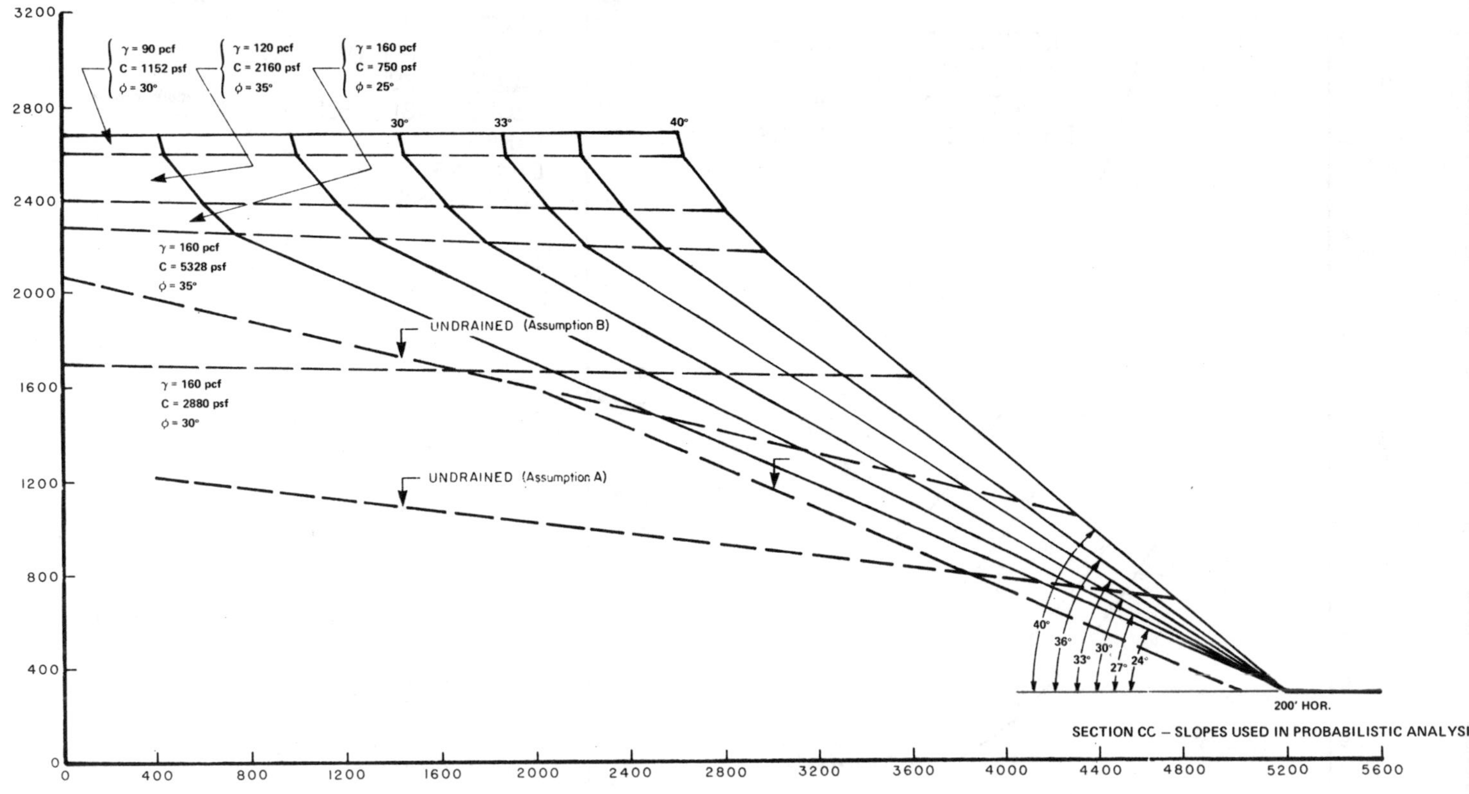

Fig. 22. Example 3, slope section used in the analysis.

varied between 24° and 40° (Fig. 22). Both factors of safety analyses and probabilistic analyses were carried out for each chosen slope angle. The results for each of the four drainage conditions are shown in Figs. 23 and 24.

The sensitivity of the factor of safety to changes in slope angle and drainage conditions is readily apparent from Fig. 23. In fact, this figure could be used as a basis for choosing the final slope angle for this design sector. However, it is still necessary to decide what is an appropriate factor of safety; this is difficult since a particular value of the factor of safety corresponds to a different probability of failure for each slope section analyzed.

The problem is largely resolved through the use of the probability of failure curves shown in Fig. 24. Using these curves, the mine planner can decide what probability of failure is acceptable to him and then look up the corresponding slope angle, without wondering what a factor of safety of, say 1.5, really means.

Let us assume that for this slope, which does not involve a haul road, that a 10% probability of failure is acceptable, i.e., 10% of the length of the slope can be allowed to fail. The corresponding slope angles then vary between 22° and 36° depending on the drainage conditions assumed. Although complete drainage might be prohibitively expensive, it appears that considerable stripping could be saved by making some improvements to the drainage (e.g., through the use of horizontal drain holes). The optimization of drainage vs. slope angle would then be one of the associated designs that the mine planners would undertake in order to arrive at an optimum design for the final slope angle. Although a detailed drainage design has not yet been completed for this pit, it appears that the condition shown as undrained (assumption A) in Fig. 24 could be achieved by one means or another, giving a final slope angle of 30° for this design sector.

ACKNOWLEDGMENT

The author would like to acknowledge the assistance given him by Graeme Major in the preparation of this chapter. He would also like to acknowledge the partial support given by the Pit Slope project of the Canada Centre for Mineral and Energy Technology in developing some of the probabilistic concepts and examples mentioned in the chapter.

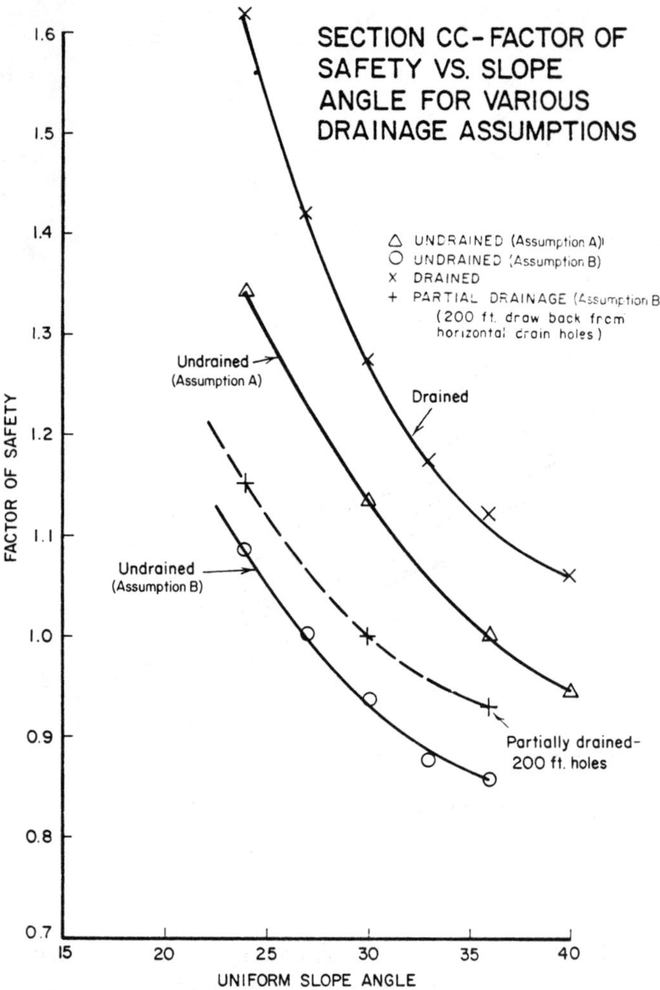

Fig. 23. Factor of safety curves for various drainage conditions.

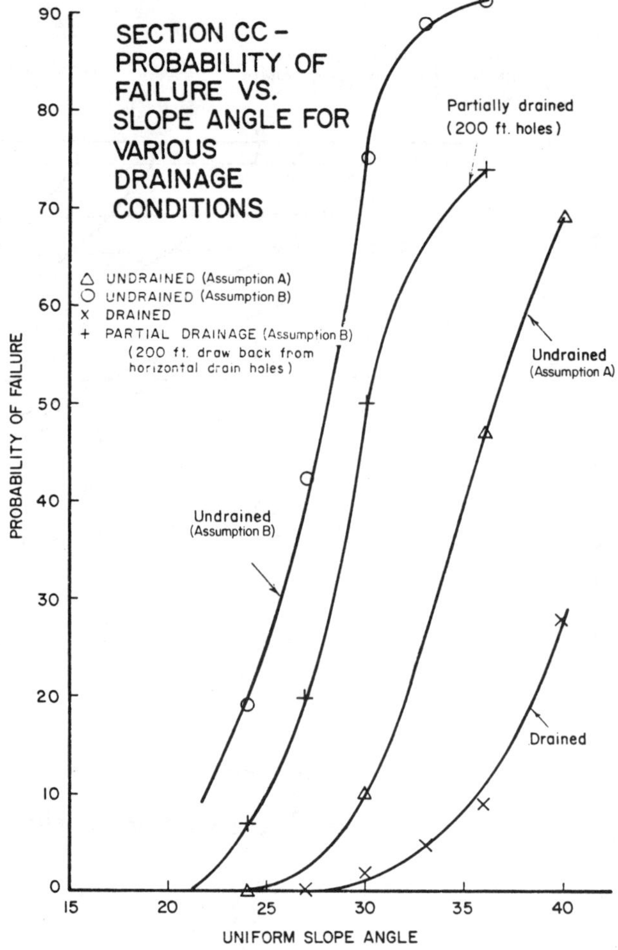

Fig. 24. Probability of failure curves for various drainage conditions.

REFERENCES

Ashby, J. P., 1971, "Sliding and Toppling Modes of Failure in Models and Jointed Rock Slopes," M.Sc. Thesis, Imperial College, London.

Bishop, A. W., 1971, "The Influence of Progressive Failure on the Choice of the Method of Stability Analysis," *Geotechnique,* Vol. 21, No. 2, pp. 168-172.

Bishop, A. W., 1955, "The Use of the Slip Circle in the Stability Analysis of Slopes," *Geotechnique,* Vol. 5, No. 1, pp. 7-17.

Bishop, A. W., and Morgenstern, N., 1960, "Stability Coefficients for Earth Slopes," *Geotechnique,* Vol. 10, pp. 129-150.

Bukovansky, M., and Piercy, N. R., 1975, "High Road Cuts in a Rock Mass with Horizontal Bedding," 16th Symposium on Rock Mechanics, University of Minnesota.

Call, R. D., 1974, "Method of Computing the Minimum Resistance Step Path for Rock Slope Design," Report, Dept. of Energy, Mines, and Resources, Ottawa, Canada.

Cundall, P. A., 1974, "A Computer Model for Rock-Mass Behavior Using Interactive Graphics for the Input and Output of Geometric Data," Report, US Army Corps of Engineers, University of Minnesota.

Cundall, P. A., 1976, "Explicit Finite-Difference Methods in Geomechanics," 2nd Int. Conf. on Numerical Methods in Geomechanics, Blacksburg, VA.

Goodman, R. E., 1975, *Methods of Geological Engineering in Discontinuous Rocks,* West Publishing Co., Minneapolis.

Hammett, R. D., 1974, "A Study of the Behavior of Discontinuous Rock," Ph.D. Thesis, University of North Queensland, Australia.

Hendron, A. J., Cording, E. J., and Aiyer, A. K., 1971, "Analytical and Graphical Methods for the Analysis of Slopes in Rock Masses," US Army Engineer Nuclear Cratering Group, Livermore, Technical Report No. 36.

Hoek, E., and Bray, J. W., 1974, *Rock Slope Engineering,* Institute of Mining and Metallurgy, London.

Hofman, R., 1974-1977, Personal Communication.

Janbu, N., 1973, "Slope Stability Computations," *Embankment-Dam Engineering,* Casagrande Vol., John Wiley & Sons, New York.

Jennings, J. E., 1970, "A Mathematical Theory for the Calculation of the Stability of Slopes in Open Pit Mines," *Planning Open Pit Mines,* South African Institution of Mining and Metallurgy, Johannesburg.

John, K. W., 1968, "Graphical Stability Analysis of Slopes in Jointed Rock," American Society of Civil Engineers, Vol. 94, No. SM2, p. 467, and *Closure,* Vol. 95, No. SM6, Nov., pp. 1541-1545.

John, K. W., 1970, "Three-Dimensional Stability Analysis of Slopes in Jointed Rock," *Planning Open Pit Mines,* South African Institution of Mining and Metallurgy, Johannesburg.

Kim, H-S, Major, G., and Ross-Brown, D. M., 1978, "Application of Monte Carlo Techniques to Slope Stability," 19th Symposium on Rock Mechanics, MacKay School of Mines, University of Nevada, Reno.

Kuykendall, L., and Goodman, R. E., 1976, "Design Appendix for the Multiblock Failure Mode," Report, Dept. of Energy, Mines, and Resources, Ottawa, Canada.

Lowe, J. III, 1967, "Stability Analysis of Embankments," American Society of Civil Engineers, Vol. 93, No. SM4, pp. 1-33.

Major, G., Kim, H-S., and Ross-Brown, D. M., 1977, "Sliding Stability Analysis," *Pit Slope Manual,* Supplement 5-1, CANMET.

Major, G., Ross-Brown, D. M., and Kim, H-S, 1978, "A General Probabilistic Analysis for Three-Dimensional Wedge Failures," 19th US Symposium on Rock Mechanics, Mackay School of Mines, University of Nevada, Reno.

McMahon, B. K., 1973, "Procedures for the Design of Slopes in Fractured Rock," Internal Report 73/101, Dept. of Energy, Mines, and Resources, Ottawa, Canada.

Morgenstern, N. R., and Price, V. E., 1965, "The Analysis of the Stability of General Slip Surfaces, *Geotechnique,* Vol. 15, pp. 79-93.

Phillips, F. C., 1971, *The Use of Stereographic Projection in Structural Geology,* 3rd. ed., Edward Arnold.

Ross-Brown, D. M., 1973, "Design Considerations for Excavated Slopes in Hard Rock," *Quarterly Journal of Engineering Geology,* Vol. 6, pp. 315-334.

Sage, R., ed., 1977, *Pit Slope Manual,* CANMET.

Sarma, S. K., 1973, "Stability Analysis of Embankments and Slopes," *Geotechnique,* Vol. 23, pp. 423-433.

Spencer, E., 1967, "A Method of Analysis of the Stability of Embankments Assuming Parallel Interslice Forces," *Geotechnique,* Vol. 17, pp. 11-26.

US Army Corps of Engineers, 1968, "Stability of Earth and Rockfill Dams," *Manual, Engineering and Design.*

Wright, S. G., 1969, "A Study of Slope Stability and the Undrained Shear Strength of Clay Shales," Ph.D. Thesis, University of California, Berkeley.

2E Pit Limit Shell Generation

B. L. Seegmiller, editor

Contents

INTRODUCTION TO SUBSECTION 2E

Final pit limit shell generation or the determination of the size and shape of the open pit mine is the final step in the development of a mine plan. Input data that are required to generate the final pit limit shell include a block model, a series of geological sections and plans, a bench profile, and the final pit limit slope angles. The sections previous to this one have described how to develop the block model and how to determine the pit slope angles. The purpose of this subsection is to bring all such data together and develop a final pit limit shell. The shell generation may be done using hand or automated methods. The two chapters in this subsection describe how to proceed using each of these methods, respectively.

13 Hand Methods

Benjamin C. Koskiniemi
AMAX Arizona, Inc.

Benjamin C. Koskiniemi is currently the manager of mining and geology for AMAX Arizona, Inc. He graduated from Michigan College of Mining and Technology with a B.S. in mining engineering and a B.S. in engineering administration. He has occupied the following positions: mining engineer for Kennecott Copper Corp. in Salt Lake City, mine evaluation engineer of AMAX in Denver, and mining engineer and later chief mine engineer of ANAMAX Mining Co.

Introduction

When evaluating any ore body, one of the first questions concerns the ore reserves. In the case of an open pit mine, this is not possible to answer reliably until the ultimate (final) pit limits have been established. Techniques used in designing an ultimate pit are classed as (1) manual, (2) computer, and (3) combined manual-computer. This chapter will describe how manual techniques can be utilized in designing an ultimate pit. Certain economic and design criteria must be established before the actual design begins. In order to begin designing an ultimate pit, it will be assumed that the engineer already has the following data available: (1) vertical sections, (2) horizontal sections for each level, (3) stripping curve, (4) bench height, (5) bank slope angle between levels, (6) level berm width, (7) roadway width, (8) pit slope angles at ultimate pit limits (estimated average including roads and ramps and between roads and ramps), and (9) minimum width of pit bottom.

Design

Hand methods of ultimate pit design usually begin with vertical sections (Soderberg and Rausch, 1968). The pit limits are first located on the vertical sections,

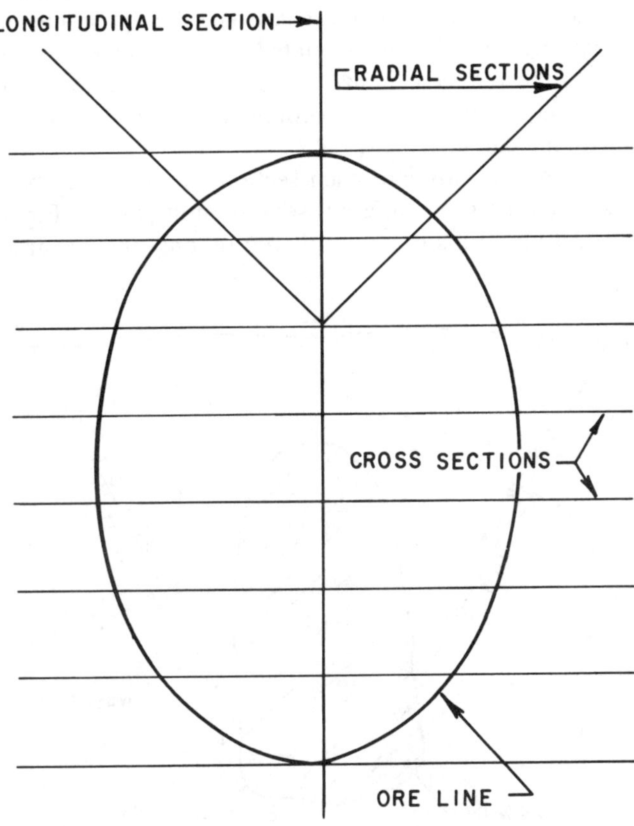

Fig. 1. Plan of ore body.

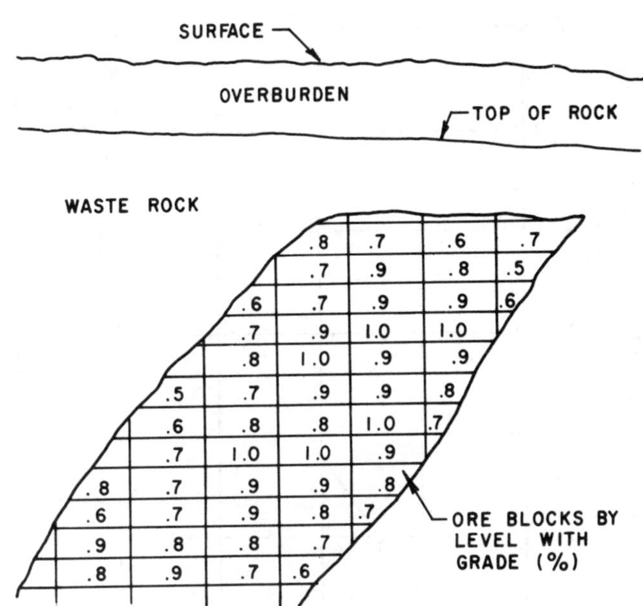

Fig. 2. Vertical section.

which consist of cross, longitudinal, and radial sections as illustrated in Fig. 1. These sections should include the mineral block inventory and surface topography as minimum requirements. An example of a vertical section is presented in Fig. 2. If there are materials of significantly different specific gravity, these areas should also be identified. This is especially important when the stripping ratio is on a tonnage basis, metric tons waste:metric tons ore (short tons waste:short tons ore).

The pit slope angles to be used when working with the vertical sections are the average angles, which include allowances for haul roads and ramps. These angles (see Fig. 3) are approximated, based on pre-

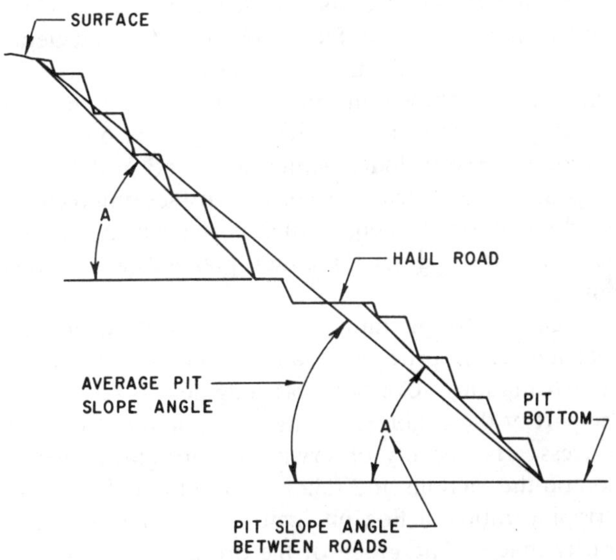

Fig. 3. Pit slope angles.

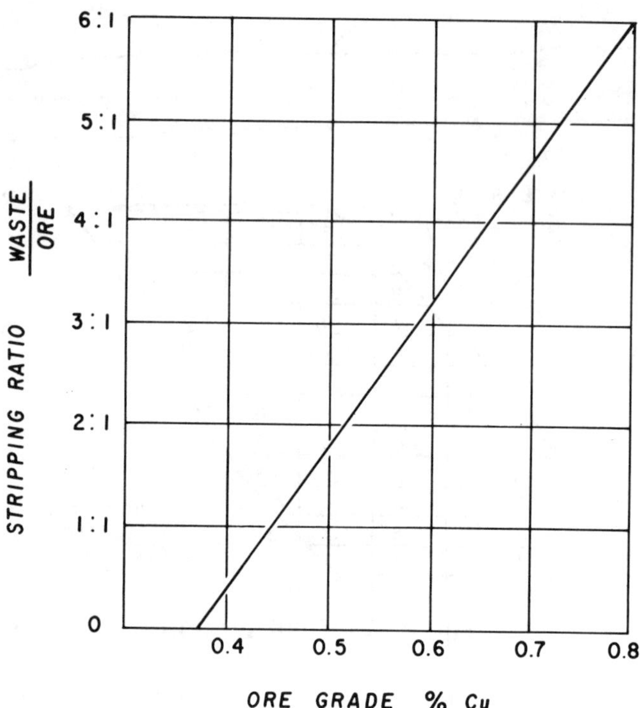

Fig. 4. Stripping curve.

liminary estimates of anticipated pit dimensions, road and ramp requirements, and pit slope stability studies.

The pit limits are located on each section so the ore grade along the pit limit line supports a stripping ratio corresponding to the break-even or allowable stripping ratio. Illustrated in Fig. 4 is the stripping curve used to evaluate the typical sections. Break-even stripping ratio signifies that the costs used include all direct costs. Depreciation is usually also included. Allowable stripping ratio usually signifies use of a profit factor in addition to direct costs and depreciation (Erickson, 1968). An illustration from Halls (1970) that depicts the pit limits based on use of these different cost assumptions is shown in Fig. 5. The design methods described apply to any of the cost assumptions. If the stripping curve includes depreciation and profit, it may be prudent to at least locate the surface intercept for the direct cost pit to ensure that permanent plant facilities and waste dumps are not planned within these limits.

Locating the pit limit on each vertical section is a trial and error process usually requiring a number of approximations. Considerable judgment is required on the part of the engineer during each phase of the design process. By visually observing the ore grade distribution on the sections and relating these to the break-even stripping ratios, a first pit limit approximation is arbitrarily made. The grade of the ore along the pit limit intercept selected is calculated, and the break-even

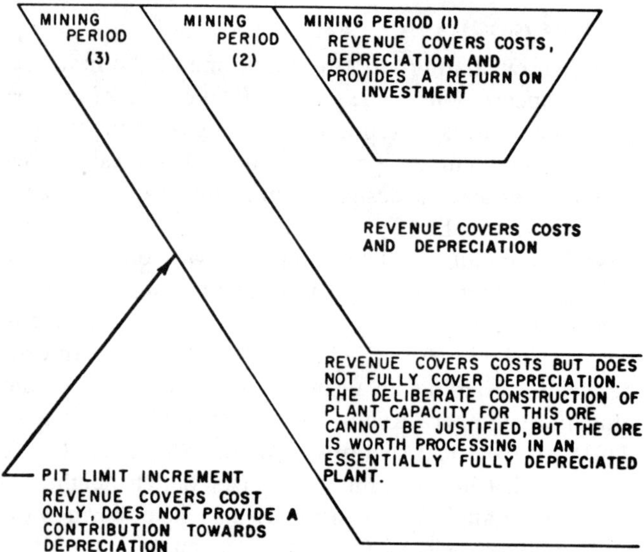

Fig. 5. Economic pit limits (Halls, 1970).

stripping ratio is determined from the stripping curve. The lengths of ore and waste along the pit limits are measured, and the stripping ratio (W:O) is calculated on the basis of these measurements (adjusting, when necessary, for changes in specific gravity). The calculated stripping ratio is compared to the break-even stripping ratio for the grade calculated and if the calculated ratio is less than break-even, the pit limits are expanded; but, if the calculated ratio is greater, the pit limits are reduced in size. These approximations continue until the pit limit is found that conforms to the stripping curve.

Each ore body and each section within an ore body usually presents a different set of conditions. In Fig. 6, the pit bottom is in waste; therefore, only the ore grade

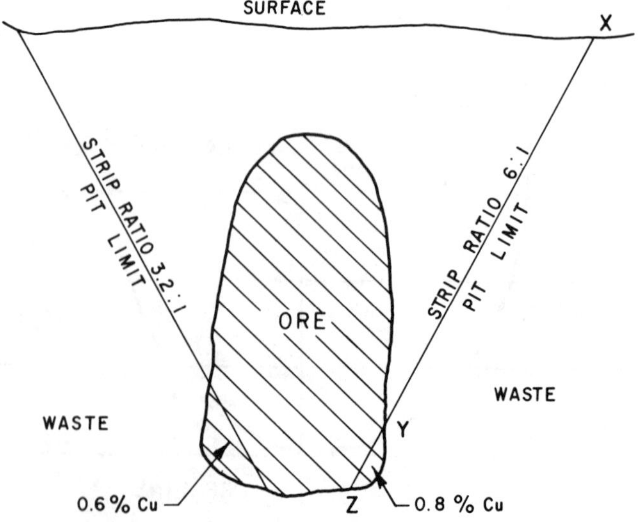

Fig. 6. Pit limits, bottom in waste.

and ore waste intercepts along the slope lines are used in determining the break-even stripping ratios and locating the final pit limits. Each slope is evaluated independently. On the right side, the grade is estimated at 0.8% Cu, and this supports a break-even stripping ratio of 6:1 (waste:ore) as determined from the stripping curve in Fig. 4. Assuming that this grade will be the same along any slope line in this area, the line is found that gives a 6:1 ratio at the designed average final pit slope angle:

$$\frac{\text{length of } X\,Y \text{ waste}}{\text{length of } Y\,Z \text{ ore}} = \frac{6}{1}.$$

On the left side, the estimated 0.6% Cu grade supports a 3.2:1.0 break-even stripping ratio, and the slope line meeting this condition is located. If the ore grade changes as the slope line is moved, the required break-even stripping ratio is also changed.

In Fig. 7, the ore extends to depth, and therefore, the pit bottom will be in ore. The pit bottom is designed at its minimum width, and the ore along the bottom is also used to calculate the break-even stripping ratios. Each slope is assumed to be influenced by one-half of the ore exposed along the pit bottom. From Fig. 4, a grade of 0.52% Cu has a break-even stripping ratio of 2:1, and the ultimate pit slope meeting this condition is located:

$$\frac{\text{length of } X\,Y \text{ waste}}{\text{length of } Y\,Z\,Z' \text{ ore}} = \frac{2}{1}.$$

In Fig. 8, the ore extends to depth on an incline with the pit bottom and one slope completely in ore.

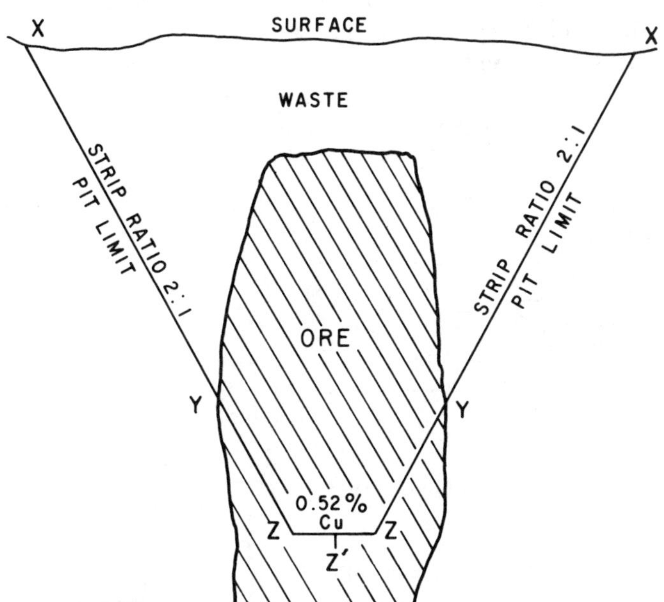

Fig. 7. Pit limits, bottom in ore.

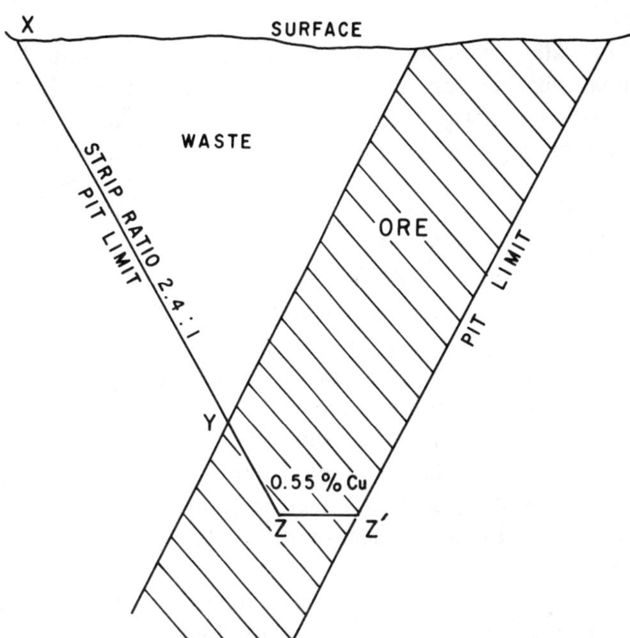

Fig. 8. Pit limits, bottom and slope in ore.

The pit bottom is designed at its minimum width, and the ore along the entire bottom is used to calculate the break-even stripping ratio for the slope in waste.

Equating pit limit distances directly is acceptable when working with parallel sections, but it can be very inaccurate when dealing with radial sections. Fig. 9

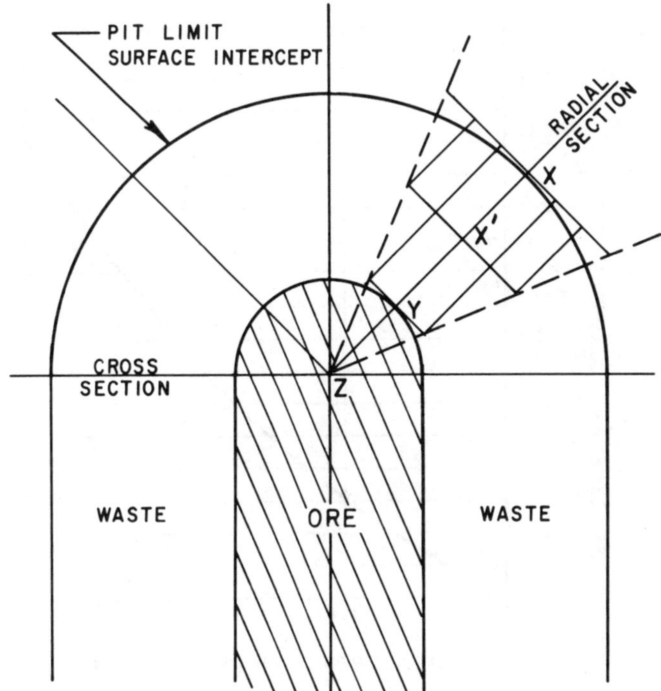

Fig. 9. Radial section, area of influence.

illustrates, in plan, why it is not feasible to locate the final pit limit on a vertical radial section simply by dividing the waste distance by the ore distance when determining the break-even stripping ratio. On the basis of simple geometric shapes, the stripping ratio in plan is approximately 7.5:1, if the waste length XY is twice the ore length YZ. The stripping ratio on the vertical section would actually be measured at 2:1. If the surface intercept was at X' with $X'Y$ being equal to YZ, the stripping ratio in plan is 3:1. On the vertical section, it would measure at 1:1. This means that the break-even stripping ratio as determined from the stripping curve must be adjusted before being applied to a vertical radial section. In this example, Fig. 10 illustrates the correction factors required to measure distances directly in locating the pit limits on the radial sections.

When the ultimate pit limits have been located on each of the vertical sections, a preliminary ore reserve can be estimated from the sections. First, the pit limit on each section must be compared to the adjoining sections to see that a logical relationship exists in regard to minability of the ore body. When calculating tonnages, parallel sections usually do not present a problem, and the area of influence is taken as halfway between adjoining sections. In the case of radial sections, the tonnages can be calculated by methods discussed by Popoff (1966) or by the sector methods first discussed by Soderberg (1959).

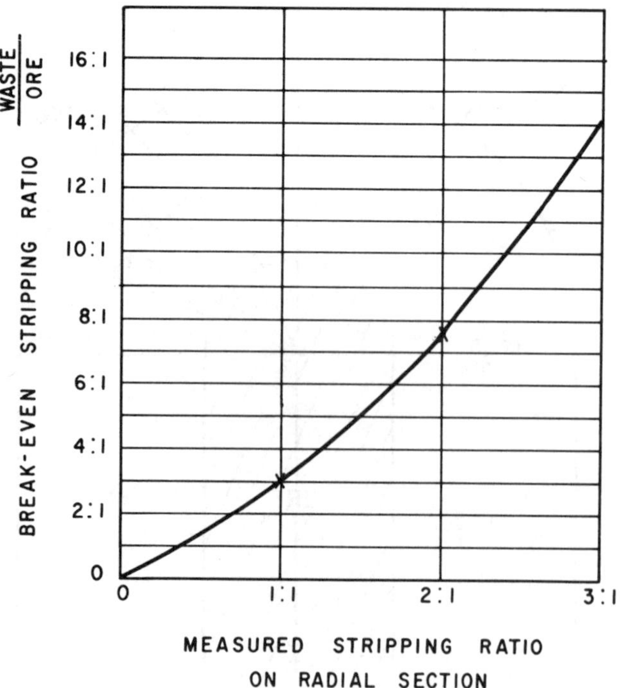

Fig. 10. Stripping curve for radial section.

The final ore reserve estimates are calculated from level plans (horizontal sections) with the pit limits for each level determined from a composite mine plan map. The mine plan map is constructed from the vertical sections. As the first step in preparing the composite, the locations of the pit bottom and the surface intercepts of the pit limits are transferred from the vertical sections to the plan map. If a vertical section does not have a single continuous slope line from the pit bottom to the surface intercept, any changes are also transferred to the plan map. The ore intercepts can also be located, if desired.

The actual designing of the composite plan generally begins with the pit bottom. The points from the sections usually present a very irregular pattern, both vertically and horizontally. In smoothing these and designing the bottom bench, the engineer has several things to keep in mind: (1) averaging the break-even stripping ratios for adjoining sections, (2) use of simple geometric patterns for ease of design, (3) location of ramp to pit bottom, and (4) watching for patterns that might lead to slope stability problems.

It should be remembered that the design is intended to optimize ore recovery and maximize profits; therefore, the design configuration must follow the ore, but because this is usually a cut and try method, the simpler the geometric shape of the bottom level, the easier it is to design the remainder of the pit.

The line plotted for each level on the composite plan is generally the median line, which is the contour elevation midway between the level elevation and the elevation of the next higher level, as shown in Fig. 11. The final pit limit design will also usually include any roads that will be in the final pit slope. In preliminary designs, the roads sometimes are not shown, and the

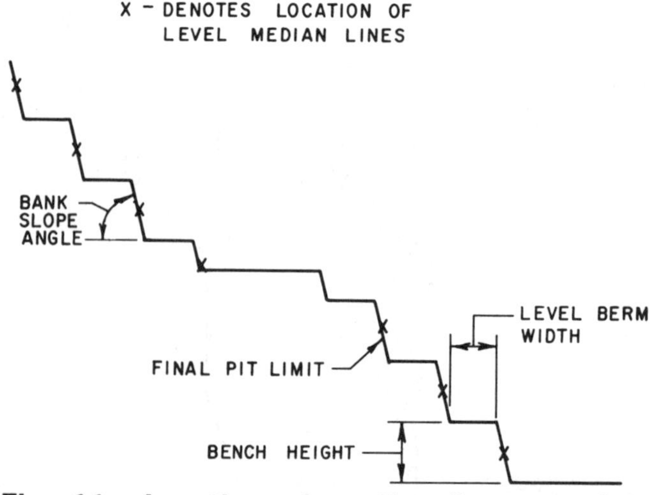

Fig. 11. Location of median lines, section A-A of Fig. 12.

median lines will be based on the flatter average overall pit slope.

Once the bottom level has been established, the pit design progresses toward the surface. Points for the level median lines are located on the plan map, and these are usually uniformly spaced and are dependent on the angles of the final pit slopes. The points for each level are connected to complete the design, as shown in Fig. 12. Care must be exercised, especially if different areas of the pit have different slope angles.

It is also important that conditions that can lead to slope failures are not incorporated into the pit design. An example of this would be an area that bulges into the pit, especially in a potentially unstable area.

When the composite ultimate pit plan is completed, the pit limits are transferred to the individual level plans (see Fig. 13). The pit can then be divided into sectors to determine whether the break-even stripping ratio requirements have been achieved. This can be done by measuring the lengths of ore and waste on each level at the pit limit within a given sector. This will provide the data for calculating the actual stripping ratio along the pit limit intercept. The ore grade is calculated by measuring the lengths of the different ore zones exposed in the sector and getting a weighted average grade. The stripping curve is checked to determine what break-even stripping ratio goes with the calculated grade, and this ratio is compared to the calculated stripping ratio. If there are any anomalous sectors, the plan can be reviewed to see how it might be affected if it were shifted in a given direction.

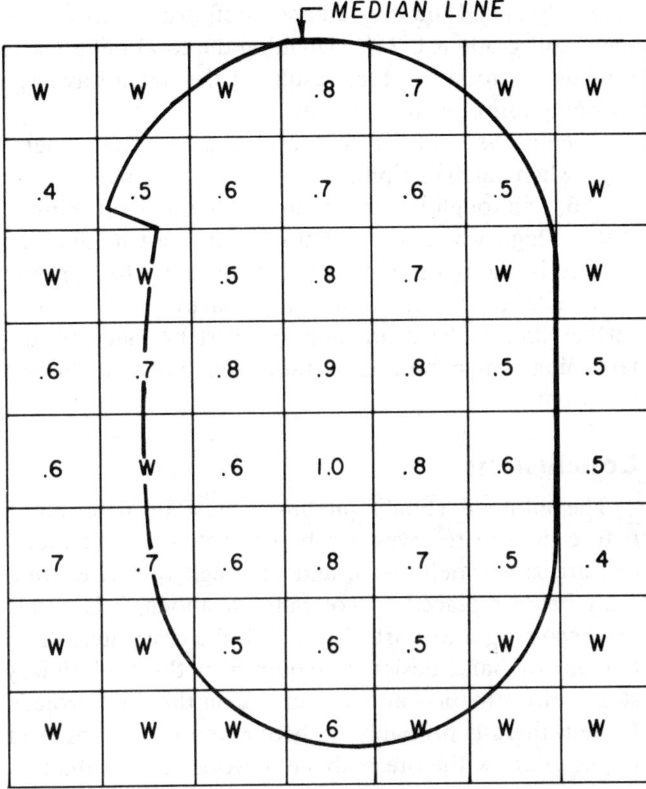

Fig. 13. Level plan with ultimate pit limit.

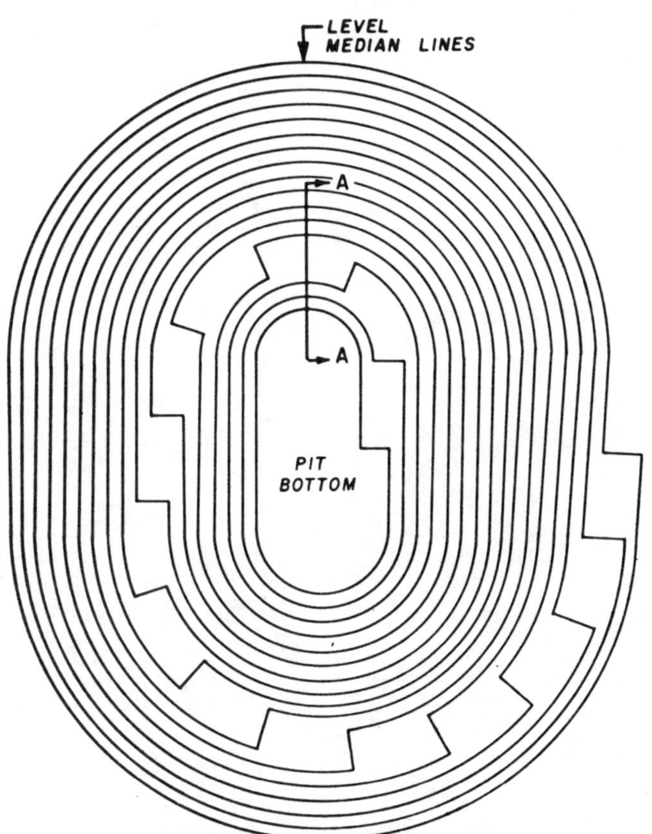

Fig. 12. Composite ultimate pit plan.

The stripping ratios at the pit limits can also be determined for each sector by overlaying the composite plan on the individual level plans and marking the ore contacts on the composite plan and planimetering the ore and waste zones. The stripping ratios can also be compared by transferring the pit limits from the ultimate pit plan to the vertical sections.

It should be kept in mind that the break-even and allowable stripping ratios are the ratios at the final pit surface and do not reflect the overall stripping ratio for a section, sector, or the entire ore body.

The final ore reserves and overall stripping ratios are determined from the level maps. The ore tons and grade, and the waste tons within the ultimate pit limits, are determined for each level, and these are accumulated to arrive at the mine totals. Ore is considered to be that material within the pit limits having a grade equal to or greater than the grade from the stripping curve at a break-even stripping ratio of zero. This is generally called the ore reserve cutoff grade. In Fig. 4, the cutoff grade is 0.3%. Dividing the total mine waste tons by the total ore tons results in the overall average stripping ratio for the ore body.

Open pit vs. underground mining is also a consideration when high stripping ratios are involved. Gill (1966), although not mentioning underground mining, states, "high waste to ore ratios should not be avoided if there is enough grade, or value, to mine the material at a suitable profit." Pana and Davey (1973) state further that, "The distinction can best be made by determining which mining method generates the largest net profit."

Conclusions

The ultimate (final) pit limits must be determined before the ore reserves can be estimated. Hand methods are still widely used, and although they are gradually being replaced by computer methods, it is surely not becoming a lost art. Even with the computer, some amount of hand design is required in the preliminary stages of evaluation and as a check on the final project. In fact, there is probably no better way for an engineer to get to know the ore body he is working with than by doing some hand design. This should give him more confidence in the results obtained from computer techniques, if that is his primary method of design.

References

Erickson, J. D., 1968, "Long-Range Open Pit Planning," *Mining Engineering,* April, pp. 75-78.

Gill, D. K., 1966, "Open Pit Planning," *Mining Congress Journal,* July, pp. 48-51.

Halls, J. L., 1970, "The Basic Economics of Open Pit Mining," *Proceedings,* Symposium on the Theoretical Background to the Planning of Open Pit Mines with Special Reference to Slope Stability, South African Institute of Mining and Metallurgy, Johannesburg.

Pana, M. T., and Davey, R. K., 1973, "Pit Planning and Design," *SME Mining Engineering Handbook,* A. B. Cummins and Given, I. A., eds., AIME, New York, pp. 10-17, 17-19.

Popoff, D. C., 1966, "Computing Reserves of Mineral Deposits: Principles and Conventional Methods," Information Circular 8283, US Bureau of Mines.

Soderberg, A., 1959, "Elements of Long-Range Open Pit Planning," *Mining Congress Journal,* April, pp. 54-58, 62.

Soderberg, A., and Rausch, D. O., 1968, "Pit Planning and Layout," *Surface Mining,* E. P. Pfleider, ed., AIME, New York, pp. 141-165.

14 Automated Methods of Final Pit Limit Determination

R. M. (Mike) Robb
United Nuclear Corp.

R. M. (Mike) Robb is senior mining engineer, United Nuclear Corp., Albuquerque, with responsibility for underground and open pit mine evaluation, design, and analysis. He has done extensive equipment evaluation and new property analysis.

Robb received a B.S., extractive industries, from the University of Idaho and has done postgraduate work at the University of Arizona. After service in the US Marine Corps as a motor transport officer, he began his career with Anaconda's Twin Buttes mine, serving successively as mining engineer, systems engineer, truck and shovel foreman, and senior mine foreman. Subsequently he was senior mining engineer for the Mined Land Reclamation Div., Dept. of Geology and Mineral Industry, State of Oregon. Prior to joining United Nuclear, he was mine superintendent for Cobre de Sonora mine, Nacozari, Son., Mexico.

Introduction

The requirements to investigate a wide range of alternatives and analyze a variety of "what if" questions make the automation of pit design a requirement in today's rapidly changing mining industry. The widespread application of computers has relieved the engineer of much of the tedious and error-prone work related to mine design. He is thus free to do the creative and innovative thinking for which he is trained.

It is now possible to investigate a large number of alternative pit designs in a reasonably short time period. In the past it was very easy for a team of engineers to consume 12 to 14 man-months in the detailed hand design of a large open pit. The pressure of today's mining industry, the desire to investigate a variety of alternative designs, and the shortage of qualified engineers make the slowness of hand design less preferred.

Prerequisites

Before any design work can be undertaken, a great deal of preparation must be completed. This preparatory work must include:

1) *Ore-Body Model*—An accurate, fully corrected, block model must be available. Each block must be identified by an X, Y, Z coordinate of some type. The grade or percentage of the minerals must be known. Other items which could be of importance in pit limit determination are often coded into the model. These factors might include rock types, mineralogy, metallurgical characteristics, trace element analysis, and potential byproduct items.

The ore-body model must reflect the true topographic surface. Provision should be made for any surface constraints such as property lines, inhabited areas, streams, etc. This is normally done by a coding system within the block of the ore-body model.

2) *Pit Slopes*—The practical pit slope angle (θ) must be determined. This is done via rock mechanics studies, previous experience, safety requirements, and contemplated equipment size. It is common for pit slopes to change from one area of the pit to another. These changes, due primarily to changing geological conditions, must be known.

3) *Bench Height*—The bench height must be previously determined. Bench height is normally a function of acceptable dilution constraints, equipment restrictions, and slope stability. The selection of bench height is a decision that will be reflected in the sizing of the truck and shovel fleet, drilling requirements, road widths, and a variety of other items. Once committed to a given bench height, it is difficult to change.

The bench height is often reflected in the block size represented by the computerized ore-body model. If the bench height is 10 m, the ore-body model would most likely be represented by a series 10-m height blocks.

4) *Maps*—A series of usable level and cross section maps must exist. These can either be the conventional geological ore reserve maps or a single digit map printed by a computer. These maps will be utilized in planning pit bottoms, laying out roads and ramps, and determining the direction of pit advance.

Fig. 1 is an example of a typical level single digit computer-created level map. Fig. 2 is an example of a single digit cross section map. Computer-generated maps will generally have a left to right, elongated scale. This elongation is due to physical limitation of the computer printer. There are printers which will produce a "to scale" map but their cost and limited application will often preclude their acquisition.

5) *Computer and Programming*—A medium size computer with either tape or disk storage must be available. Most mine design systems will require a machine with a central memory of at least 64,000 bits, and a FORTRAN compiler. Most major mining companies possess an in-house computer of sufficient size and capability to do pit design work.

Computer-Aided Design

The approach taken by most companies is for the engineer to utilize the computer as a tool to speed examination of various pit design options. This is usually done by the engineer studying the ore reserve maps and conceptualizing a logical way to exploit the ore body. The concept is then translated into a development sequence. Development is normally done through a series of pushbacks. The words *pushback, phase,* or *increment* are often used interchangeably. A pushback is normally defined as "a logical subsection or division of the ultimate or final pit which contains enough ore to sustain production over a definable time period." The time period is normally years (i.e., one year or five years). Sustained production must at least match the nominal rated mill capacity. The size of a pushback must be known.

The size of the equipment in use, or contemplated, is a major constraint in determining pushback size. For example, if the pit is utilizing 150-t trucks, 11.25-m^3 shovels, and operating under the normal requirements of two-side loading by the shovel, a minimal operating width of 50 m will be required. Under these circumstances a pushback with a width of 25 m, no matter how optimal, would not be acceptable.

```
                                                   LEVEL MAP
                                 NO CU COPPER COMPANY              PUSHBACK = 1-3120
                                                                   LEVEL = 1620

                                                            COLUMN
  0    5   10   15   20   25   30   35   40   45   50   55   60   65   70   75   80   85   90   95  100  105  110  115  120
  +    +    +    +    +    +    +    +    +    +    +    +    +    +    +    +    +    +    +    +    +    +    +    +    +
ROW......................................................................................................................... ROW
50+.......................................................................................................................+ 50
49+.......................................................................................................................+ 49
48+.......................................................................................................................+ 48
47+.......................................................................................................................+ 47
46+.......................................................................................................................+ 46
45+.......................................................................................................................+ 45
44+.......................................................................................................................+ 44
43+.......................................................................................................................+ 43
42+............................1..........................................................................................+ 42
41+.......................................................................................................................+ 41
40+............................1111111111.................................................................................+ 40
39+...........................111.........................................................................................+ 39
38+...........................11114454....................................................................................+ 38
37+.......................11111444444.....................................................................................+ 37
36+.....................11111111111111....2222............................................................................+ 36
35+.....................11111111111.......................................................................................+ 35
34+.....................1111333...........5555......44454.................................................................+ 34
33+.....................***111111.........55.......3334.55555.............................................................+ 33
32+.....................******..........................222222............................................................+ 32
31+.....................*****...........................115112222.........................................................+ 31
30+.....................*****..........................555555.............................................................+ 30
29+.....................33333.............................................................................................+ 29
28+.....................222222222.........................................................................................+ 28
27+.......................***.............................................................................................+ 27
26+.......................***.............................................................................................+ 26
25+.......................***.............................................................................................+ 25
24+......................*****............................................................................................+ 24
23+........................***............................................................................................+ 23
22+.....................4444****..........................................................................................+ 22
21+.....................4444..........1111................................................................................+ 21
20+.....................3333..............................................................................................+ 20
19+.........................2222.........................2222.............................................................+ 19
18+.........................33............................................................................................+ 18
17+..........................*****........................................................................................+ 17
16+.......................................................................................................................+ 16
15+.......................................................................................................................+ 15
14+............................***........................................................................................+ 14
13+.............................***.......................................................................................+ 13
12+.......................................................................................................................+ 12
11+.......................................................................................................................+ 11
10+.........................................***...........................................................................+ 10
 9+.......................................................................................................................+ 9
 8+.......................................................................................................................+ 8
 7+.......................................................................................................................+ 7
 6+.......................................................................................................................+ 6
 5+.......................................................................................................................+ 5
 4+.......................................................................................................................+ 4
 3+.......................................................................................................................+ 3
 2+.......................................................................................................................+ 2
 1+.......................................................................................................................+ 1
```

Fig. 1. Single digit level map.

Automated Mine Design

The investigation and analysis of pushbacks is most easily done by utilization of a computerized mine design system. While the details of a mine design system will vary a great deal, they will all follow the general outline shown in Fig. 3.

Pit Generation Program

The pit is created by generating a right normal cone, with the apex of the cone being the pit bottom and the top of the cone being at the topographic surface. The complement of the half angle is equal to the pit slope. The pit generation program computes which blocks, or portions thereof, are inside the pit cone. The creation of the cone and its relationship to an actual pit are shown in Fig. 4.

The pit bottom is defined by use of an X, Y, Z coordinate system. The coordinate system can take the form of: levels, rows, and columns; elevations, northings, and eastings; line segments, center points, and circles, or any plausible combination. Most pit generation programs will allow for changes in the pit slope angle (θ), different block sizes, and irregular-shaped pit bottoms.

Matching Program

The function of the matching program is to match the blocks that are in the computer-generated pit to its counterpart in the ore-body model. This is done by creating the mathematical pit file and comparing it to the ore-body model. If a match is found, the block information (i.e., tons, grade, etc.) is set aside for use in the ore reserve calculations.

Ore Reserve Output

The actual format of the ore reserve output will depend on the material being mined, ore grade cutoffs,

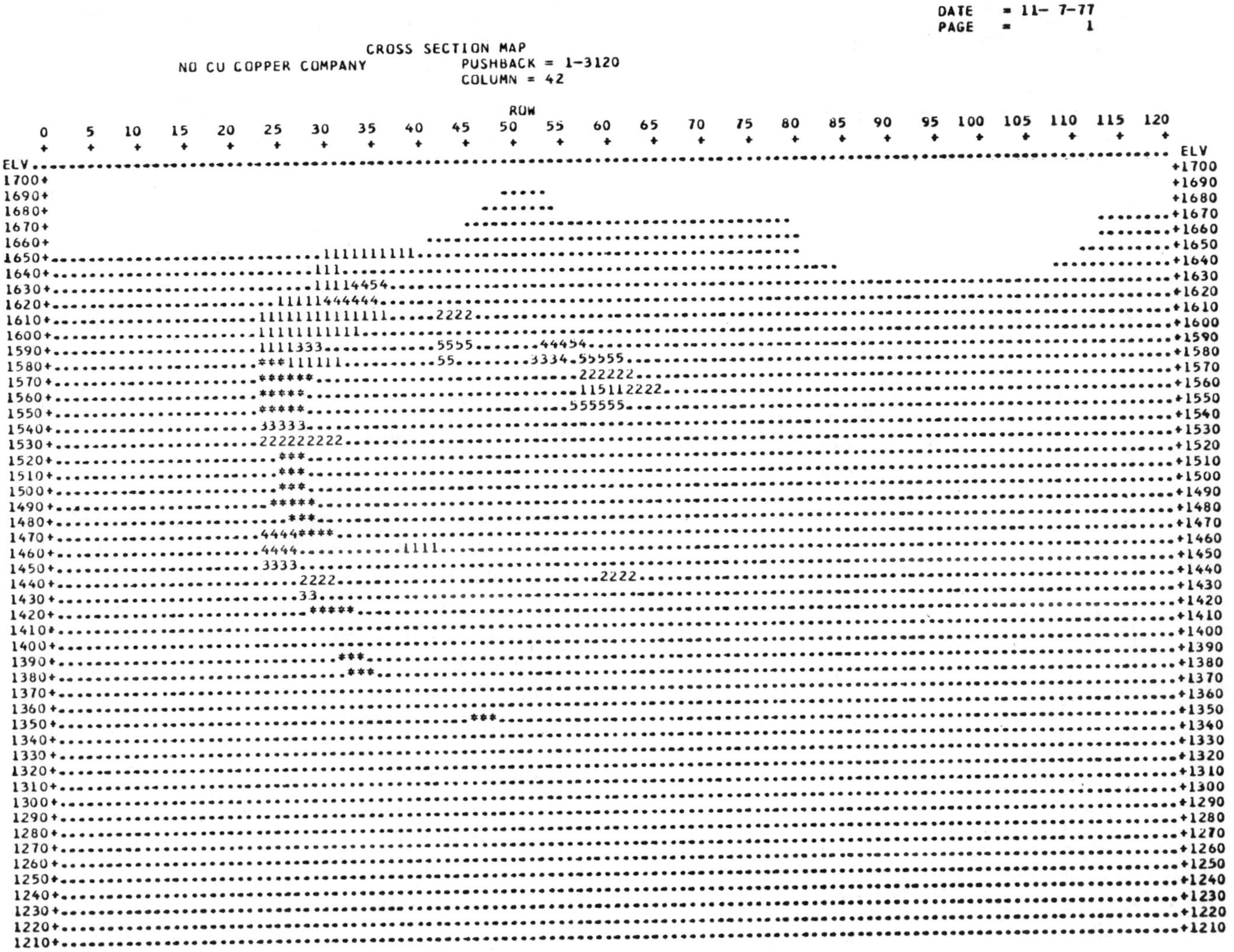

```
                                                                          DATE   = 11- 7-77
                                                                          PAGE   =        1
                              CROSS SECTION MAP
              NO CU COPPER COMPANY          PUSHBACK = 1-3120
                                            COLUMN = 42
```

Fig. 2. Single digit cross section map.

and most importantly, what the design engineer wants. An example of an ore reserve output is shown in Fig. 5.

Analysis

Utilizing the ore reserve output the engineer must now analyze the results of the pushback. The economic viability is the ultimate factor in an analysis. The relative economic worth of a pushback is a function of: waste to ore ratio, operating costs, mill requirements, grade mill recovery, and selling price of the product produced. If the pushback is not acceptable, the model is restored to its previous status and a different pushback is run. If the pushback meets the criteria of an acceptable pushback, the engineer then designs and runs another pushback. This process is repeated until the ore-body model is mined out. The pushbacks are then scheduled as a function of time, and a detailed financial analysis is undertaken.

Break-even Analysis

A widely practiced method of determining final pit limits is the break-even analysis. Break-even analysis may be defined as "the point in the life of a mine where the cost of operation equals the value of the product sold."

One technique that has proven very effective is to divide the ore body into segments such as those shown in Fig. 6. Each section is analyzed by a series of pushbacks to determine the economic break-even point. Based on grade cutoff constraints, known or anticipated operating costs, and operating parameters, the point where cost equals income is determined. After each segment has been analyzed the break-even pushbacks are plotted on a map. An example is shown in Fig. 7.

The pit outline thus obtained must then be engi-

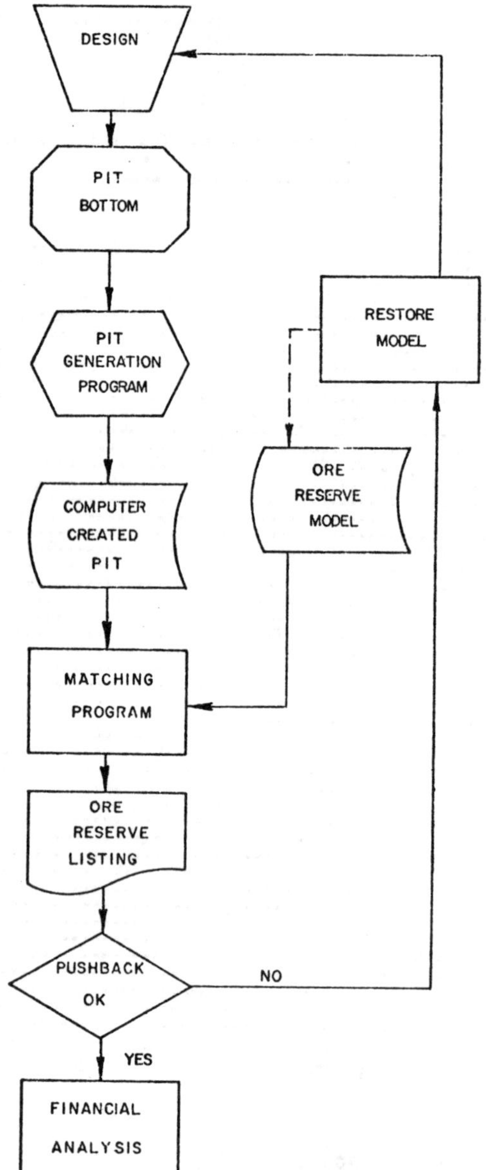

Fig. 3. Generalized mine design system.

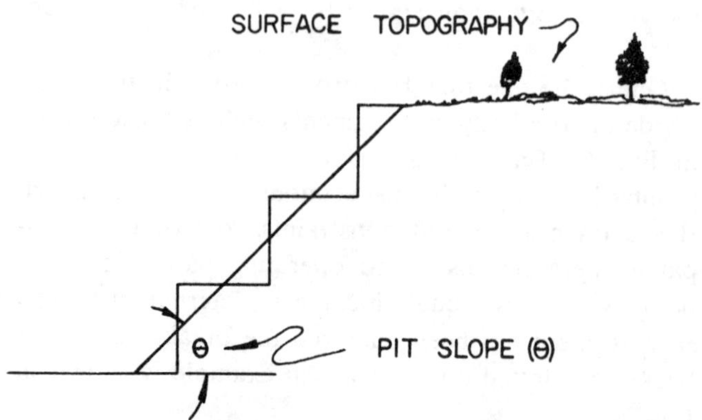

Fig. 4. Computerized pit generation.

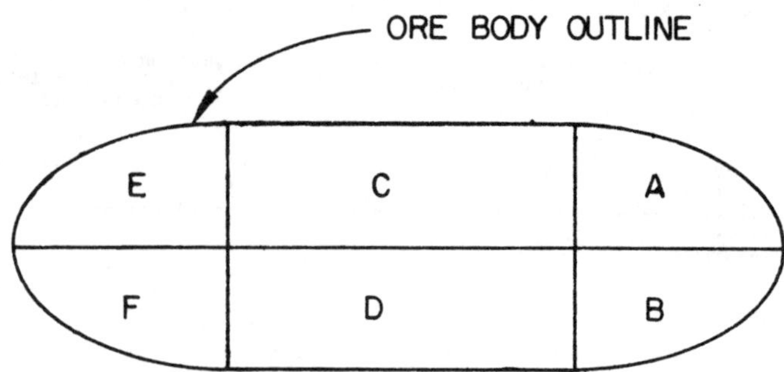

Fig. 6. Plan view break-even analysis.

neered into a minable, realistic shape, and the final financial analysis undertaken.

Fully Automated Ultimate Design

There are, in theory, several ways to determine the ultimate pit limits. The Lerch-Grossman technique and critical path analysis of either grade or financial return are but a few of the techniques mentioned in professional publications. These techniques can be of great assistance in determining the final ultimate pit outline. They do not, in general, provide any means for determining pushback size or location. Only the most general of operating constraints can be accommodated.

Another real constraint of a fully automated pit limit determination system is the cost. A block model will have at least $(X \cdot Y)^n$ possible block combinations to be analyzed. Even on today's fast computers this can easily involve several hours and a high cost. While this is a small percentage of the total capital cost of a mine, it is an upfront cost which could substantially add to the cost of a feasibility study.

Summary

Computer-aided mine design and optimization is a tool that is only as good as the input data and capability of the person utilizing it. It must be remembered that judgment, caution, practicality, and a great deal of

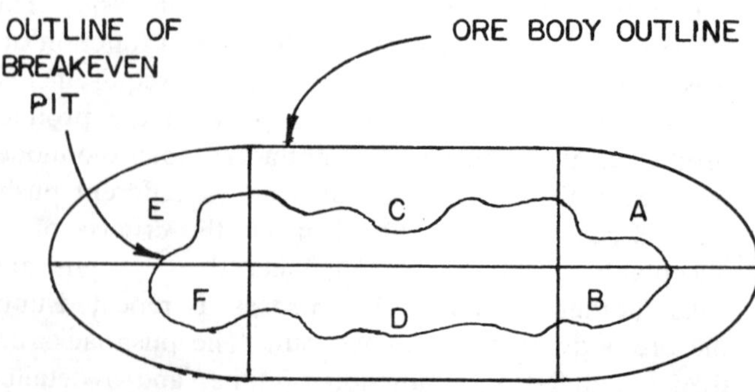

Fig. 7. Pit outline after break-even analysis.

DATE = 11- 7-77
PAGE = 1

ORE RESERVE LISTING
NO CU COPPER COMPANY
PUSHBACK = 1-3120
TONNES X 1,000.

	.800 -100.		.600 - .799		.400 - .599		.200 - .399		.000 - .199		TOTAL	
LEVEL	TONS	GRADE	TONS	GRADE	TONS	GRADE	TONS	GRADE	TONS	GRADE	TONS	GRADE
1700	0.	0.0	0.	0.0	0.	0.0	0.	0.0	10.	0.0	10.	0.0
1690	0.	0.0	0.	0.0	0.	0.0	0.	0.0	62.	0.0	62.	0.0
1680	0.	0.0	0.	0.0	0.	0.0	0.	0.0	333.	0.0	333.	0.0
1670	0.	0.0	0.	0.0	0.	0.0	0.	0.0	894.	0.0	894.	0.0
1660	0.	0.0	0.	0.0	0.	0.0	0.	0.0	852.	0.0	852.	0.0
1650	0.	0.0	0.	0.0	0.	0.0	0.	0.0	811.	0.0	811.	0.0
1640	0.	0.0	0.	0.0	0.	0.0	0.	0.0	1622.	0.0	1622.	0.0
1630	0.	0.0	0.	0.0	0.	0.0	0.	0.0	1112.	0.0	1112.	0.0
1620	0.	0.0	0.	0.0	0.	0.0	0.	0.0	1361.	0.0	1361.	0.0
1610	0.	0.0	0.	0.0	0.	0.0	0.	0.0	1839.	0.0	1839.	0.0
1600	0.	0.0	0.	0.0	0.	0.0	0.	0.0	2152.	0.0	2152.	0.0
1590	0.	0.0	0.	0.0	375.	0.408	0.	0.0	2380.	0.0	2755.	0.056
1580	0.	0.0	947.	0.798	0.	0.0	0.	0.0	2348.	0.043	3295.	0.260
1570	1008.	1.089	480.	0.750	0.	0.0	0.	0.0	2328.	0.017	3816.	0.393
1560	2108.	0.938	0.	0.0	0.	0.0	0.	0.0	2818.	0.021	4926.	0.414
1550	1278.	0.876	500.	0.609	0.	0.0	0.	0.0	2777.	0.020	4555.	0.325
1540	2497.	0.930	0.	0.0	0.	0.0	0.	0.0	3544.	0.024	6041.	0.399
1530	2006.	0.926	0.	0.0	990.	0.501	0.	0.0	1809.	0.013	4805.	0.495
TOTAL	8897.	0.941	1927.	0.737	1365.	0.475	0.	0.0	29052.	0.013	41241.	0.262

STRIPPING RATIO AT 0.200 CUT OFF = 2.383

Fig. 5. Ore reserve output.

common sense must play an important part in determining the actual pit design.

References

Connell, J. P., 1969, "Computer Reports with a Purpose," AIME Intermountain Minerals Conference, July.

Gauthier, F. J., and Gray R. G., 1971, "Pit Design by Computer at Casper Copper Mines, Ltd.," *The Canadian Mining and Metallurgical Bulletin,* November, pp. 95-102.

Lerch and Grossman, 1965, "Optimum Design of Open Pit Mines," *The Canadian Mining and Metallurgical Bulletin,* January, pp. 47-54.

Lipkewich, M. P., and Borgman, L., 1969, "Two and Three Dimensional Pit Design Optimization Techniques," *A Decade of Digital Computing in the Mineral Industry,* A. Weiss, ed., AIME, New York, pp. 505-523.

Mirani, A. M., 1969, "Mine Planning Systems," *A Decade of Digital Computing in the Mineral Industry,* A. Weiss, ed., AIME, New York, pp. 525-537.

3

PRODUCTION PLANNING

Contents

Introduction to Sections 3A, 3B

Once the ore body has been defined, an economic evaluation has been made of the model, and the final pit shell has been generated, the engineer is ready to "detail" the method and rate of extraction. Numerous problems including slope geometry, working faces, roads, dumps, drainage, and equipment constraints are analyzed through the use of maps and drawings showing different phases or time periods during the life of the mine. In conjunction with these phase plans, the engineer will develop production schedules. The objective of this production planning, whether short or long range in nature, is to develop an economically viable plan that is realistic, flexible, and concise.

3A Operating Layout and Phase Plans

S. P. Winkelmann, editor

Contents

15 Selected Aspects of Production Planning—Berkeley Pit

Don Hendricks
Alan Dahlstrand
The Anaconda Co.

Donald A. Hendricks graduated from Montana School of Mines with a B.S. in mining engineering. From 1958 to 1970, he held the following positions at Cominco American, Inc.: surveyor, safety and personnel officer, production engineer, mine foreman, mine engineer, and chief engineer. He is currently the chief operations engineer of the Anaconda Co. in Butte, Montana.

Alan Dahlstrand graduated from the University of New Mexico with a B.S. in geology. He held the following positions at the Anaconda Co.: sampler and assistant mining engineer for underground mines and exploration activities, assistant mining engineer, planning engineer, and engineering and planning superintendent for pit mining. He is currently director of mine planning for Anaconda's Butte operations.

Introduction

The subjects discussed in this chapter deal with production planning, or more specifically, how the results of production planning may relate back to long-range planning and final pit design. The items discussed here are: operating slopes, geometry, bench height, ore and waste faces, relationship to equipment selection, and computer methods. Many different ideas can be presented on these subjects, but this material is a recap of mining experience in the Butte ore body.

Operating Slopes

Since the early years of the Berkeley pit, there have been problems with designed slopes in some areas while mining in other areas has progressed as planned. The final design and interramp operating slopes for most areas of the pit have varied between 0.66 rad (38°) and 0.78 rad (45°). The 0.66 rad (38°) slope has been used primarily in wet alluvial material and some extremely unstable areas in rock, while most other areas have been mined up to 0.78 rad (45°). When interramp operating slopes have been modified from the designed slope, it has been because of slope instability due to water, adverse structure, or a change in ore location from that projected. The overall operating slopes may vary widely from one area of the pit to another and are a function of ore location, bench operating widths, and the position of haul roads.

Changes in operating slopes in unstable areas have been made in several ways. Buttressing of the toe by leaving a wide berm may be required, but this can result in a temporary loss of ore reserves and an increase in stripping ratio. If the pit wall is being mined to final position, the ore loss is permanent. When a decision is made to leave a wide berm, a minimum width is usually left as a first step. If slope failure persists, more berms are left on benches below. Operating slopes have also been changed when ramp and road locations are changed due to slope problems or when a change in projected ore location requires a change in pushback design.

During the early stages of the Berkeley pit, very few changes in operating slopes were necessary since most slope failures were local bench failures. Over the last ten years, as a pit increased in depth and width, the frequency and severity of slope problems required changes of planned operating slopes in increasing numbers. Berm widths left below unstable areas have varied from 15 m (50 ft) to 60 m (200 ft). Extra berm width has also been allowed in some areas for water collection and pumping.

Having the flexibility to adjust operating slopes as problems arise is very important and is directly dependent upon the amount of operating room which has been designed into the mine (Figs. 1a and 1b). Lack of operating room (Fig. 1a) not only limits bench development, equipment size, and adjustments for geological changes, but narrow pushbacks are especially vulnerable to pit wall slides. Slope failures cannot be absorbed within narrow pushbacks but must be mucked up, or stripping must be started on a new pushback, thereby raising stripping ratios and costs. With adequate mining width, adjustment of the pushback for a slide can still allow plenty of mining room (Fig. 1b). Changes in projected vs. actual ore position may not cause as much ore loss if the pushback is wider.

A slope monitoring program requires the establishment of a network of monitoring stations which will provide adequate information about the stability of the pit walls. The scope of the program will depend upon the size and depth of the pit, the geological and physical setting, the estimated monetary consequences of a major slide, and the availability of capital to fund the program.

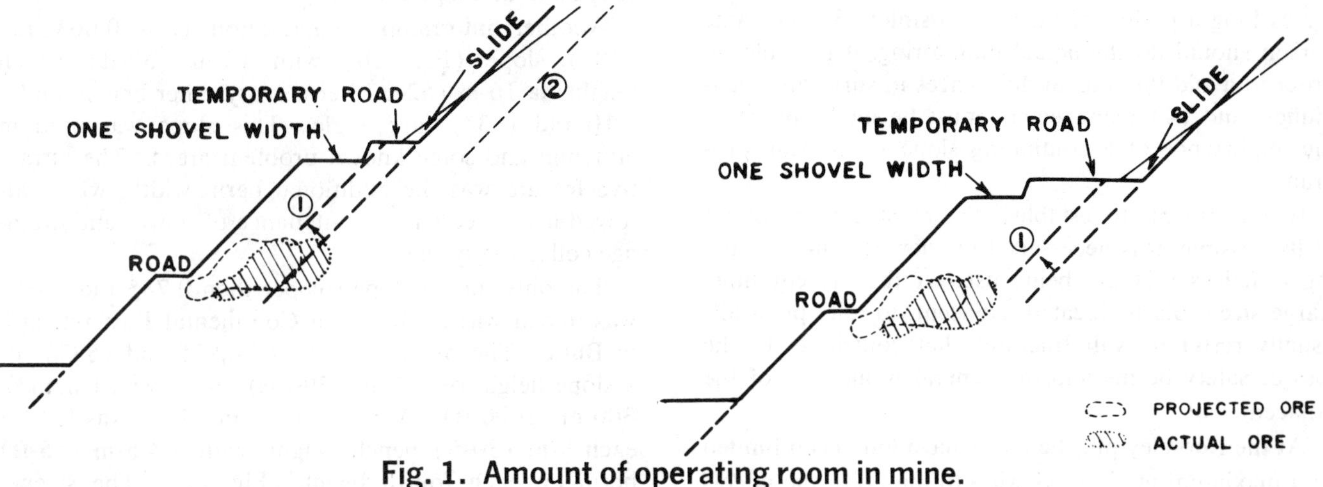

Fig. 1. Amount of operating room in mine.

The program in present use at the Berkeley pit utilizes extensometers, ground-water measurements, and a network of 60-odd survey points. The extensometers are used in areas in which visual examination reveals that movement has already occurred. Readings are taken every other day, and graphs of these measurements are kept to show when acceleration of the movement occurs. Continued acceleration of subsidence gives advance warning of an impending slide. In one critical area, a switch has been installed and connected to the weight so that a vertical displacement of the weight of 152 mm (6 in.) or so will trip the switch and light a red, flashing light mounted on a nearby pole. The light is in view of the truck dispatcher, and he can give warning, when necessary, to trucks in the area. A constant downslope movement in this area has made it necessary to periodically adjust the linkage between the weight and the switch to prevent false alarms.

Ground-water levels are taken weekly in alluvial areas by measuring the water column in a network of cased, vertical holes. This information is recorded and graphed for possible correlation with slope movements in these areas.

The monitoring of survey points is done with a 1-sec theodolite and a laser distance-measuring device. The distance from the control points to the monitoring points varies from 457 m (1500 ft) to 2438 m (8000 ft). Considering the equipment used and distances involved, the resultant data is accurate to within 9-12 mm (0.03-0.04 ft). To calculate and report numbers that are carried out beyond 4 mm (0.01 ft) would be misleading under the circumstances.

Some of the problems encountered in monitoring points are: (1) survey points should be located away from operating equipment but within areas which may affect the operations if there is a failure; (2) survey points should be located where they will be accessible for as long a period of time as possible; (3) the same person should do the actual monitoring, if possible, in order to avoid the human differences in surveying techniques; and (4) management must be made aware of the importance of a continuing slope monitoring program.

In an attempt to establish the most competent pit walls possible, it is necessary to review the in-pit blasting practices that are being used at the present time. Large-sized blasts located adjacent to the pit walls usually result in wall fractures that contribute to the loss of safety berms and the general weakening of the adjacent wall.

At the Berkeley pit, the blasts have long been limited to a maximum of six rows wide to minimize the effects of blasting upon the adjacent city of Butte. To protect the walls, the patterns are now designed so that a final, free-faced blast is made adjacent to each planned crest. The hole spacing and loading of the final row are varied to fit the different rock types encountered throughout the pit. The biggest problem encountered so far has been hard toes resulting from the misjudging of the rock hardness in the blast area and/or the inability to drill deep enough to get the desired subdrilling, both of which result in unbroken ridges between the individual drill holes and between the holes and the planned toe. To overcome this problem, the final row hole spacing has been tightened correspondingly. At the present time, all final, single-row blasts are spaced 4.5 m (15 ft) between holes and 1.8 m (6 ft) away from the final toe. The amount of explosive varies from 68 kg (150 lb) per hole to 136 kg (300 lb) per hole for soft and hard rock zones, respectively.

Geometry

Slope geometry is one facet of pit design and planning which can be quite variable, depending on bench height, structural controls, rock type, blasting practices, and ground-water conditions.

Several slope cross sections have been used at the Berkeley pit in the past; not one has been a complete failure or success since conditions are always so variable.

The most common interramp slope configuration, as shown in Fig. 2a, is a 0.785 rad (45°) overall slope used with 10-m (33-ft) benches. Berms are collected every other bench with a width of 10 m (33 ft). The bench face angle is 1.10 rad (63°). The success of the 0.785 rad (45°) slope varies widely from one area of the pit to another. In most cases, the 10-m (33-ft) berm width was adequate to catch loose rock, but one of the big drawbacks was lack of access for power drops and drainage control.

Another interramp configuration is a 0.663 rad (38°) slope (Fig. 2b) with 10-m (33-ft) bench heights, a 16-m (52-ft) berm every other bench, and a 1.10 rad (63°) face angle. This slope was used in alluvium and some known problem areas. The attractive feature was the additional berm width, which allowed later access for maintenance of power and drainage collection systems.

The only time a slope steeper than 0.785 rad (45°) was mined was in the small Continental East pit, also in Butte. The overall slope was 0.925 rad (53°) for a slope height of 119 m (390 ft) and a wall length of 300 m (1000 ft). A berm of 1.5 m (5 ft) was left for each 9-m (30-ft) bench height, with a 4.5-m (15-ft) berm on each fourth bench (Fig. 2c). The steeper

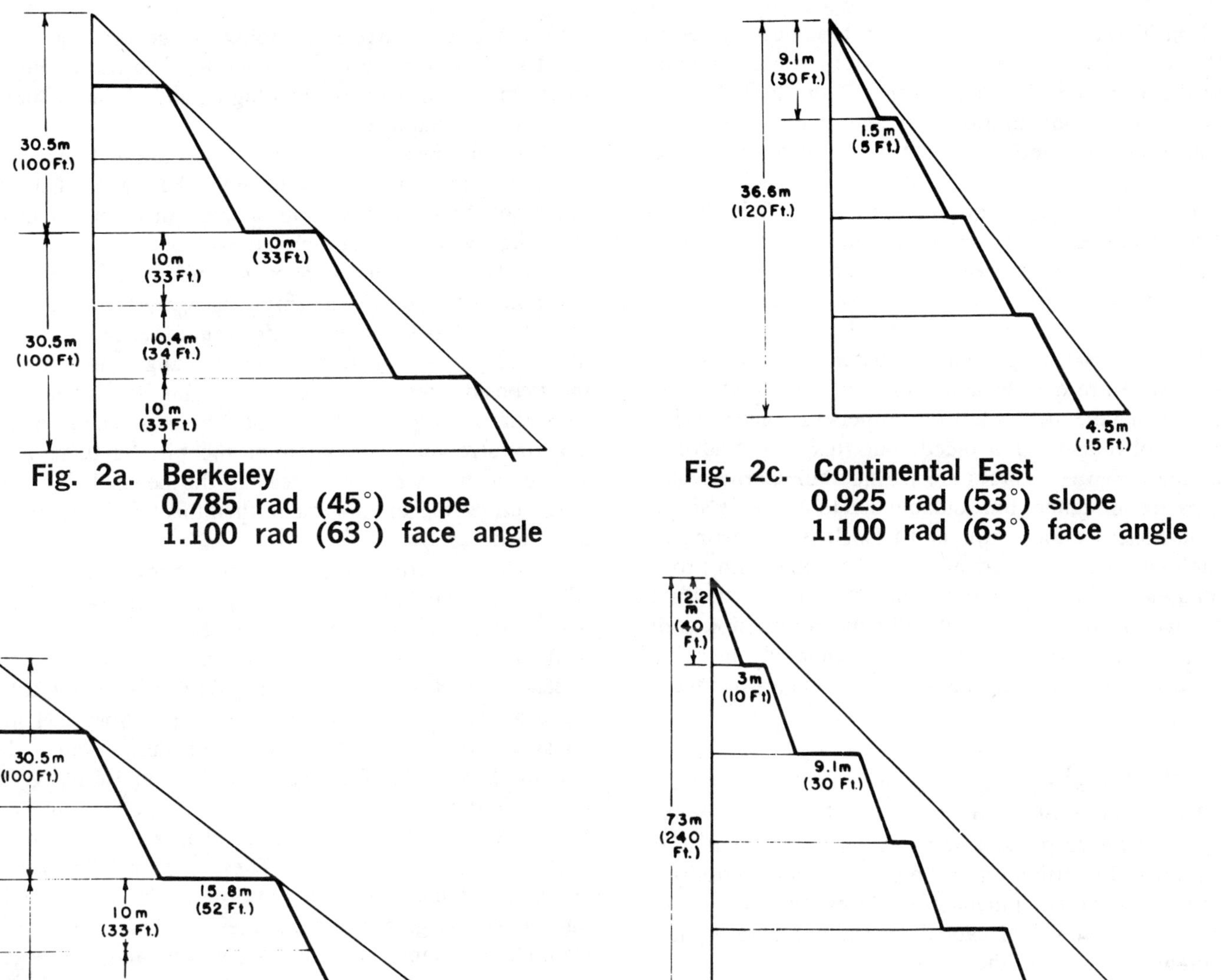

Fig. 2a. Berkeley
0.785 rad (45°) slope
1.100 rad (63°) face angle

Fig. 2c. Continental East
0.925 rad (53°) slope
1.100 rad (63°) face angle

Fig. 2b. Berkeley
0.663 rad (38°) slope
1.100 rad (63°) face angle

Fig. 2d. Berkeley
0.785 rad (45°) slope
1.222 rad (70°) face angle

Fig. 2. Interramp slope configurations.

slope was attempted due to the small size of the final wall and apparent lack of oblique structures. Success was achieved in the central portion of the wall, but slides developed in the east and west ends of the wall in fault zones, thereby causing the loss of some ore.

During 1976 at the Berkeley pit, the bench height was changed to 12 m (40 ft) for economic reasons, and the 12-m (40-ft) bench was still within the capabilities of present equipment. The wall configuration was also changed for overall slope angles. A face angle of 1.222 rad (70°) was attempted, and a berm was left on each bench 3 m (10 ft) and 9 m (30 ft), alternately. For a 0.785 rad (45°) slope, a 19-m (63-ft) berm would be left every 73 m (240 ft) vertically (Fig. 2d). This would allow access for pit utilities. The berm on each bench was intended as an additional safety feature, but due to adverse structure, overbreak from blasting, overdigging, and inability to achieve the 1.222 rad (70°) face angle, the 3-m (10-ft) berms would often fall off. This left a 9-m (30-ft) berm which was full of rubble and that did not allow access or protection for power poles or drainage ditches.

Consideration is now being given to collecting berms every other bench with a 12-m (40-ft) berm and retaining the 19-m (63-ft) berm every 73 m (240 ft) vertically. This configuration may improve the required protection from sloughs, and it would also give added berm access.

A complete analysis of pit slopes is now being made under the direction of a consulting firm. The results of this study should establish a wide variety of slope recommendations based on physical conditions and economics.

The geometry of pit plans is based upon the shape of possible ore deposit and economics. However, certain pit shapes, such as outside corners or convex walls, should obviously be avoided, but this is not always possible when economics are considered. Corners and noses are sometimes unavoidable due to ore-body shape or pushback configuration and ties. It is interesting to note that outside corners or noses developed into problem areas less often than the walls between for the first 15 years of the Berkeley pit. The overall pit size, wall length, and adverse structure encountered probably added to the increasing number of problems in recent years.

Bench Height

The selection of the optimum bench height depends upon a number of considerations, among which are: (1) vertical distribution of the ore, (2) production requirements and equipment size, (3) existing equipment and the availability of capital for the purchase of new equipment, and (4) other factors.

A brief discussion of each point is warranted for clarity.

1) In the case of an ore body which is irregular, both vertically and horizontally, there could be one or more bench heights that would optimize the ore tons and/or grade. A computer file of the projected mineral grades throughout the mine area makes it possible to simulate the mining of the deposit at varied bench heights and reference elevations. Comparison of the ore tons and grade, bench by bench as well as totally, will show the optimum bench height(s), from which an intelligent selection can be made.

2) The planned production requirements of the mine will determine the size of the equipment that will be used in the pit. The desired bench height will be determined within certain limits by the size and type of equipment selected for the desired production. For instance, it would be impractical to choose a 15.2-m (50-ft) bench interval if front-end loaders were selected for the loading units. Conversely, a 7.6-m (25-ft) bench height would not be economical if large electric shovels are to be used for loading. In general, savings can be realized if the bench height can be made equal to the maximum vertical working height of the loading and drilling equipment.

3) Oftentimes a mine must be designed around existing equipment that will limit the bench height selection. In such cases, the planner must look ahead to a time when this equipment will be replaced with new units. The trend in open pit mining is to reduce operating costs by purchasing larger more productive shovels and trucks. If it is judged that the future availability of capital will support such a change, the planner must consider the possibility of changing bench heights. This change could be throughout the entire mine or it may involve only a portion of the pit. If there are savings to be made by increasing the bench height to better utilize bigger equipment, then this feasibility must be considered.

4) Other factors must also be considered in the selection of an optimum bench height. Among these are safety and weather conditions. Safety, of course, is a major consideration in that operators cannot be expected to work under a towering bank that might collapse at any moment and bury or injure them. This consideration must be applied to both the broken rock and the final wall. Weather conditions in Butte, MT, are such that extended subzero cold spells result in frozen layers on the top of the broken muck that may be up to 2.1 m (7 ft) thick. These frost slabs must be reached by the shovel bucket in order to break them off and prevent large overhanging crests. These frost slabs eventually collapse of their own weight and are large enough to do considerable damage to a shovel or loader. Even if no damage is done, the size of the slabs causes problems and delays in loading them into trucks.

Ore and Waste Faces

The complexity of the ore deposit will affect the mine planning to a greater or lesser degree. The Berkeley pit ore deposit falls into the more complicated category, as can be seen in Fig. 3. The ore varies from soft zones of secondary enrichment, to hard primary veins, to old square-set stopes backfilled with underground waste and loaded with old timbers, to old fire areas which sometimes reignite upon exposure to air. Under such varied conditions, the daily scheduling of ore becomes much more complex than that in many of the copper deposits being mined today.

In addition to the variety of ore types, the ore grades fluctuate vertically and horizontally throughout the mine. The zone of secondary enrichment is quite irregular and often badly diluted with oxide wastes which trace irregularly through the ore areas. In gen-

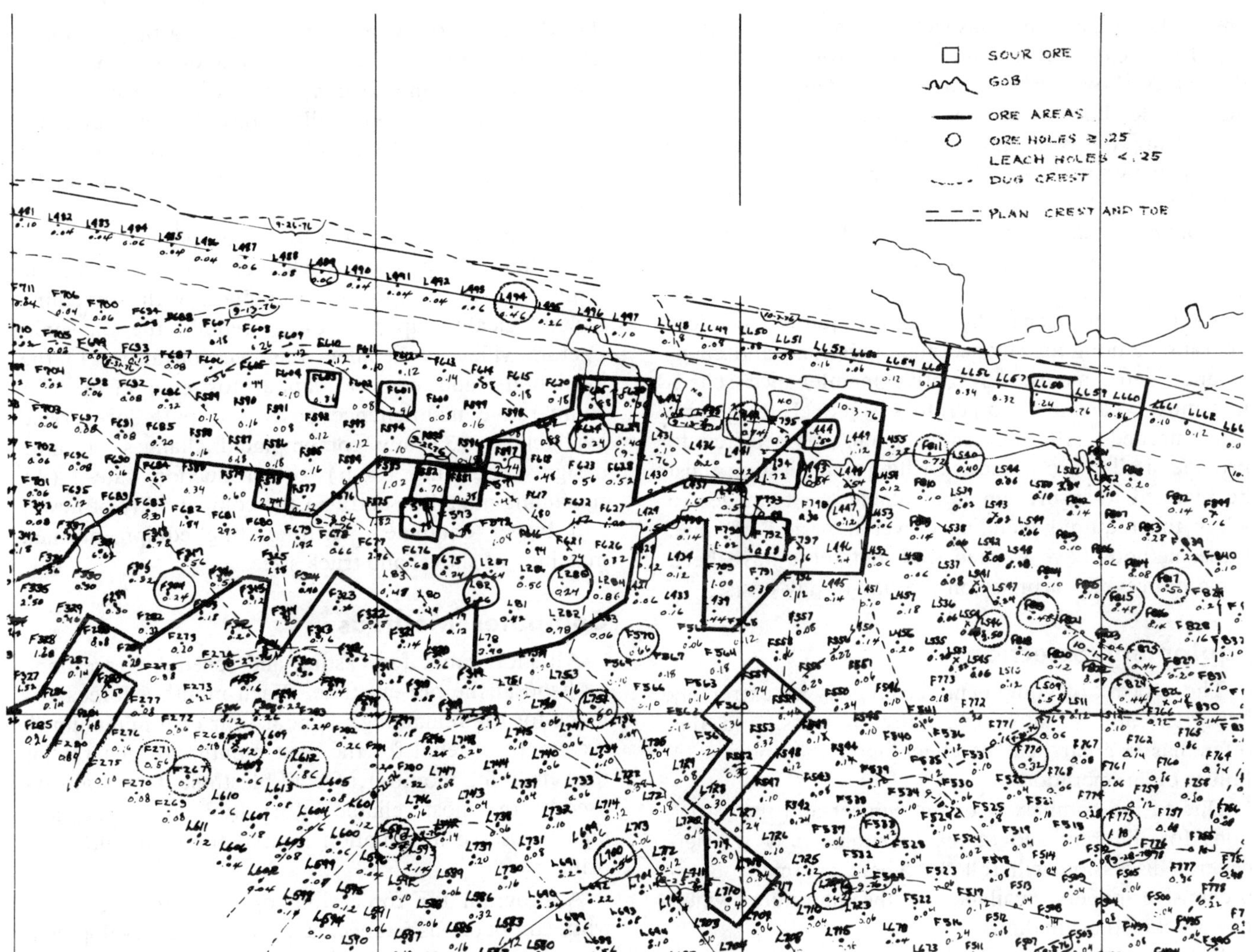

Fig. 3. Berkeley pit blasthole map.

eral, the ore from the enriched zone is soft, but it is also more difficult to concentrate due to mineralogical problems.

Vertically, below the zone of secondary enrichment, the primary vein system is encountered. The larger veins were mined years ago by underground methods, but there are sufficient small veins left to justify the mining of certain areas. In addition, the waste used to backfill the mined-out stopes often runs over 1% Cu. The combination of acid-soluble copper, ferrous sulfate, and rotten timber from these areas creates a condition locally referred to as sour ore. In sufficient quantity, this type of ore causes excessive frothing in the flotation circuits, which can actually inundate the sump pumps and shut down the plant.

The ore minerals are chalcocite, enargite, and chalcopyrite, with much lesser amounts of a half dozen or so other copper minerals, all mixed with zinc ore in certain areas. The primary host rock is quartz monzonite, but dikes of quartz porphyry are found in certain areas. These quartz porphyry areas are extremely hard and greatly reduce the concentrator tonnage when mined.

The local problem is to combine all of these elements, secondary enriched ore, sour ore, vein ore, high zinc ore, and hard ore, into a daily schedule that will provide a relatively uniform feed to the concentrator (Fig. 3). This is accomplished by making a daily schedule for each shovel, shift by shift, showing where the shovel is to mine, what tonnages of ore and/or discard should be produced, and combining these shovel schedules (colored accordingly as staked in the field) into booklets which are given daily to production foremen, shovel foremen, truck foremen, and truck dispatchers.

Blastholes are sampled and surveyed. The areas suspected of zinc associations are also sampled for zinc, and assays over 1% ore plotted on the bench work maps. Likewise, areas suspected of sour ore are assayed for acid-soluble copper and ferrous sulfate, and the values above certain limits are plotted on the work maps. The shovel scheduler then plans the daily mining, trying to avoid grade fluctuations greater than 0.10% between shifts, allowing a maximum of 3 629-000 kg (4000 st) of high zinc bearing ore and/or approximately 25% of sour ore per shift and blending the ore hardnesses to provide a uniform grinding rate at the concentrator.

To make this scheduling work requires close cooperation among the engineering, operations, and maintenance departments. Preventive maintenance schedules must be arranged to permit key shovels to operate when needed. Alternate shovels must be selected to provide similar ore grades. Drilling and blasting must be planned to insure ample available ore tonnages to allow the required blending. Over a period of time, these methods have worked well, but there have been some noticeable surprises on the day-to-day basis.

Equipment Selection

The process of equipment selection is an extremely important and sensitive evaluation process, and this often has a great deal of effect upon the success or failure of an operation.

Due to the complexity of the subject and the variations for different geographic locations, only a brief, general discussion will follow. There are many comprehensive papers available in the literature which cover detailed evaluations.

From a long-range planning viewpoint, the basic process involves choosing the types of equipment, the sizes of equipment, and the numbers of units required. Factors which affect the choice of equipment are: (1) required production, (2) haul distance, (3) operating room within the mine design, (4) availability and costs of power, fuel, etc., (5) weather conditions, (6) material type (rock, alluvium, etc.), (7) mine life vs. capital required for a specific mining system, and (8) operating characteristics of the equipment.

The size of equipment is influenced by a variety of factors, such as mine design, production rates, comparative operating costs, and capital costs. The total numbers of units needed are determined from haul distances and production rates. The outcome of equipment selection may dictate changes in mine design and rates of mining if the benefits of one type of mining system place it far ahead of other systems. Both planning and operating personnel contribute to the efforts of equipment selection.

The subject of short-range production planning is generally involved with equipment replacement, although it is closely related to the long-range efforts. The most important step in evaluation of equipment replacement is the keeping of accurate records for maintenance costs and availabilities. These figures are primary input for determining how unit costs increase with operating hours and when a machine should be replaced.

After a decision is made to replace a piece of equipment, the next question is whether to replace with the same type of equipment, to replace with something technologically superior, or to rebuild the old unit. Factors which might affect this decision are: (1) the availability and cost of capital for equipment, (2) comparative maintenance costs and parts or service availability, (3) any contemplated changes in mine design (operating room) or production rates, (4) remaining mine life, and (5) the compatibility of newer and bigger equipment with existing equipment, i.e., matching of shovel and truck sizes.

Computer Methods

All short-range forecasting and scheduling are done by hand at the present time. Each month a forecast is made for the remaining calendar year or a minimum of six months (detail for the next month and general for the remaining months). Detail blasthole data is always used as the most reliable information, while the geological reserve information is used for areas ahead of drilling.

A program for forecasting on a short-range basis with computer assistance is now in the development stage in Butte. The data base for forecasting will be a combination of all available blasthole information, along with some geological reserve blocking for several benches in each pushback. Since the geological reserve information is carried on a file composed of 15.2× 15.2×12.2 m (50×50×40 ft) blocks, the minifile to be used for forecasting will be the same.

Current computer capabilities allow the blasthole grades to be coded, averaged, and compared to original projected block data. Individual block variances and ore gain and loss maps can be run. New blasthole data will be coded on a regular basis using a digitizer and added to the minifile along with remaining reserve data for each bench of each active pushback. The minifile can then be mapped on 200-scale by plotter for use by an engineer in forecasting alternatives. The map of each bench will also have the pushback toes plotted on it, as indicated by B and C in Fig. 4.

The engineer can then begin the physical layout of bench development alternatives, taking into account shovel capacities and shovel maintenance schedules. One of the alternatives should give the mine the neces-

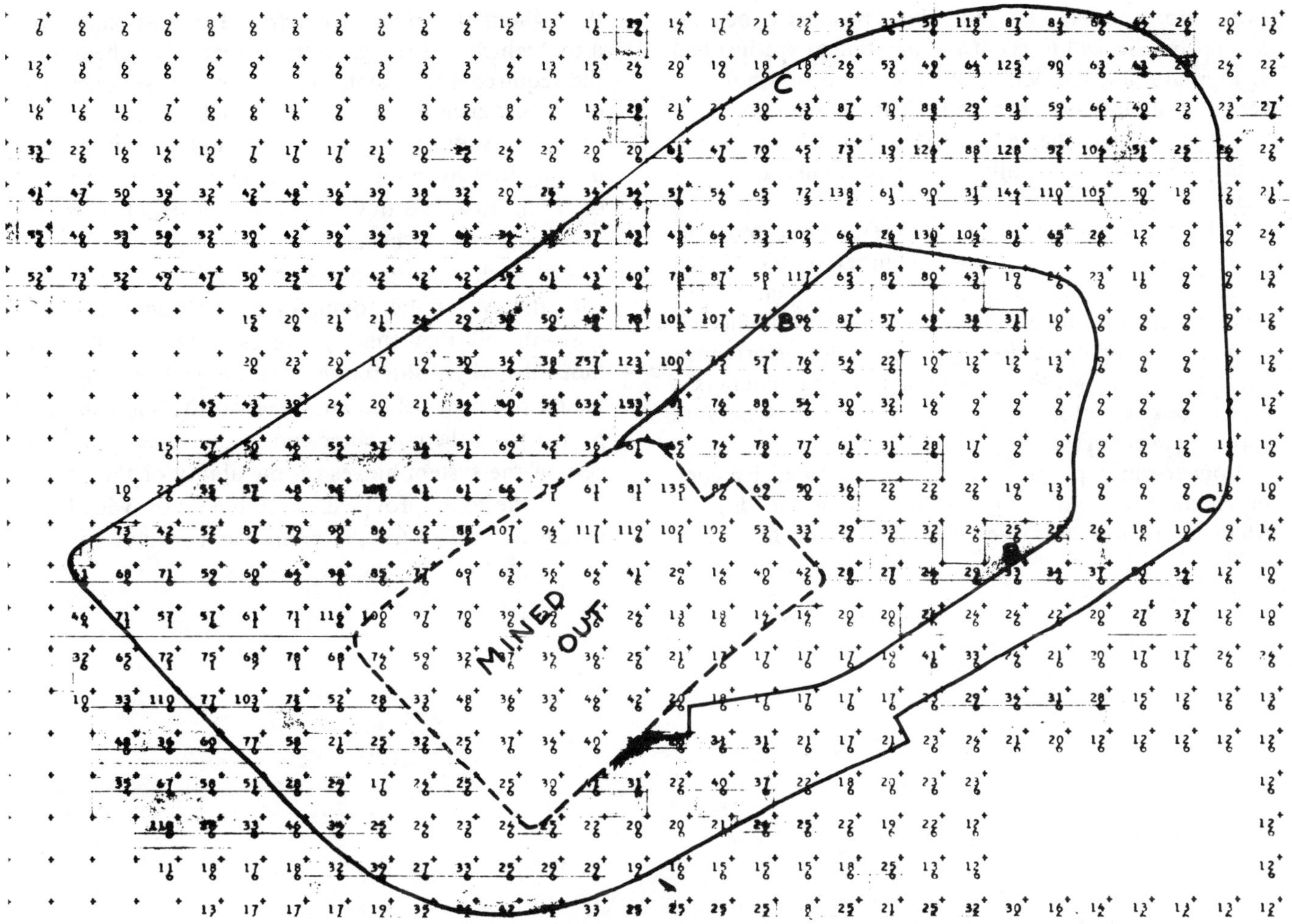

Fig. 4. Reserve bench map.

sary ore, discard, and the most even grade throughout the six-month forecast period. The mining layout by month will be digitized and the reserve figures tabulated by computer.

When the best forecast is chosen, the tonnages and grades should be compared with the long-range forecast to check that there are no major deviations between forecasts during the same time period. If large devia-

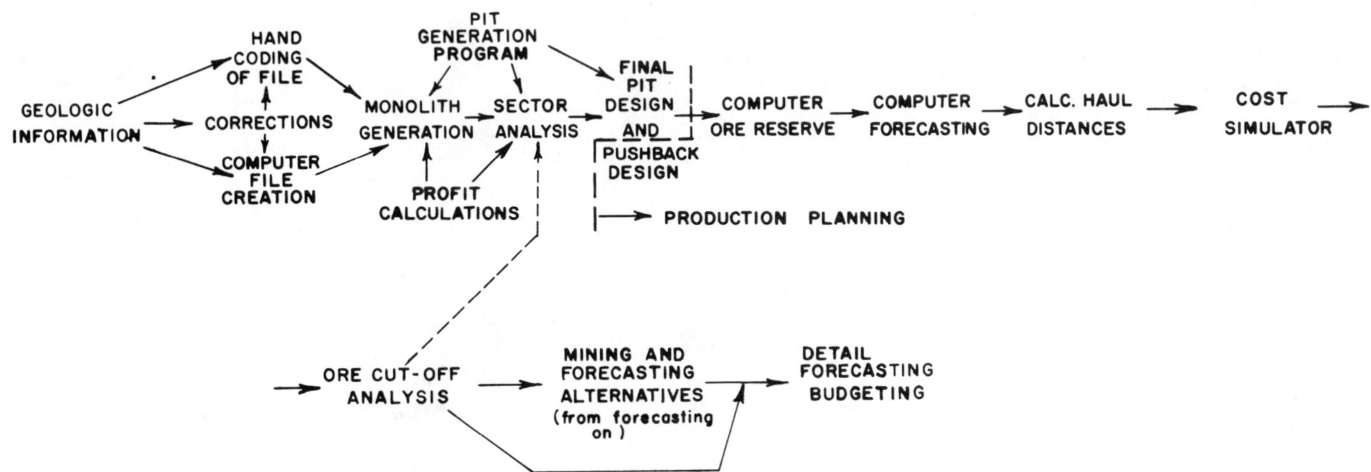

Fig. 5. Mine planning process flowsheet.

tions are present, the individual pushback benches should be checked to see if the differences are justified. If they are not, the performance data on blastholes vs. reserves should be examined. There might be some tonnage or grade changes in certain areas, which would justify development drilling, reserve changes, and/or plan changes.

It is anticipated that the use of the digitizer and computer will greatly increase the number of short-range forecasting alternatives that can be run each month. The current blasthole information on the minifile will also give a good up-to-date figure on mine performance, thus giving the geologists and long-range planners a quick look at information requirements and possible future plan changes.

Longer-range production planning is aided by computer methods and is generally outlined in Fig. 5. Many computer aids used are similar to those used throughout the mining industry. The cost simulator is a tool which was designed to simulate the scheduling of the required equipment, manpower, and supplies based on the production forecast for a truck and shovel operation. It is presently used as a long-range decision-making tool by comparing alternatives, but the process of trying to refine this tool to aid in detail forecasting and budgeting has begun.

Although many planning problems have been solved with the aid of the computer applications, it must be remembered that this system is only a tool. Much thought and consideration must be given to each individual problem before anything worthwhile can be expected from the computer end of the system. The success of the system hinges on the ability of the engineer to use the tools provided, in combination with his experience and feel for the mining and planning process.

16 Aspects of Production Planning: Operating Layout and Phase Plans

Thomas R. Couzens
Pincock, Allen & Holt, Inc.

Thomas R. Couzens has over 20 years of experience in open pit operations in the southwestern United States. He was senior pit engineer at the Questa, NM, mine of the Molybdenum Corp. of America. He has direct working experience with the varied components of successful pit operations, with premining activities including planning of exploratory drilling, ore reserve estimates, mine access, and waste disposal. He also has considerable experience in applying computer techniques to the needs of the mining engineer. He is presently the manager of mine planning for Pincock, Allen & Holt, Inc., Tucson, AZ.

He received his B.S. in mining engineering from the University of California and his A.B. from Stanford University. He is a member of AIME and of Tau Beta Pi.

Introduction

This chapter covers certain aspects of open pit production planning based on experience, pointing out approaches to planning that may help produce a realistic picture for evaluation and operation. The subjects discussed are roads, preproduction stripping, dump planning, dewatering, the relationship of production planning to equipment selection, and computer methods for the short-range or operational pit planning process. Actual determination of ore limits, pit design, and production schedules are covered in other chapters.

This discussion is somewhat general. Specific problems that occur in mine planning are extremely varied, and it is not possible to cover many of them in a chapter; therefore, only some of the principles that are important in the approach to open pit production planning problems will be discussed.

Certain characteristics of open pit planning should be kept in mind even though they may seem elementary. First, we must keep our objectives clearly defined while realizing that we are dealing with estimates of grade, projections of geology, and guesses about economics. We must be open to change. Second, we must communicate. If planning is not clear to those who must make decisions and to those who must execute the plans, then the planning will be either misunderstood or ignored. Third, we must remember that we are dealing with volumes of earth that must be moved in sequence; geometry is as important to a planner as is arithmetic. Fourth, we must remember that we are dealing with time, that volumes must be moved in time to realize our production goals, and that the productive use of time will determine efficiency and cost effectiveness. Finally, we must seek acceptance of our plans such that they become the company's goals and not just the planner's ideas.

Long-Range and Short-Range Production Planning

Open pit production planning is of two kinds. First, there is the short-range production planning that is a necessary function of an operating mine. It can be called operational planning. The second kind of production planning is connected with long-range feasibility type planning. This long-range production planning supplements the pit design and reserve estimation work that is usually done for feasibility or budget studies and is an important element of the decision-making process.

At an operating property, it is common to have certain regularly scheduled planning procedures and reports. These are often related to an annual ore reserves estimation and commonly include a number of years of yearly planning for long-range guidance and perhaps quarterly or monthly plans for short-range planning. The operating staff may develop weekly or daily plans within this framework. The whole process is tied together by production reporting and regular comparison in order to guide future estimating. The mine operators and their associated planners have the task of making the operation work. The operations planner may feel that he is caught in the middle between the mine superintendent, who is saddled with many problems, such as cost control, labor relations, and equipment availability, and the long-range planners, who have indicated that things should go smoothly if the long-range plan is followed. It is often not so simple, however, and the problems discussed in this chapter may indicate how difficult and demanding operations planning can be.

Much of the emphasis in published works on planning has been from the long-range planning and feasibility study approach. However, it is important for people in long-range planning to think like operational planners at certain stages of their work. Obviously, there are many details that a long-range planner cannot go into as deeply as the operational planner, but anyone in long-range planning should frequently ask himself, "How would I plan this if I had to be the mine superintendent and actually make it work?"

At operating mines, there is usually reasonably good interaction between long- and short-range planning, but in new property studies or feasibility studies, production type planning may be neglected. By the time the production planning stage of the feasibility study is reached, many calculations have probably been made on the backs of envelopes and on scratch paper and perhaps in a preliminary feasibility report. In addition, some of the executives have probably roughed out their ideas of what is required to make a worthwhile investment. Considerable time and money may have been spent assembling the basic deposit model data, perhaps involving several months of intensive computer work.

At this point, there is a temptation to take some shortcuts through the production planning stage. Someone has made some operating cost calculations on the basis of average grade, average waste ratio, and total tonnage, and these have entered into the economics of defining the ultimate pit. At this stage, there may be several ideas about operating rates, operating costs, and equipment. Also, the feasibility study may be exceeding the original schedule both in time and cost, and there may be pressure on the planners to supply numbers. Often the mill has been partially designed on the basis of certain throughputs, and everyone is somewhat surprised at the way capital costs are mounting.

It is important that production-type planning not be slighted at this stage because it will influence a number of decisions that can be crucial to the success of the evaluation of the operation if it goes into production. Most of us have seen examples where hasty decisions at this point have been extremely costly in the long run or have led to operating conditions that later proved unworkable. Though the production planning in a feasibility study is not truly a detailed operational plan, it should be made as close to an operating plan as it can be in order to estimate realistically how the operation will go. The planner should have the attitude that he will not be confined to certain operating slopes, waste ratios, or preproduction stripping numbers until he is satisfied that he has a truly workable mine plan. Such a plan must allow for access, haulage, enough operating room to work mining equipment efficiently, and enough ore exposure to assure a proper mill feed even in the face of some uncertainty.

Use of Phase Plans

This subject may be a partial repetition of material covered in other chapters but it is reviewed here because of its relevance to equipment selection, dump planning, and other aspects discussed in this chapter. In order to break up the overall pit reserve into more manageable planning units, planners usually make some kind of phase designs. Ideally these are made for time periods, probably years, for the first few years, and then perhaps for multiple-year periods on to the exhaustion of the reserve. In the beginning, however, it may be useful to take a look at the pattern of mining in a coarser division than years so phases or stages are developed. Phase plans are a preliminary attempt to relate the geometry of mining the ore to the geometry of the distribution of the ore in the deposit. They are the framework within which more detailed time period plans can be made.

Ordinarily, there is some amount of stripping, waste, or low-grade material above the ore which must be removed before the ore is exposed to mining. Unfortunately, it is not always a simple matter to schedule this waste removal. Consequently, the planner needs to develop a picture of the pit at the end of each phase in order to start to visualize the waste/ore relationships. We usually start with an approximate tonnage of ore to be developed in each phase and then, using whatever slope criteria we have, we determine the waste that must be removed to release this ore.

It is important to show at least one haul road in each phase to be sure that enough volume has been allowed to provide a way out. The phases need to be wide enough for the equipment to work its way down in an efficient operating manner. The practical width is not a fixed dimension but must be related to the size of the equipment and to the amount of equipment used in a stage or pushback or slice. For example, if there will be two shovels working on the stripping in the same slice, there must be enough width to allow room for drilling to the bench below as well as for haulage, power cables, and whatever else is required while a bench is being mined.

Once the phases are designed and measured, the planner can determine the approximate total material movement required to provide a continuing ore supply with a reasonable ore exposure. When the level of annual tonnage output is decided, he can then go back and draw up the yearly ore time period plans, measuring them, and tabulating results. It may be necessary to go through this process two or three times before an adequate plan is developed. Even then, after each pass, a planner may see things he would probably change if he were going to do it again, unless the deposit and the mining are so regular that he is sure that a given amount of stripping will automatically release a given amount of ore. I don't know any good substitute for this process and it may take considerable time, especially when the plans are balanced to equipment capability.

Development of phases also makes it possible to determine the amount of preproduction stripping required and to get a more exact determination of the actual best fleet of equipment. We commonly develop a graph that shows time relationships through phases with total tonnage movement and waste/ore ratio. On such a graph, the planner can see what has to be done to smooth out or otherwise adjust the production. It is not usual to operate with one waste ratio for the entire mine life, but within certain periods, a somewhat even production rate will be maintained. Adjusting production in this way avoids the need for sudden buildups in equipment or periods of surplus equipment; it also lessens problems of labor supply.

When a mine is working several large shovels, certain combinations appear to be best from the standpoint of utilizing both equipment and manpower. This may affect the waste ratio or indicate when an operation should go to a certain waste ratio. An example of this is included in the discussion of equipment selection.

Two examples of production graphs are shown. Fig. 1 illustrates a plan where the milling rate was constant and the plan was worked out to achieve a good blend of ore, good ore exposure, and good operating conditions. The waste was determined as required to achieve these requirements. Consequently, the yearly waste/ore ratios fluctuated. An average waste/ore ratio

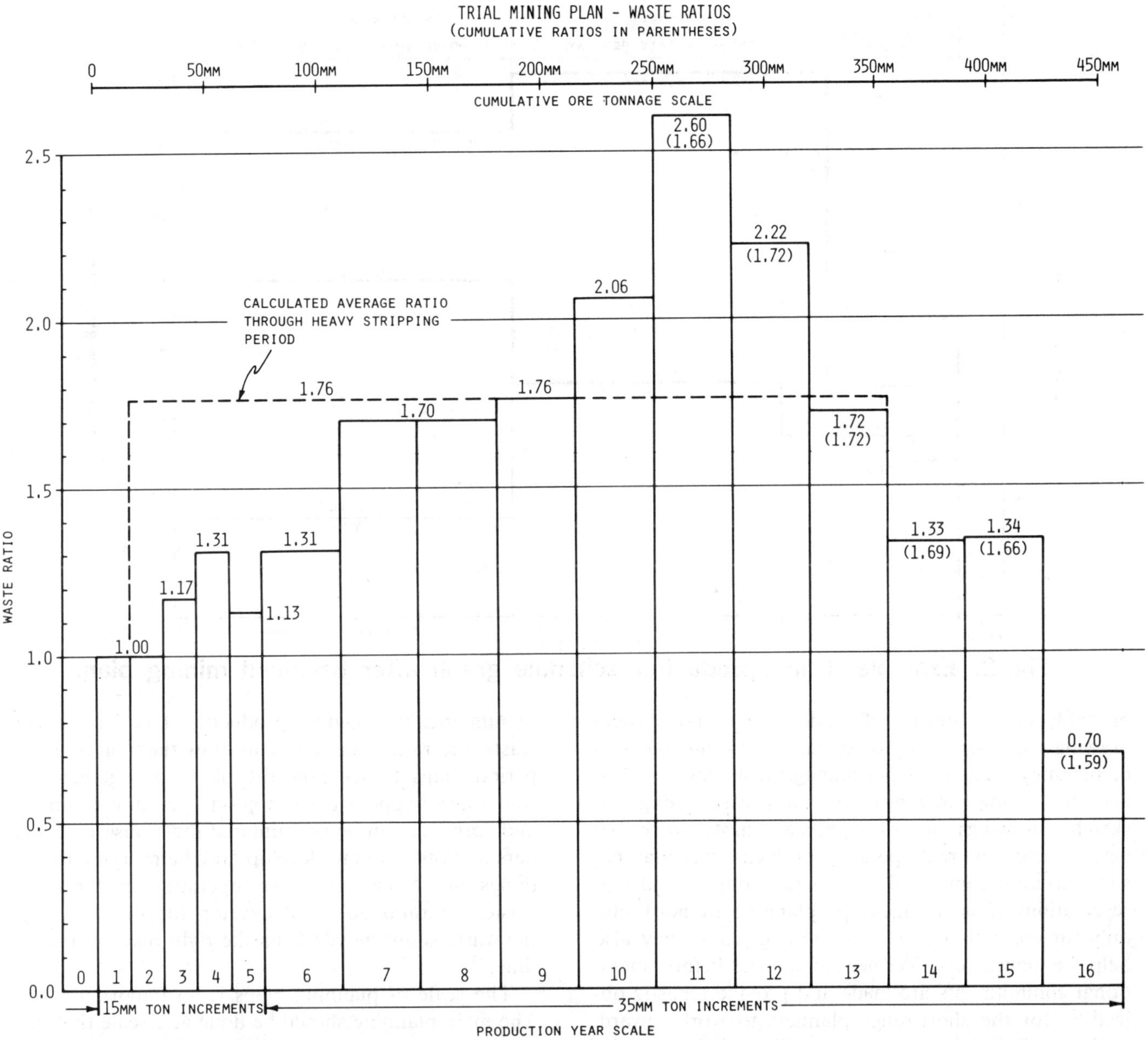
TRIAL MINING PLAN - WASTE RATIOS
(CUMULATIVE RATIOS IN PARENTHESES)
0
50MM
100MM
150MM
200MM
250MM
300MM
350MM
400MM
450MM
CUMULATIVE ORE TONNAGE SCALE
WASTE RATIO
2.5
2.0
1.5
1.0
0.5
0.0
CALCULATED AVERAGE RATIO
THROUGH HEAVY STRIPPING
PERIOD
1.00
1.17
1.31
1.13
1.31
1.76
1.70
1.76
2.06
2.60
(1.66)
2.22
(1.72)
1.72
(1.72)
1.33
(1.69)
1.34
(1.66)
0.70
(1.59)
0
1
2
3
4
5
6
7
8
9
10
11
12
13
14
15
16
15MM TON INCREMENTS
35MM TON INCREMENTS
PRODUCTION YEAR SCALE

Fig. 1. Waste/ore ratios vs. time in a trial mining plan before smoothing.

was then shown with a dashed line through the heavy stripping period (end of year 13). Replanning with this ratio in mind would enable a better equipment and manpower estimate to be made.

Fig. 2 illustrates a different type of production schedule graph. It illustrates the various relationships in a finished plan between total tonnage movement, ore requirements, waste ratio, and shovel shifts. This kind of presentation to management makes it possible to communicate the mining schedule better than just bare statements of tonnage and waste/ore ratios.

It is important to make the transition from phase plans to time period plans as soon as the phase designs are sufficient to set the overall pattern. The yearly plans enable definite production goals to be set in space as well as in quantities of material to be moved, and they allow better economic evaluations than phase averages can provide. Also, the yearly plans give a better definition of the relationship of the phases to each other as they overlap in the complete mine operation, thus showing actual operating slopes and haulage routes.

Graphic Methods

One of the most important aspects of any pit planning is graphic representation. More misunderstandings

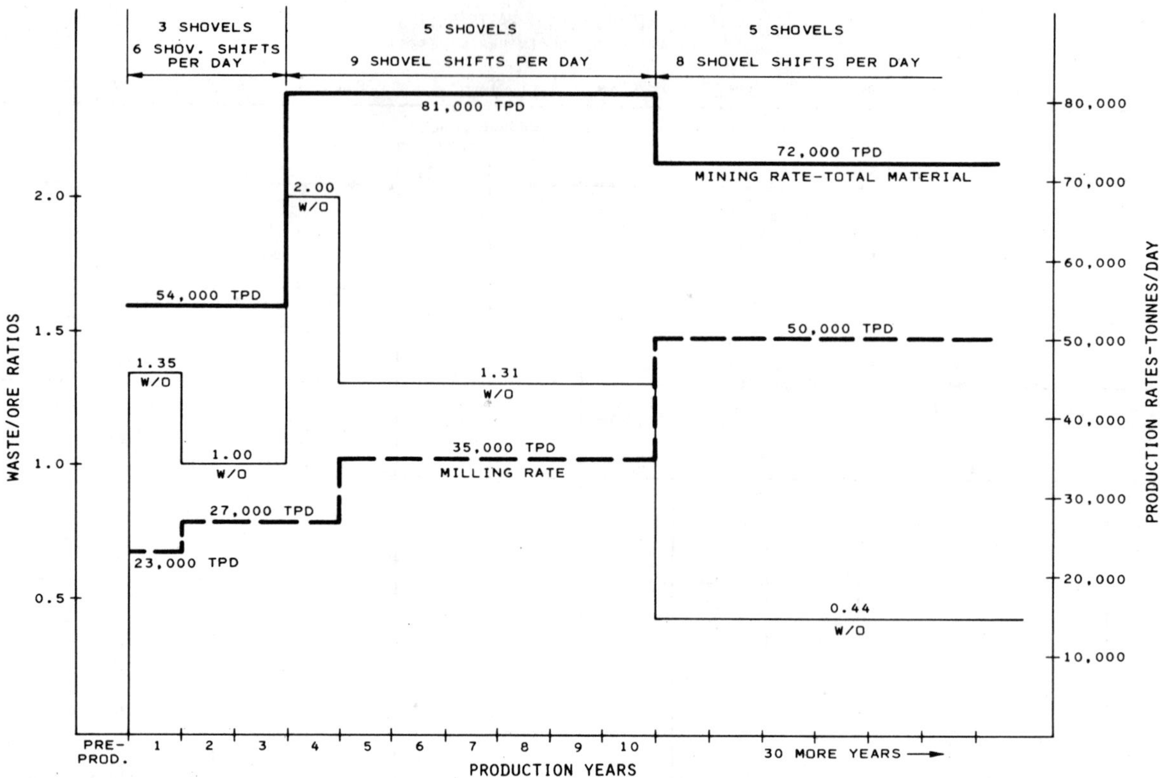

Fig. 2. Example of final production schedule graph after balanced mining plan.

probably occur because of inadequate or nonexistent drawings than from any other factor. Communication of planning goals to upper management requires clear and presentable drawings. In many cases, rough or sketchy drawings not only indicate hasty work but actually can hide real operating problems that may not show up until committed work is far along. Good representations of each stage of pit planning are not useful only for cosmetic effect or for selling plans; they also help the planner discover needed changes before operational commitments are made and provide realistic objectives for the short-range planners to work toward.

Except for illustrative sections or sections used for special derivations, such as pit wall placement, most of the pit planner's work is done on plan maps. A set of bench maps showing the topography or surface contour, location of ore, geologic boundaries, and design limits for each bench is essential. Pit composite maps showing the shape of the mine at the end of each planning period should be kept up as the plan is developed. These enable the planner to avoid conflicts between features of the plan, provide a picture of the access at each stage of development, and illustrate the actual working slopes, operating room, and spatial relationship between ore and waste.

These maps are often neglected. We all tend to take shortcuts under pressure, and one of the easiest shortcuts to take is to develop a general, phased mining plan to numerically schedule production based on average waste/ore ratios, and never to draw the final mine plan period maps to see how the plan fits together. This sometimes happens even in good operating companies, and surprises can be encountered that cause both planners and operators to develop gray hairs trying to make things fit. Sometimes even operating at the proper waste/ore ratio does not develop the ore properly if the waste is not moved from the right place at the right time.

The scale of planning maps is an important choice. The main planning should be done at a scale that keeps the whole pit on one sheet, if possible, and yet permits sufficient detail to be realistic. For the average medium to large size metal or industrial minerals pit, common planning scales are 1 in. = 100 ft (1 mm = 1.2 m) or 1 in. = 200 ft (1 mm = 2.4 m) or 1:1000, 1:1250, and 1:2000 in metric ratios. A metric scale of 1:1250 is very close to the English 1 in. = 100 ft (1:1200). Adequate road design can be done at these scales for most planning purposes, and if specific detailed drawings have to be made for construction of complicated intersections or drainage structures, these exceptions can be handled at whatever scale is required.

Geologic detail mapping is commonly carried at a larger scale such as 1:500, but this is usually too large to keep the entire mine on one sheet. For pit planning, it is usually better to reduce the necessary geologic outlines to fit a convenient planning scale than to obscure the planning by having it on multiple sheets of paper that are difficult to relate to each other.

Finally, there is the question of how to depict the bench locations on the mine plan. Personally, I have found the use of bench centerlines or midbench contours to be the simplest and most straightforward way to represent the shape of the mine. There is seldom the necessity to depict toes and crests of each bench except for illustration to people who are unused to looking at planning maps or for certain other specific purposes. Usually, it is easy for people working with the maps to get accustomed to visualizing the toe and crest locations, and the mine map draftsmen can easily develop toe and crest lines for making field notes for layout work. The drawing in Fig. 3 illustrates a mining plan composite map. It shows the bench centerline contours at the end of a year, indicating the haul roads, stripping area, and part of the waste dumps. Outside of the pit area, the contours are labeled with their true elevations; inside the pit, the elevations refer to the bench toe elevations, and the bench centerlines are one-half the bench height above the labeled elevation. In other words, it is the flat areas between centerlines that are labeled. On ramps, the bench centerlines cross the ramp halfway between benches, and the labels are at the actual bench elevation on the road. To those who are accustomed to using toes or crests, this may at first seem confusing, but in the long run, the ease of relating these composite maps to the bench maps for measuring and the simplicity of the picture offset any initial problem with labeling.

Roads

The planning of roads is one of the most important aspects of open pit planning. Because of their effect on everything to do with the pit, road considerations need to be worked into the planning at as early a stage as possible. Roads are difficult to include in some of the computer pit generations. For this reason, they are sometimes left out of the early economic evaluations. Pits can be designed without consideration of roads but it has been my experience that even after an economically optimum pit is designed, if roads are absent, the changes required to bring the pit into a realistic mining configuration are often drastic in terms of tonnage as well as in the shape of the pit.

In rail pits, which were common in the previous generation, a great deal of attention was given to the layout of rail haulage. The fact that railroad operations are not as flexible as truck operations forced this kind of planning to be dominant. Now, with the advent of truck haulage, some things can be done more easily in less room with more force fitting of haulage than was possible with railroads. As a result, road design is sometimes neglected in long-range planning. Someone has to come to grips with roads at some point, and the long-range planner is really letting down on the job if he leaves it to the mine superintendent and his staff. He is reducing his own credibility and is forcing the decision-making process into the short-range phase where the decisions sometimes do not adequately reflect the long-range needs. If haulage and access are provided for in the long-range planning, a lot of the other problems take care of themselves; otherwise, haulage and access changes may force operations to depart from the long-range plans to the point where nobody considers such planning worthwhile. The ultimate pit design may change several times as new knowledge, additional drilling, and changes in economics force constant redesign. Nevertheless, the final road should be shown because it does give an estimate of the tonnage necessary and prevents an uncomfortable awakening to the fact that the actual mining is going to be more than initial forecasts called for.

The first thing in the layout of a new pit is to decide where the road exit or exits from the pit will be. This is dependent on the location of crusher or dump points and is greatly influenced by topography. Considerable thought should be given to selecting these exits. Depending on need, there may be one or more such exit points.

In the intermediate stages, the roads should also be thought out carefully. It is nice to develop the final road layouts as early as possible, but many roads are temporary, lasting for a period of a few months to a few years, and then they are replaced by other roads that serve new pushbacks or stages in the pit. Often it seems attractive to try to design some kind of an external haul road that will not be disturbed by the mining, but in many cases this is not practical because the connections to such a road often cause more trouble than just moving a road in the mining phases.

Mine superintendents understandably like to have more than one way in and out of the pit in case something happens to a haul road. A slide or some other disruption or just the problems of operating a road when there is mining above may cause traffic delays. Sometimes a second road is not feasible if it requires a lot of stripping, but the pit designer and long-range planner should always keep in mind a way to get the ore out if there is an interruption in the main haulage system.

In laying out roads in long-range plans, there is usually a question whether to spiral the road around the pit or to have a number of switchbacks on one side

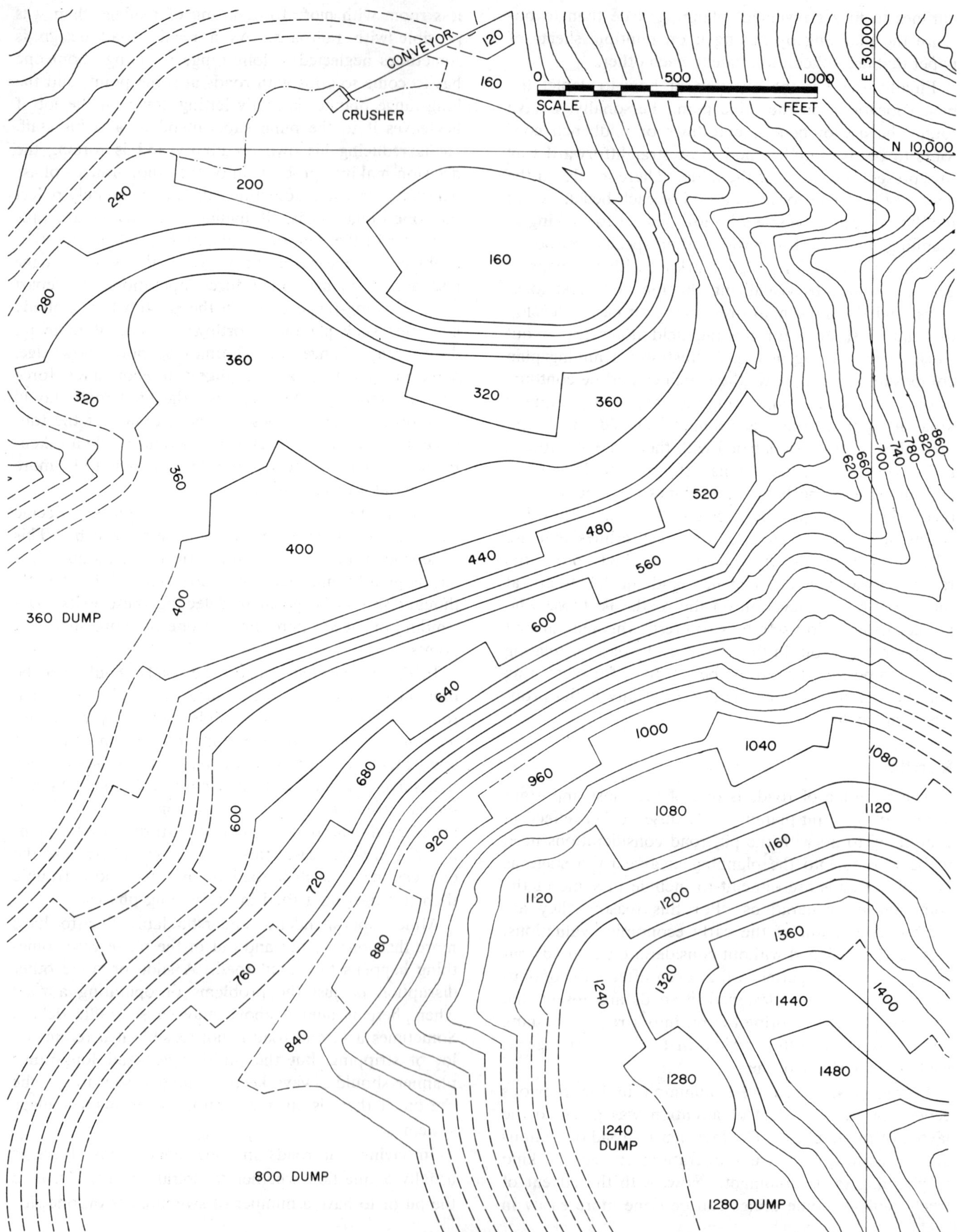

Fig. 3. Example of a mining plan composite map for a specific time; pit lines are midbench contours.

or a combination of both. Sometimes the geometry of the deposit leaves little choice, for example, when there is a gently sloping ore contact in some area that provides room to work in switchbacks at little stripping cost. The planner must take advantage of these things and always design the pit to fit the shape of the deposit. Generally speaking, it is desirable to avoid switchbacks because they tend to slow down traffic, cause greater tire wear and various maintenance problems, and are probably more of a safety hazard than spiral roads. However, if there is a low side to the pit, it may be better to have some switchbacks on that side than to accept a lot of stripping all the way to the top of the high side to provide room for a road or series of roads on that side. If switchbacks are necessary, it is important to leave enough length at the switchbacks for a flat area at the turns so that trucks don't have to operate on extremely steep grades at the inside of curves and so on. The planner should also think about the direction of traffic and some of the problems the drivers may have with visibility on switchbacks.

One of the important considerations in laying out the haul roads is width. There is a tendency either to design roads somewhat narrow to save stripping or to go the other way and design a great highway that may be too costly. Naturally, wide roads are desirable but we have to balance these against other factors. A common rule of thumb is that the design width should be no less than four times the width of the haul trucks. This allows for two-way haul truck traffic and room for an outside berm and an inside ditch if necessary. Large pits commonly have 25- to 30-m roads and there may be stretches of haul road, such as where several streams of traffic come together near a crusher, where a greater width will be desirable. Fig. 4 shows a minimum haul-road cross section for an actual truck.

Table 1 shows some typical haul-road widths as related to trucks. The design width will probably be adjusted to a rounded standard figure for basic design, subject of course to variations as at turns, switchbacks, or high traffic density areas.

Table 1. Minimum Road Design Widths for Various Size Rear Dump Trucks

Truck size *	Approx width, m	4× width, m	Design width m	Design width ft
35 ton †	3.7	14.8	15	50
85 ton	5.4	21.6	23	75
120 ton	5.9	23.6	25	85
170 ton	6.4	25.6	30	100

* Nominal size in short tons.

† Metric equivalents: 1 st×0.907 184 7=t; 1 ft×0.304 8=m.

Some mines have two lanes in one direction over part of the haul road. For example, they allow passing for uphill loaded traffic and keep the downhill empty traffic in a single lane. These things have to be worked out in some detail by the designer after he knows something about the shape of the pit, the equipment, and the traffic density.

The second basic consideration is road grade. In a pit where there is a considerable vertical component to the haulage requirement, the grade will have to be fairly steep to reduce the length of the road and the extra material necessary to provide the road length. However, the practical maximum is usually considered to be 10%. A number of pits operate quite well at 10% both favorable and unfavorable to the loads. If it does not cause too much extra stripping or unduly complicate the road layout, 8% is probably preferable because it gives a bit more latitude in building the road and fitting in bench entries without having some locally

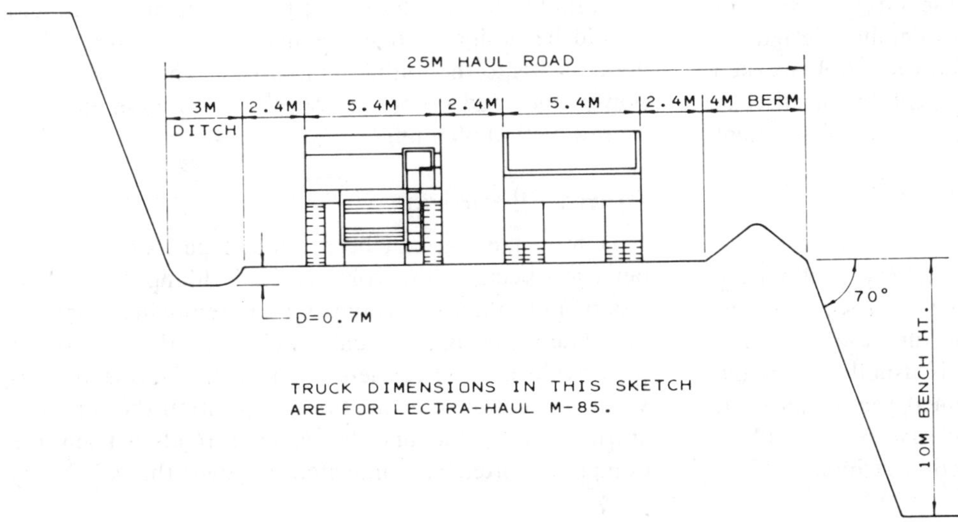

Fig. 4. Typical design haul-road width for two-way traffic using 77.11-t (85-st) trucks.

oversteep places. Unless there is a long distance to travel without requiring much lift, there is normally nothing to be gained by flattening the road below 8%; the extra length on the grade and the complications of fitting the road into the available room or doing extra stripping would probably offset any increase in uphill haul speed. The geometry of the pit is the main consideration, and the roads must be designed to fit the particular situation. Often there will be a number of different grade segments in haul roads for this reason.

Safety features should always be kept in mind, particularly in the case of downhill haulage; some means must be provided to reduce speed or handle the truck that loses its brakes. Probably the most successful thing in this area is the center muck berms called whopper stoppers, straddle berms, and various other slang expressions. These should be high enough to impinge on the undercarriage of the truck if the driver needs to use the berm to reduce his speed. It is not necessary to build a big barrier that forms another crash hazard. The important thing here, as much as designing something to stop the trucks, is training the drivers to get on the berm or into the bank immediately if they start to lose control of the truck, before they develop such speed that whatever they do is hazardous. If the driver tries to turn into a berm or a bank at too high a speed, the truck will probably turn over.

Other safety features that may be considered are runaway ramps and turnouts. In time these may be required, but again, intelligent operation, good training, and thorough equipment maintenance are the main contributors to safety.

In the actual construction of main haul roads, superelevation on curves, calculated vertical curves, special surfacing, and such highway type design features may be indicated, particularly with large equipment and high tonnage operations. In ordinary mine planning, however, the planners are more concerned with estimating bulk volume removal in the pit and will not spend too much time in this kind of detailed design. If the basic room is left and the routes are well worked out, the mining department and field engineers can usually build excellent haul roads to promote efficient haulage.

Preproduction Stripping

One of the biggest challenges in any preliminary feasibility planning is determining the proper amount of preproduction stripping. By the time the feasibility mining schedule is being made, it is usually surprising how much capital has been estimated for the project, and there is often a temptation to save a few million dollars by reducing the preproduction stripping. This is a natural place to think of cutting back. Some of the suggestions that are brought to the planner are difficult to cope with because they deal with uncertainties, such as slopes. Someone may suggest steepening the slopes to save stripping expense, assuming that it can be made up later when the operation starts to show a better cash flow. The planner may not have enough information to know how steep the walls will stand, and yet he is reluctant to take too big a risk. We sometimes see mines where decisions have been made to open the mine with an insufficient amount of preproduction stripping, causing severe problems later.

The important thing to remember in preproduction stripping (and this of course is true in any subsequent phase as well) is that enough work has to be done to expose a sustaining ore supply and to keep the mine in a condition that allows it to be operated efficiently at all times. Setting the amount of preproduction stripping involves determining the volume that must be moved to open up the ore and to provide enough developed ore to last until more is uncovered. It may be something of a juggling act to fit this in with production rates and the equipment buildup schedule and to determine the amount of time required. Scheduling the preproduction stripping is very important. Sometimes the preproduction work is contracted but in a large tonnage new mine, which will use shovels and trucks for both stripping and mining, it is probably better to use the preproduction phase to build an organization and to gain operating experience. These are all factors that the planner must evaluate.

One aspect of preproduction work that should not be overlooked is the development of initial roads. This may involve considerable excavation and construction work before the mine is even in condition to get the stripping started efficiently. This usually requires considerable dozer work, perhaps some scraper excavation, and careful planning of the initial access road. Ideally, the initial road should become part of the haul road and should be designed to serve this purpose by providing good access to the initial dump areas. The provision of power lines and other services may also enter into the preproduction planning.

Dump Planning

Dump planning can be a difficult part of the planner's job because the volumes he is filling tend to be less definite than the volumes he is removing from the pit. The pit design is usually well worked out in order to provide a good sequence, and considerable time is spent on economic evaluations to position the pit walls at the right place and to develop roads. Then the dumps are often just indicated as "over there." Many factors enter into dump planning and the planners have to balance a variety of factors such as the length of hauls, the required lifts, and the relationship of dumps to property constraints, other installations, drainage, and, in recent years, reclamation or environmental requirements.

Pit planning should include an estimate of where the dumps are going to be at each stage in connection with a haulage study and the necessary haul and lift examination. The planner can look at the trade-off between an additional lift and a longer haul. This is valuable knowledge, which most often requires an actual map layout of the dumps.

Several years ago, I worked as a mine planner at a mine where we had a canyon to dump into along one side of the mine, and we were pushing a ridge back a few hundred feet. As we came off the ridge, we could dump around the end of the ridge straight into the canyon. The canyon was almost parallel to the ridge. By doubling back and hauling up, we could get more into the canyon than by dumping straight out, but at the cost of a longer haul and lift. Obviously, the temptation to get production efficiency led us to want to dump straight off and let the dump run into the canyon. The only trouble was that the dump toe would soon get out to where we would have to stop advancing in that direction because of a major haulageway below. If we took the easy dump haul while stripping the upper part of the ridge, when we got down lower on the ridge, we would find that the easy haul was used up, and we would have to lift waste farther up the canyon. This would mean adding considerable cycle time because we would be mining from a lower elevation and dumping at a much higher elevation. So when we looked ahead at the truck requirements, they tended to balloon in the later phases of stripping.

To convince ourselves and our superiors, we ended up building a physical model of the topography out of various materials such as egg cartons and modeling clay. We measured out volumes of sand representing the waste volumes from each bench, and we actually built the dumps in several different ways, photographing the layout and keeping track of the haulage and lift involved. This was used to predict the number of trucks needed to keep the pushback going. Because of a tight time schedule for ore development and room for only two shovels (one part of the time), we had to keep the stripping moving. We discovered that we would have really hurt ourselves in the long run if we had taken the easy dump haul in the beginning. By using this model, we were able to demonstrate to everyone's satisfaction that we should discipline ourselves to haul to the upper end of the canyon in the early phases, leaving room down below for later when it would be needed for the lower waste.

Unfortunately, in feasibility studies there may not seem to be time to do such studies. The important point is to face the fact that dump planning can have an important bearing on pit planning, particularly in the haulage layout, scheduling, and equipment estimating areas. Haulage studies are discussed in another section of this book so we will not go into detail here, but we need to recognize their importance. For example, the same waste/ore ratio in different time periods in a pit may not mean much if the haul distances differ widely because cost and truck requirements may be quite different at the same ratio. Ideally, we should have enough planning to be able to optimize the waste/ore ratio on a relative cost basis within the ore development constraints.

Dewatering

There are two reasons for dewatering in an open pit mine. First, it is very difficult to operate with much water in the pit. Second, the presence of water in the pit walls usually reduces slope stability. Consequently, the planner is faced with the problem of considering water at nearly every step of his work. Water affects blasting, equipment operation and maintenance, road construction, and even ore quality. There are experts in the field of rock mechanics who will discuss the effects of water on slope stability and pit design in this book, so I will not cover this to any great extent except to acknowledge the importance of both water and structure in pit design.

In new pit designs, it is important to estimate the amount of expected water. This means looking at rainfall records, drill logs, and hydrologic reports, if they exist, as well as talking to geologists on the job. It is often tempting to ignore the geology but since we are working in a geologic medium, attention should be paid to faults, aquifers, or underground workings which could indicate potential water problems. Even arid areas may require considerable water management planning. In wet climates, a constant coping with water becomes part of normal operating methods. However in dry areas, a lot of water may fall in two or three storms and though the total precipitation may be low, the effect of precipitation is concentrated into a short period and causes much trouble.

In many places, feasibility studies must be related to environmental planning, and the planner may have to consider what happens to water that is removed from the pit. Can it be discharged into natural drainage or must it be impounded, treated, or recycled? Of course, recycling may have economic benefits to offset the cost consequences.

In planning a pit, if it appears that a significant amount of ground water or surface water will be encountered, plans must be made to remove it. This may mean peripheral ditches at the pit bottom, as in the case of some of the uranium operations that collect drainage from the pit walls, and then channel it to a sump from which it is pumped out of the pit. Horizontal drain holes may be needed, or in some mines, it may be necessary to drill and blast a collecting sump and install a well-type pump that will draw the water down enough to allow drilling and mining in the bottom. Sometimes underground workings can be used, or they may even be driven in order to drain part of the pit walls. Most mines at some time use one or more pumps to move water out of a sump to the mill or to a drainage channel for discharge.

Perched water bodies can exist in the walls of the pit. These may have an effect on blasting, or they may create ground-water problems even in an area without slides. Equipment delayed on wet or spongy roads can create problems and add to operating costs. Sometimes these perched water bodies may be carried down the side of the pit and encountered repeatedly.

In long-range planning, if it is known that there are wet areas, cost projections may be influenced by indicating different explosives in certain areas or by dictating extra room for sumps, ditches, or pipelines. In a preliminary study, it is not usually necessary to design in detail the drainage features in the pit, but some allowance should be made for their handling, and some money should be budgeted to cope with water.

Streams or even dry arroyos coming into the pit area may have to be diverted to avoid bringing surface water into the pit. Another aspect of drainage to consider is the drainage of dumps in order to prevent erosion or dump instability. In many cases, this involves diversion of natural drainage or runoff from collecting areas created by the mine itself. The dumps may have to be sloped upward toward the crest to prevent rapid gullying and crest erosion of inactive dumps.

Most of this can be handled adequately in the planning stage by the mine planners. In cases where major stability questions, hydrology studies, or water quality questions are involved, the services of specialists may be required.

Relationship to Equipment Selection

The level of material movement in an open pit mine is generally set during the planning process by looking at the requirements for ore and the waste needed to make that ore available. Before a mine comes into operation, of course, a major purpose of the feasibility study is to determine equipment requirements and to guide equipment selection. Several factors enter into the selection of the kind of equipment to be used, including the rock characteristics, the shape and continuity of the mineralization, the selectivity needed in the mining of the ore, the size and geometry of the mine, and the distances required to move the material. The amount of equipment is dictated partly through the efficient size of units that can be applied; other factors are the need for geographical distribution in the pit to provide ore blending or coverage of various required working areas, the need for backup, and the mobility of the equipment.

Attention has to be paid to the work schedule and the labor force. Though equipment productivities are often given in such units as tons per hour, actual crew scheduling is done by shifts so it is usually best to figure equipment productivities in terms of output per shift. It is then easy to relate production requirements, equipment requirements, and labor requirements.

Usually for expensive pieces of equipment, like shovels and trucks, a three-shift operation is the most economical. Therefore, it is good to have a total shovel shifts requirement, for example, in multiples of three so equipment can be evenly distributed among the shifts. If you have four shovel shifts scheduled on one shift and three on each of the others, then you must own enough trucks to cover four shovels during the one shift. Otherwise, you may be limiting the production of the shovels on the four-shovel shift. This can usually be best worked out on a trial table of various combinations, perhaps for each of several sizes of equipment. The planner can then look at the utilization that he thinks is most realistic. An example of this is shown in Table 2 which gives several options for an actual case study. In this case the 6-shovel, 12-shovel shifts per day situation would probably give the best unit cost effectiveness for the equipment if the 152 407 tpd (168,000 stpd) rate fits the material movement needs. Five shovels at 12 shovel shifts per day would be difficult to continue over a long period of time since it would require working four of the five all the time.

An example of relationships of shovel utilization to required shovel shifts for a given size shovel and tonnage requirement is shown in Table 3. In this case, the shovel size was already determined and the question was whether two or three shovels should be used. The numbers indicated that two shovels would require unusually high utilization and that three would be more realistic during one period. Incidentally, it was unlikely in this case that two shovels could handle the geographical distribution required to keep separate areas

Table 2. Possible Shovel Combinations to Meet Tonnage Requirements

Number 15-Yd * shovels in fleet	Tons * per shovel shift	Shovel shifts per day	Utiliza-tion, %	Tpd	Million tpy at 350 days	Million tpy at 360 days
5	14,000	10	67	140,000	49.0	50.4
5	14,000	11	73	154,000	53.9	55.4
5	14,000	12	80	168,000	58.8	60.5
6	14,000	10	56	140,000	49.0	60.5
6	14,000	11	61	154,000	53.9	55.4
6	14,000	12	67	168,000	58.8	60.5

* Metric equivalents: 1 yd × 0.914 4 = m; 1 st × 0.907 184 7 = t.

moving, which was an overriding factor. This kind of study can be put together for different sizes of equipment for comparison.

Long-range production planning may be done in operating mines either for expansion or to modify the existing operation. This may involve selection of additional equipment, if targets are not being met and so on. In this case, there is considerable influence from the equipment already being used. The planner must consider the problems of mixed fleets both in sizes and manufacture and in the applicability of maintenance facilities and such considerations.

Computer Methods

Computer usage is perhaps one of the most difficult areas to comment on in short-range or production planning. There has been a tremendous amount of work in the last several years on application of computer methods to mine planning. Much significant work has been done in this area. Some companies have spent large amounts of money applying computer technology to planning problems, and most people are using computers to some extent to help handle large volumes of data and to speed up some of the classification and sorting problems. But there are still widespread weaknesses in the applications of computers to the production planning level. Production planning contains elements of judgment and subjectivity, and it is subject to the interaction of many variables. It is still difficult to work this into a fully computerized system.

One of the main shortcomings of computer work in short-range planning is the lack of flexibility of most computer systems. This includes lack of flexibility in programs and the difficulties in tailoring or altering programs to specific needs which may frequently change in some detail. Also, planners often have difficulty in working with the computer processing system. This includes availability of the computer (not the overall availability, but availability to the engineer at the time he needs his answers), and also communications between the planner and the computer. If the communication between the planner and the computer has to go through various echelons of specialists, systems people, or operators, there is often a kind of fatal delay that drives the planner back to manual methods, usually with a feeling of frustration.

My own experience has been that the most helpful computer methods do simple things over and over. It

Table 3. Shovel Utilization at Various Schedules—Case Study

No. of shovels	7 days 3 shifts	6 days 3 shifts	5 days 3 shifts	7 days 2 shifts	6 days 2 shifts	5 days 2 shifts
Case A: First Period, Total Tonnage Requires 30 Shovel Shifts Per Week						
2	30/42=0.71	30/36=0.83	30/30=1.00	30/28=n.g.	30/24=n.g.	30/20=n.g.
3	30/63=0.48	30/54=0.56	30/45=0.67	30/42=0.71	30/36=0.83	30/30=1.00
Case B: Second Period, Total Tonnage Requires 36 Shovel Shifts Per Week						
2	36/42=0.86	36/36=1.00	36/30=n.g.	36/28=n.g.	36/24=n.g.	36/20=n.g.
3	36/63=0.57	36/54=0.67	36/45=0.80	36/42=0.86	36/36=1.00	36/30=n.g.

Notes: Fractions indicate that required shovel shifts per week/total shovel shifts for fleet=utilization (in decimal); n.g. means no good; required shovel shifts exceed available shovel shifts for this work schedule; most promising work schedules are underlined.

is important that the work is under direct control of the planning engineer. For this reason, I have had my best success in short-range planning with the programmable calculators or desk-top computers. It is possible to do a lot of work on such a system while keeping it totally under one's own control. This includes handling large data files on cassette tape and getting formatted, printed, or plotted output. The execution speed may be much less than a large memory computer, but the actual elapsed time for the job may be less because of continuous application and lack of delays.

Discussion Questions

A few questions are included to stimulate discussion. There are so many questions that can present themselves to the mine planner that it would not be possible to anticipate them all. Also, many questions have no simple or analytical solution, and the planner is forced to use trial solutions and compare alternatives.

The solutions to many of the problems are dictated by various physical, topographic, or legal constraints. A basic rule in mine planning is that the planner must visualize and simulate using his best estimates of what will happen and then design what appears to be necessary. It is hoped that a discussion of these questions will bring forth ideas as to how to approach problems and sharing of experience on how such problems have been handled by others.

Haulage is one of the most important aspects of mine planning. This has been discussed in handbooks and textbooks and is also the subject of another section of this book, so it is not dealt with in detail in these questions.

Roads

1) How much stripping tonnage increase can we expect when adding a haul-road segment to a pit? Isolate an area with 15-m benches where an average of 20 benches must be stripped back 30 m for an average length of 500 m and a rock tonnage factor of 2.6 t/m^3. That piece of excavation will amount to $(20\times15)\times500\times30\times2.6$ or 11 700 000 t. This, of course, is oversimplified but it gives a feeling for the effect of adding a segment of road to a pit area.

2) List some of the influences road width can have on pit production rates and operating costs.

3) How does the planner balance additional stripping cost against maintenance and operating cost in designing roads?

4) How should road ramps be built in a pit? Should drilling be done to ramp grade or should full bench drilling be done and ramps built of broken material?

Preproduction Stripping

1) How should the timing of preproduction stripping be worked into planning? List some of the factors that must be taken into account in a feasibility study in determining the time and cost of preproduction stripping.

2) What factors enter into the decision for or against contract stripping?

Dump Planning

1) How does the planner figure the trade-off between extension of a lower dump and starting a dump at a higher elevation? (This problem and solution were submitted by R. R. Leveille of Chino Mines Div., Kennecott Copper Corp.).

Assumptions: (1) truck operating cost per minute is the same on level hauls as on a grade, (2) common starting point, and (3) road to higher dump exists.

Definitions: L is incremental height between higher and lower dump [meter (feet)]; SLG is speed loaded on upgrade [km/h (mph)]; SLO is speed loaded on level road [km/h (mph)]; SEG is speed empty return on downgrade [km/h (mph)]; SEO is speed empty return on level road [km/h (mph)]; G is grade of road to higher dump; F is factor derived as $\frac{1609 \text{ m/km}}{60 \text{ min/hr}}=27$ $\left(\frac{5280 \text{ ft/min}}{60 \text{ min/hr}}=88\right)$; SDG is slope distance to higher dump [meter (feet)]; TTG is round trip travel time on grade (min); TTO is round trip travel time on level road (min); and HDO is economic distance from common starting point for advance of lower dump ($TTO=TTG$).

Equations:

$$SDG=\sqrt{L^2+\left(\frac{L}{G}\right)^2}$$

$$TTG=\frac{SDG\times\left(\frac{1}{SLG}+\frac{1}{SEG}\right)}{F}$$

$$HDO=\frac{TTG\times F}{\left(\frac{1}{SLO}+\frac{1}{SEO}\right)}$$

$$HDQ=\frac{\sqrt{L^2+\left(\frac{L}{G}\right)^2}\times\left(\frac{1}{SLG}+\frac{1}{SEG}\right)}{\left(\frac{1}{SLO}+\frac{1}{SEO}\right)}$$

Example: Given L is 80, SLG is 9, SLO is 28, SEG is 23, SEO is 28, and G is 7%,

$$HDO = \frac{\sqrt{80^2 + \left(\frac{80}{0.07}\right)^2 \times \left(\frac{1}{9} + \frac{1}{23}\right)}}{\frac{1}{28} + \frac{1}{28}}$$

$HDO = 756$ m (2479 ft).

2) What factors influence dump location?
3) How should dumps be drained?

Dewatering

1) This situation was submitted by John H. Lucas of the Permanente Plant, Kaiser Cement & Gypsum Corp. It covers a multitude of questions and is a good example of the kind of problems that face a planner.

The Permanente quarry is a surface mining operation encompassing an area approximately 1372 × 762 m (4500 × 2500 ft) in size covering a faulted-fractured limestone deposit. The general mining plan is to mine the limestone back to the footwall of the Franciscan shale formation. There is some trapped water within the footwall and final wall of the quarry. This water bleeds out continuously and causes a weakness in the final pit slope. Also, the present mining plan will take the final quarry limits to within 30.48 m (100 ft) of a small running stream lying on the hanging wall side of the quarry.

Some flatter than 45° back sloping has been done on the footwall side, where the top of the final slope is 91.44 m (300 ft) above the present top working bench. Ultimate planned mining is to go 304.8 m (1000 ft) deeper than this top bench. Work on horizontal drain pipes and vertical holes with piezometers has been performed to help prevent major sliding of the final pit slope.

The main problem is to be sure that planning and designing of the quarry creates a safe final pit slope to enable maximum limestone extraction at a minimum cost without hydrostatic pressures causing the footwall to slide or the hanging wall to collapse due to entrapped water and the close presence of the small running stream.

2) What are some of the factors that enter into road and dump drainage?

Equipment Selection

1) What are the physical factors to be considered in equipment selection?

2) What are the economic factors to be considered in equipment selection?

3) What are the operating and maintenance factors to be considered in equipment selection?

4) What is the starting point in equipment selection?

5) How do you determine productivities for equipment selection?

6) Select the optimum truck fleet configuration in a porphyry/skarn copper pit with a mix of 5.5-m (6-yd), 13.7-m (15-yd), and 18.2-m (20-yd) shovels. Consider the increased cost of effectiveness of larger hauling units, limitations on road width, limited loading height of smaller shovels, and the need for a multiple of small shovels in the ore areas to provide a reasonable blend of mill ore. (This was submitted by R. R. Leveille of Chino Mines Div., Kennecott Copper Corp.).

3B Production Schedules for Different Ore Production Rates

S. P. Winkelmann, editor

Contents

17 Production Schedules

Michael B. Kahle
Phelps Coal Co.
Fred J. Scheaffer
Morrison-Knudsen Co., Inc.

Michael B. Kahle is manager of development for Phelps Coal Co., Dallas, TX. He has a B.S. in geological engineering and an M.S. in mining engineering. Mr. Kahle has over 15 years of professional experience in the mining industry, in copper, uranium, oil shale, coal, and lignite projects involving truck-and-shovel operations, scraper operations, and dragline operations. Prior to joining Morrison-Knudsen, Mr. Kahle held various positions with Kennecott Copper Corporation's Bingham Mine at Bingham Canyon, Utah.

Fred J. Scheaffer is a project engineer for the Industrial and Mining Engineering Div., Morrison-Knudsen Co., Inc. He has an Engineer of Mines degree from the Colorado School of Mines. Mr. Scheaffer has about nine years of professional experience in the mining industry in copper, oil shale, coal, and lignite projects involving truck-and-shovel operations, scraper operations, and dragline operations. Prior to joining Morrison-Knudsen, Mr. Scheaffer held various positions with Kennecott Copper Corporation's Nevada Mines Div. at McGill, Nevada.

Introduction

Production scheduling, along with production planning, provides projections of future mining progress and time requirements for the development and extraction of a resource. These schedules and plans are used by management as a means of attaining the following objectives: (1) maintaining or maximizing expected profit, (2) determining future investment in mining, (3) optimizing return on investment, (4) evaluating alternative investments, and (5) conserving and developing owned resources.

The first four goals are generally concerned with mining costs, both capital and operating requirements, and as such, play an important role in production planning. However, this chapter is concerned with the fifth management objective of resource development in order to conserve and perpetuate the corporate entity. The following discussion is based on the premise that detailed economic evaluations and market surveys have been performed and analyzed and that the results indicate a viable project.

Relationship of Production Scheduling to Mine Design

Mine Design

The development of a mine design for an open pit mining operation occurs in three stages. The first stage is the development of a long-range mine plan based on a mineralization inventory of the resource. This mineralization model is built from borehole data collected during exploration and development drilling programs and the geological interpretation of data. The major goal of this stage is to examine and evaluate the mineral deposit in sufficient detail to define economic tonnages and grades/quality of the resource, quantities of waste, and the geometry of the mine. These parameters are used to establish ore reserves, economic pit limits, stripping ratios, and initial investment planning.

The second stage in the design of a mine is intermediate-range planning. The intermediate-range plan establishes the five to ten-year resource and waste production requirements for obtaining optimum or near-optimum cash flows within the total reserves as outlined in the long-range plan. This planning technique allows the removal of material in large increments while maintaining the required pit slopes and providing for operational and legal constraints. Mine management is also provided with sufficient time for analyzing capital requirements, specifically equipment units with long delivery times.

The third stage in mine design is short-range mine planning. This phase of the mine design is concerned with daily, weekly, monthly, and yearly mine schedules and plans. These short-range mining activities are dependent on three basic activities: (1) production schedules, (2) operating equipment, and (3) material handling procedures.

This chapter discusses the first activity, production schedules, and presents some of the methods and procedures used in production scheduling for various production rates.

Production Schedules

Production scheduling is important to the overall mine design because of the substantial costs associated with labor, supplies, and equipment which are affected by the production schedule.

The generalization of production scheduling is difficult. Most mines vary in size, mining method, geometry, and management philosophy. Consequently, scheduling procedures used for optimum results at one mine may be completely different at another. Some of the more universally accepted concepts used in many mining operations are discussed in the following section.

The production schedule is a plan relating to (1) production rate and (2) operating layout. These factors establish the main criteria for the development of a production schedule. The production rate determines the limits of production capacity for a production unit such as a shovel and a fleet of haulage trucks. A series of these production units establishes the overall production of the mine. The operating layout establishes the physical constraints which will be encountered by the production units. Time, a finite constraint, establishes the duration or length of the schedules.

Production Rate: The production rate is material per unit of time for an equipment unit or a series of equipment units. The material factor of the production rate can be described as follows: (1) metric tons (short tons) per hour, shift, day or year, and (2) cubic meters (cubic yards) per hour, shift, day or year. Care must be used when describing these rates because of the major confusion associated with the time element. This confusion usually occurs because of the difference between an operating hour and a scheduled hour. A scheduled hour relates usually to the time paid the operator or time scheduled for the operator on the equipment unit. An example of scheduled time would be 60 min to an hour or 8 hr per shift.

An operating hour usually refers to the productive time of the production unit. An example of an operating hour would be 60 min (scheduled hour) minus normal operating delay time, such as fueling, lubrication, coffee break, etc.

The time factor of the production rate can also be

described as: (1) hours per shift, (2) shifts per day, and (3) days per year. These criteria are usually established by a management decision based on socioeconomic conditions such as holiday or vacation schedules at other surrounding mines, labor contracts, and total plant utilization philosophies.

Operating Layout: The operating layout element of production scheduling is the establishment of the physical or operating constraints of the mine design. Some of the key factors that must be taken into account when developing an operating layout are: (1) established pit operating procedures, (2) expected ore grades, (3) planned operating slopes, (4) designed haul roads, (5) planned dump development, (6) planned backfilling and reclamation sequences, (7) designed surface and ground-water controls, (8) required equipment size and maneuverability, and (9) planned bench development.

The main objective of operating layout in production scheduling is to determine how far in advance a certain resource must be stripped to maintain the required production rate and resource grade or quality.

Varying Production Rates

An economic analysis reflecting typical corporate philosophies of most open pit mines requires a high present value which favors mining of the best grade first and the general desire to recover capital expenditures by mining the lower stripping ratio areas. The analysis of different mine production schedules and rates requires the simultaneous examination of the existing physical constraints (operating layout) and the present mine equipment capabilities (production rates).

A major task in production scheduling associated with the operating layout is to avoid high stripping requirements for short periods of time. High stripping requirements for short time durations result in high equipment investment and excess equipment to meet the demand. However, as soon as the high stripping requirement has passed, the excess equipment is no longer needed and results in lower equipment utilization. Production scheduling allows the scheduling of waste stripping and ore production to keep equipment requirements constant using a metric ton-kilometer (short ton-mile or billion cubic yard-mile) basis.

Another task in production scheduling is to provide sufficient operating room and mining faces to permit economical mining practices. Limited cut widths and mining faces decrease production stripping requirements but result in a more costly mining operation through decreased equipment efficiency and utilization.

Production scheduling also analyzes the production capacity of the existing equipment. This evaluation usually includes an examination of previous equipment performance levels and a projection of expected equipment performance. The major items reviewed by the engineer are: (1) expected equipment fleet sizes; (2) projected equipment availability; (3) projected equipment utilization; (4) planned haulage profiles and conditions; and (5) anticipated mining conditions, digging, development work, weather, water, and labor constraints. These items allow the engineer to schedule or adjust a short-range mine plan and to develop a production plan which the mine operator can use in meeting operating goals and objectives.

Methods for Production Scheduling

There are various methods and techniques available to the engineer for use in production scheduling. The two major methods employed are manual methods and/or computer methods. Various degrees of computer utilization supplementing the manual method of production scheduling exist. However, the methods and techniques employed depend largely on the equipment available to the engineer. It should be stressed that the computer serves as a tool for the engineer and usually provides the required data quicker and more accurately than manual methods. Since the concepts for both methods are basically the same, this chapter will discuss computer methods used in production scheduling.

Available Computer Systems

During the previous decade many types of computer models have been developed describing in a numerical or geometrical form an ore body and mine. These models have provided a valuable tool to the engineer for evaluating the long, intermediate, and short-term planning goals. However, these models do not provide the detail necessary for close control and planning required for shovel cuts and grade control on a week-to-week or month-to-month basis.

The application of graphic digitizers used in conjunction with the computer has provided the tools necessary to handle large masses of data for defining individual shovel cuts and cut sequences as well as shift or daily ore grade control. This production planning system is presently being used to: (1) forecast pit area geometry by shovel cut, (2) display ore grades by cut and to guide how to blend ores for grade control, (3) prepare production schedules, and (4) compare mining development progress to production schedules.

Another major computer method used in production planning is the truck simulation model. To schedule and evaluate haul truck requirements properly, computerized truck simulation models have been developed and are used almost universally by mining companies

Table 1. Result for a Haul Cycle Simulation

Morrison-Knudsen Co., Inc. Haul Cycle Summary by Bid Item EM120 Date 11/18/77 Page 1

Job- Example -Excess Overburden

Item Bid	Haul #	Haul Unit	Cap. BCY	Volume BCY	Load Rate	Spread Hr	Theoret Unit Hr	Assum. EFF.	Number of Units			Calc EFF	Actual Unit Hr	Haul Dist Ft	M.P.H. with Fixed Time	M.P.H. with-out	Prod./Unit with Round up	Prod./Unit w/o Round up	Cycles per Hr
									Theor.	w/Eff	Use								
1	791	TS-24	21.00	202000	1000	202	1589	0.800	7.87	9.83	10	0.787	2020	4240	9.7	14.7	100	102	4.76
	1	Total		202000		202	1589					0.787	2020	4240	9.7	14.7			
BCY Mile				162212															

and equipment manufacturers. This simulation model is based on a stochastic process based on the performance curve supplied by the equipment manufacturer. Stochastic simulation is particularly valuable when analyzing the effect of scheduling trucks of significantly different size or speed to meet a desired production rate. Although this technique has been used extensively for haul trucks, it is being used for scrapers, trams, and highway trucks. Table 1 shows the typical result for a haul cycle simulation.

Application of Model Results

Although the computer provides an essential tool for production planning, the results must be evaluated, adjusted, and applied by the engineer to determine the effects on production scheduling. Depending on which computer method is employed—truck simulation models, mineralization models, geometrical generators, or slice and cut sequence summaries—the end result is an application of the data to determine equipment requirements. This determination, as described previously, requires the evaluation of the existing equipment to determine its production capability. In addition to the evaluation of the existing equipment, additional equipment requirements must be determined and evaluated. This task consists of reviewing existing equipment performance levels, planned haul profiles, and operating conditions and constraints.

The computer serves as a tool that provides quick analysis of various production schedules and rates. This capability provides the engineer with a tool that enables him to respond to management's request for timely evaluations of various production levels to optimize the mining plan and the overall mine economics.

Economic Considerations: Once total equipment requirements are determined, the results are provided for management's review. In addition to the technical evaluation, a financial evaluation should be performed and presented to management. Based on the results of the economic evaluation, management will make a decision as to whether the engineer should proceed with production planning. If the results of the financial evaluation meet and satisfy corporate criteria and requirements, then the decision will be to proceed. If the results are negative, management may decide to discontinue the evaluation or revise the original criteria established for the evaluation.

Production Scheduling Considerations: After management has reviewed and approved the equipment evaluation, the engineer must proceed with the tasks of production scheduling and developing sequential mine plans for the proposed development.

Based on the selected production level, production rates are established for the various equipment spreads planned. Once these production rates are established, operating layouts and plans can be developed in the detail necessary for operations to implement. These plans should be of sufficient detail to delineate by level where the production units are scheduled and the scheduled volumes, tonnages, and grades to meet required mine output. These plans and schedules should incorporate blending, if necessary, and provide constant equipment requirements and high utilization of the available equipment.

Production Schedule Example

The example discussed herein outlines a typical problem faced by engineers and operators. Mine management has requested an increase in the mine production level for the remaining portions of the planned mining period. The engineer is faced with the problem of determining the number of trucks and shovels necessary to achieve the requested production.

The individual shovel cuts have already been determined by short-range computer simulation. The haulage cycles have also been determined for truck assignment requirements. The mining plan has 171 operating

Table 2. Truck Requirements Calculated by Truck Haulage Simulation Program

Shovel No.	Truck Assignment, ton *	Computer Projected Shovel Productivity, ton * per shovel shift	Experienced Shovel Productivity, ton * per shovel shift
A	4- 65	9,600	9,400
B	4- 65 2-100	11,600	10,600
C	3- 65	5,000	6,000
D	5-150	11,600	12,000
E	6-150	11,600	11,500
F	6- 65	11,600	12,000
G	4- 65	5,000	5,000
H	2- 65	5,000	5,000
I	8- 65	11,600	9,800
J	3-150 1-100 1- 65	11,600	11,500
K	5-150	11,600	10,600
L	8-100	11,600	12,500
M	5-150	11,600	11,300
N	4- 65	5,000	5,500
Average productivity		9,600 =	9,400

* Metric equivalent: 1 st×0.907 184 7=t.

Table 3. Number of Shifts Multiplied by the Ratio of the Shovel's Planned Productivity to the Experience Productivity

Shovel No.	Cut	Cut		Required Shovel Shifts
A	$\frac{24{,}600}{28{,}200} \times 93 +$	$\frac{24{,}200}{28{,}200} \times 420$	=	442
B	$\frac{23{,}600}{31{,}800} \times 513$		=	378
C	$\frac{10{,}000}{18{,}000} \times 513$		=	291
D	$\frac{25{,}750}{36{,}000} \times 51 +$	$\frac{22{,}300}{36{,}000} \times 462$	=	319
E	$\frac{24{,}600}{34{,}500} \times 93 +$	$\frac{24{,}200}{34{,}500} \times 420$	=	360
F	$\frac{23{,}700}{36{,}000} \times 513$		=	338
G	$\frac{8{,}000}{15{,}000} \times 513$		=	277
H	$\frac{12{,}000}{15{,}000} \times 45 +$	$\frac{12{,}400}{15{,}000} \times 468$	=	409
I	$\frac{25{,}000}{29{,}400} \times 234 +$	$\frac{30{,}000}{29{,}400} \times 279$	=	481
J	$\frac{30{,}000}{34{,}500} \times 189 +$	$\frac{25{,}000}{34{,}500} \times 324$	=	399
K	$\frac{25{,}000}{31{,}800} \times 513$		=	402
L	$\frac{30{,}000}{37{,}500} \times 513$		=	410
M	$\frac{30{,}000}{33{,}900} \times 513$		=	452
N	$\frac{14{,}200}{15{,}000} \times 153$		=	436
				5394

Table 4. Truck Requirements Needed to Meet Production Requirements

Shovel No.	Planned Profile Distance, ft *	Ratio of Shovel Shifts	Schedule Trucks			Average Trucks		
			65	100	150	65	100	150
A	4524	442/513×	4	0	0	3.45	0	0
B	6447	378/513×	4	2	0	2.95	1.47	0
C	4500	291/513×	3	0	0	1.71	0	0
D	9176	319/513×	0	0	5	0	0	3.11
E	13,589	360/513×	0	0	6	0	0	4.22
F	3640	338/513×	6	0	0	3.95	0	0
G	4027	277/513×	4	0	0	2.17	0	0
H	500	409/513×	2	0	0	1.59	0	0
I	10,314	481/513×	8	0	0	7.52	0	0
J	9431	399/513×	1	1	3	0.78	0.78	2.34
K	9366	402/513×	0	0	5	0	0	3.92
L	8895	410/513×	0	8	0	0	6.39	0
M	10,418	452/513×	0	0	5	0	0	4.41
N	3939	436/513×	4	0	0	3.41	0	0
			41	12	24	27.53	8.64	18.00

* Metric equivalent: 1 ft × 0.304 8 = m.

days or 513 shovel shift days remaining in which to accomplish the production requirements.

Table 2 shows the truck requirements for each shovel as calculated by the truck haulage simulation program. Also shown is the planned shovel productivity required to meet the increased production requirement. The experienced shovel productivity represents the actual productivity realized by each shovel to date. The shovel productivity statistics are presented in tons per shovel shift. As can be noted, the average shovel productivity required for the increased productive level is close to the experienced shovel productivity. Consequently, it can be said in general that the existing shovel fleet size will accommodate the desired production rate.

The next calculation that must be analyzed is scheduling the increased rate with the original plan and cut sequence. This calculation determines the number of shifts each shovel must work to achieve the new mining schedule and cut sequence.

Table 3 shows the calculation which multiplies the number of shifts originally planned for each cut by the ratio of the shovel's previously planned productivity to the experience productivity.

The required 5394 shovel shifts in 171 operating days is equivalent to 31.5 shovel shifts per day or to scheduling an average of 10.5 or 11 shovels per shift. Since there are 14 shovels in the shovel fleet, the shovel schedule will require a shovel utilization of 11/14= 79% on the average.

Table 4 summarizes truck requirements needed for the scheduled shovels to meet the production requirements. The table indicates that a total of 77 trucks should be scheduled. However, the average number of trucks and truck type is based on required shovel shifts which are shown under average trucks. Thus, on the average, the total number of trucks needed is 55 trucks: 28 59-t (65-st) trucks, 9 90-t (100-st) trucks, and 18 136-t (150-st) trucks for a ten-shovel schedule.

This method of calculation can be computerized for timely evaluation by the engineer, operator, and maintenance shop personnel. The results allow the engineer and operator to quickly schedule and make changes for operating conditions and equipment availability. It also provides the maintenance shop personnel with the criteria and priorities for repairing the necessary equipment to meet production requirements as identified in the production schedule.

3C Equipment and Facilities

J. T. Crawford, III, editor

Contents

INTRODUCTION TO SUBSECTION 3C

Equipment and facility evaluation and selection are vital links between exploration, ore reserve estimation, and mine planning, and a financially viable producing mine. The equipment and facilities are the tools with which the production goals of a mining plan are achieved. All material that has been presented thus far in this book could be meaningless or at best, the intended mining plan results difficult to achieve, unless this phase is given proper attention.

Production equipment and facilities have a strong bearing on the feasibility of bringing a deposit into production and on the actual financial results. They represent the majority of the capital required and influence importantly the operating costs. The evaluation and selection process must seek to optimize economically the equipment and facilities needed to accomplish a specific mining plan, while ensuring production reliability and minimizing financial risk.

This subsection presents two chapters on equipment evaluation and one on maintenance and ancillary facilities, and support equipment. The process of evaluating equipment alternatives is illustrated by case examples. Some important considerations in designing maintenance and ancillary facilities and selecting support equipment are presented.

Because the proper selection of production equipment is so important financially to the viability of a property, the evaluation process requires an investment analysis approach. The discounted cash flow technique used is presented in detail by the case examples.

Numerous computerized assists have been developed to aid shovel and haulage truck evaluation. A haulage truck profile simulation technique and a type of operating performance monitoring are included in the discussion of shovels and haulage trucks.

18 Drill Evaluation

R. H. Heinen
Dravo Corp.

R. H. Heinen currently holds the position of senior mining engineer of Dravo Corp. in Denver, CO. Prior to this, he was plant engineer for the Nevada Mines Div. of Kennecott Copper Corp. He has been active in AIME serving as secretary of the Eastern Nevada section. Mr. Heinen holds a bachelor of science degree in Chemical Engineering from the University of Nevada.

Introduction

Drilling is the initial operating step in open pit mining. It goes hand-in-hand with the blasting operations to ensure adequately broken material for the excavation equipment employed. The drilling effectiveness is highly dependent on the quality of the drill evaluation. This discussion will center on the parameters that are needed to evaluate drill requirements for a new mine. The procedure that is generated can, however, be used to evaluate whether to expand existing operations and replace existing equipment. Methods of breaking ground, other than blasthole drilling and blasting, such as ripping, will not be discussed.

The importance of correct drill evaluation extends to the other facets of open pit mining. Inadequate fragmentation, the direct result of poor drilling and blasting techniques, results in oversized, hard digging for the excavating equipment. The result is higher secondary blasting costs, reduced excavation efficiencies, higher excavation costs, higher road repair costs, and higher haulage equipment costs. The main objective of evaluating drill fleet requirements, therefore, is to determine the equipment needed to minimize the operational constraints, caused by blasthole drilling, on other facets of open pit mining at the lowest cash flow costs.

Drill Evaluation

Effective drill evaluation starts with a complete evaluation of the excavation equipment to be used. The drilling fleet is developed for the excavation operation to be employed, not vice versa. The excavation equipment evaluation must include the bench height and width to be mined, the physical arrangement of the excavating equipment (all equipment on one or mutliple levels), and the projected annual operating rate that the excavation equipment would be required to meet.

The example used to illustrate drill evaluation is based on a hypothetical open pit mine that will operate 356 days per year, 3 shifts per day. The production rate will be 90 718 t (100,000 st) of material mined per day, of which 31 751 tpd (35,000 stpd) is ore. The excavation equipment to be employed will consist of four 15.3-m^3 (20-cu yd) shovels. These shovels will normally operate on different levels and mine 12.2-m (40-ft) high benches with widths between 30.5 and 152.4 m (100 and 500 ft).

There are three basic elements which must be considered in evaluating a drilling system; they are:

1) Ore and waste production schedules, operating conditions, and rock types encountered.

2) Equipment productive capacities: (a) pattern size; (b) metric tons (short tons) of material affected per hole drilled; (c) drill production rate, meters (feet) drilled per shift or hour; and (d) drill availability and utilization, %.

3) Capital ($) and operating costs ($/hr).

Only rotary blasthole drilling will be evaluated in this discussion because of the normally superior efficiency of rotary blasthole drilling over other drilling methods. The evaluation procedure reviewed is applicable to other drilling methods with some slight modification.

The evaluation will start with the review of some of the drills available as grouped into four classes based on the drill-hole size capability (Hoppe, 1976). Table 1 lists representative rotary blasthole drills that

Table 1. Representative Rotary Blasthole Drills by Class

Class	Manufacturer's name & model	Maximum down pressure capability, kg (lb)	Single pass capability, m (ft)	Typical bit size mm (in.)
229 mm (9 in.)	BE-40R	22 680 (50,000)	8.2 (27)	229 (9)
270 mm (10⅝ in.)	GD-80	36 288 (80,000)	16.8 (55)	270 (10⅝)
	BE-45R	36 288 (80,000)	16.8 (55)	270 (10⅝)
	IR DM-6	40 824 (90,000)	9.8 (32)	270 (10⅝)
311 mm (12¼ in.)	M-4	49 896 (110,000)	16.8 (55)	311 (12¼)
	GD-120	49 896 (110,000)	19.8 (65)	311 (12¼)
	BE-60R	49 896 (110,000)	19.8 (65)	311 (12¼)
	IR DM-7	40 824 (90,000)	9.8 (32)	311 (12¼)
381 mm (15 in.)	M-5	49 896 (110,000)	16.8 (55)	381 (15)
	GD-130	58 968 (130,000)	20.1 (66)	381 (15)
	BE-61R	49 896 (110,000)	19.8 (65)	381 (15)

are available from various manufacturers. This is not a complete list but is meant to show the four classes or groups that will be evaluated. Table 1 also includes some of the pertinent information for the drills listed. Only crawler-mounted drill rigs are listed because of their universal acceptance in open pit mining and their ability over truck-mounted drills to support a heavier rig and to move more easily on rough or inclined surfaces.

Rock Types

One of the major variables in evaluating the drill fleet requirements for an open pit mine is the rock types (Williamson, 1968) that will be encountered. For a new mine, the rock type information is usually generated from exploration drill holes that were used to outline the ore body. Table 2 is a listing of the rock types to be encountered in this hypothetical case.

Table 2 shows the rock types in a drillability chart based on experience from existing operations. If no actual operating data are available, a similar listing can be derived by subjecting rock samples obtained from the exploration drilling program to compressive strength, abrasive or scratch hardness, impact hardness, and toughness tests. Most drill bit manufacturers have the facilities to conduct these tests at very nominal costs.

Table 2. Comparison of Rock Types

Formation drillability	Formation compressive strength, kg/m²* (psi)	Rock type	% Material mined in each formation (in example)
Soft	0–29 300 (0–6,000)	Altered sediments Altered porphyry Altered shale Diabase Schist Altered quartzite	30
Medium	29 300–97 700 (6,000–20,000)	Monzonite porphyry Quartzite Shale Porphyry Rhyolite Biotite-argillic porphyry	35
Hard	97 700–293 000 (20,000–60,000)	Limestone Magnetite Marbalized limestone Garnetized limestone Jasperoid	35

* kilogram per square meter (kg/m²) × 9.807 = pascal (Pa).

The importance of knowing the rock types to be encountered cannot be stressed enough. Rock types will dictate the blasthole pattern size to be used, the drill pulldown requirements, the bit type to be utilized (steel tooth or carbide insert bits), and the production rate at which the drill will perform.

Other factors such as faults, shear planes, and saturated formations play a very small part in the evaluation of the drill fleet. To simplify this discussion, their effects on drill performance will not be included.

Drill-Hole Pattern and Tonnage Configurations

Drill-hole patterns and the subsequent tonnage affected by each hole are governed by the following variables:

1) Bench widths which generally dictate whether a single- or a multiple-row drilling pattern would be applicable. In this example the bench width is between 30.5 and 152.4 m (100 and 500 ft) so a multiple-row drilling configuration would, most likely, be used.

2) Excavation equipment type and size which dictate the bench height to be mined; 12.2 m (40 ft) in this example.

3) Hardness of the rock. The harder the rock the closer the spacing of the patterns to achieve the desired fragmentation.

4) Haulage system. Rail versus truck haulage systems may impose constraints in terms of utilizing the blasting efficiency of a given drill pattern. Throw of material and back-break become important constraining factors.

5) Stability of pit walls limiting the size of the drill pattern and the explosive charge per hole.

6) The characteristics and economics of available explosive agents.

7) The physical dimensions of the mining operation.

The evaluation of the variables listed will lead to an approximate pattern size and tonnage configuration per hole for each of the drill fleets to be evaluated. Tables 3 and 4 list typical pattern sizes and tonnages affected by each hole based on the rock drillability and the loading and haulage configuration given. Pattern sizes and tonnage figures can best be estimated from actual operating mines with similar operating and rock type conditions.

Production Rates

Rotary blasthole production rates are dependent on the type of rock formation encountered, the type and size of the bit used, operator skill, the thrust exerted by the drill on the bit, the drill compressor size, type of

Table 3. Typical Pattern Size for 12.2-M (40-Ft) Bench Height

	In m (ft)			
	229 mm (9 in.)	270 mm (10⅝ in.)	311 mm (12¼ in.)	381 mm (15 in.)
Soft formation	8.2 × 9.1 (27 x 30)	9.1 x 10.0 (30 x 33)	10.0 x 11.0 (33 x 36)	11.0 x 12.2 (36 x 40)
Medium formation	7.3 x 8.2 (24 x 27)	8.2 x 9.1 (27 x 30)	9.1 x 10.0 (30 x 33)	10.4 x 11.0 (34 x 36)
Hard formation	6.4 x 7.3 (21 x 24)	7.3 x 8.2 (24 x 27)	8.2 x 9.1 (27 x 30)	9.1 x 10.0 (30 x 33)

terrain, distance moved between holes, and the drill's capability of drilling the hole with or without changing drill steel (single-pass capability). Most drill manufacturers have estimated production rates for varying rock types, but the best estimates come from information derived from existing open pit mine operations. The typical production rates for the four drill fleets in varying rock formations are listed in Table 5.

Drill Availability and Utilization

The procedure for determining drill availability and utilization is outlined in Table 6. This information is based on a typical open pit mining operation. The following equations can be used to obtain the percent availability and utilization for the drill fleets to be evaluated:

$$\%\ \text{Availability} = \frac{\text{Possible avail. time—M\&R outages}}{\text{Possible available time}} \times 100\%.$$

$$\%\ \text{Utilization} = \frac{\text{Avail. oper. time—oper. restrictions—long drill moves—personnel time—other}}{\text{Available operating time}} \times 100\%.$$

For the example given, the percent operated of the total possible time is equal to the percent availability times the percent utilization, or 53%.

The figure of 53% is lower than other operating equipment such as shovels and trucks primarily because of the following operational restraints:

1) Shutdown of drills during blasting.

2) The need for drill fleets to have excess capacity to ensure a steady supply of blasted material for loading.

3) Scheduling problems.

4) Unscheduled maintenance failure of the loading equipment.

In actual practice, the amount of time lost due to

Table 4. Material Affected Per Hole*

	In met (ton)			
	229 mm. (9 in.)	270 mm (10⅝ in.)	311 mm (12¼ in.)	381 mm (15 in.)
Soft formation	2351 (2592)	2874 (3168)	3449 (3802)	4264 (4700)
Medium formation	1882 (2074)	2351 (2592)	2874 (3168)	3556 (3920)
Hard formation	1463 (1613)	1882 (2074)	2351 (2592)	2874 (3168)

$$\text{Tons/hole} = \frac{\text{Pattern size} \times \text{bench height}}{\text{Density}}$$

Example: Soft formation and 9 in. fleet size:

$$\text{Metric tons/hole} = \frac{8.2 \times 9.1 \times 12.2\ \text{m}}{0.390\ \text{m}^3/\text{mt}} = 2351$$

$$\text{Tons/hole} = \frac{27 \times 30 \times 40\ \text{ft}}{12.5\ \text{cu ft per ton}} = 2592$$

Table 5. Typical Production Rates

	m/operating shift (ft per operating shift)			
	229 mm (9 in.)	270 mm (10⅝ in.)	311 mm (12¼ in.)	381 mm (15 in.)
Soft formation	91.4 (300)	137.2 (450)	167.6 (550)	173.7 (570)
Medium formation	83.8 (275)	121.9 (400)	155.4 (510)	161.5 (530)
Hard formation	68.6 (225)	103.6 (340)	134.1 (440)	140.2 (460)

operational restraints can vary from 5 to 40% of the available operating time.

Long drill moves and other major interruptions are defined as occurrences when the drill must be moved far enough on one level to require additional trail cable in the case of electric drills, approximately 229 m (750 ft), the drill mast must be lowered, or the drill is moved to a new level. The figure listed in Table 6 for this category reflects a minimum number of drill moves such as when a drill is servicing one or two shovels very close together which routinely does not require long drill moves.

Table 6. Determination of Drill Availability and Utilization, Typical Drill Operating Time Table

	Hr	Days
Total calendar time	8760	365
Less holidays per year	216	9
Possible available time	8544	356
Less M&R outages	1440	60
Available operating time (equipment availability)	7104	296
Less operational restrictions	624	26
Less long drill moves and other major interruptions	216	9
Less personnel time		
travel time	432	18
lunch	432	18
other	72	3
Less other nondrilling time		
lubrication & inspection	288	12
short moves	72	3
running repairs	216	9
other	216	9
Net drilling time	4536	189

Example calculations:

$$\% \text{ Availability} = \frac{356 - 60}{356} \times 100\% = 83\%$$

$$\% \text{ Utilization} = \frac{296 - 26 - 9 - 39 - 33}{296} \times 100\% = 64\%$$

$$\% \text{ Operated of total time} = 83\% \times 64\% = 53\%$$

$$\text{Opr. hr. per shift} = \frac{296 - 26 - 9 - 39 - 33}{296} \times 8 \text{ hr/shift} = 5.1 \text{ hr/shift}$$

It is assumed for the initial evaluation that all four drill fleets will have the same availability and utilization figures.

Minimum Drill Fleet Requirements

Now that all the drill productive capacities are known, the minimum number of drills in each drill class required to maintain broken material for the four 15.3-m³ (20-cu yd) shovels can be calculated by the equation:

$$\text{Minimum drill fleet} = \frac{\text{Total drill hr}}{\text{Total poss. hr/drill} \times \text{avail.} \times \text{util.}}$$

These calculations and the calculations for determining the total drill hours are listed in Table 7. The minimum drill fleet size varies from a high of four drills for the 229-mm (9-in.) fleet to a low of one drill for the 381-mm (15-in.) fleet.

Drill Operating Costs

Drill operating costs (Chitwood and Norman, 1977) vary widely throughout the mining industry. A typical set of operating costs without equipment depreciation is listed in Table 8.

The labor costs listed are based on a drill crew consisting of a driller and a helper as well as the indirect labor costs such as supervision, vacation, sickness, and accident replacement labor. Some operations today that utilize drills and single-pass capabilities have eliminated the drill helper from the drill crew which would reduce the labor cost per hour in Table 8 by $5.70. It should be noted that the labor costs are based on an 8-hr operating shift per crew and must be adjusted to the drill operating hours per shift, as shown in the table.

The use of either electric or diesel-electric generator power for operating the drill varies with the nature of

Table 7. Drill Fleet Number Evaluation

Drill fleet size	Form. Tonnage, tpy (stpy)	Material affected/hole mT/hole (ton/hole)	No. of holes drilled	M drilled/hole (ft drilled/hole) (bench height + subgrade drlg.)	Total m (ft) drilled, m(ft)	Drilling rate, m/shft. (ft/shift)	Drill oper. shifts	Op. hr. per shift	Drill oper. hr	Min. no. of drills needed
229 mm (9 in.)	9 688 896 (10,680,000)	2351 (2592)	4,120	13.7 (45)	56 510 (185,400)	91.4 (300)	618.0	5.1	3,152	
	11 303 712 (12,460,000)	1882 (2074)	6,008	13.7 (45)	82 406 (270,360)	83.8 (275)	983.1	5.1	5,014	
	11 303 712 (12,460,000)	1463 (1613)	7,725	13.7 (45)	105 956 (347,625)	68.6 (225)	1545.0	5.1	7,880	
Total	32 296 320 (35,600,000)		17,853		244 872 (803,385)		3146.1		16,046	4
270 mm in. (10⅝ in.)	9 688 896 (10,680,000)	2874 (3168)	3,371	13.7 (45)	46 237 (151,695)	137.2 (450)	337.1	5.1	1,719	
	11 303 712 (12,460,000)	2351 (2592)	4,807	13.7 (45)	65 933 (216,315)	121.9 (400)	540.8	5.1	2,758	
	11 303 712 (12,460,000)	1882 (2074)	6,008	13.7 (45)	82 406 (270,360)	103.6 (340)	795.2	5.1	4,056	
Total	32 296 320 (35,600,000)		14,186		194 576 (638,370)		1673.1		8,533	2
311 mm in. (12¼ in.)	9 688 896 (10,680,000)	3449 (3802)	2,809	13.7 (45)	38 528 (126,405)	167.6 (550)	229.8	5.1	1,172	
	11 303 712 (12,460,000)	2874 (3168)	3,933	13.7 (45)	53 945 (176,985)	155.4 (510)	347.0	5.1	1,770	
	11 303 712 (12,460,000)	2351 (2592)	4,807	13.7 (45)	65 933 (216,315)	134.1 (440)	491.6	5.1	2,507	
Total	32 296 320 (35,600,000)		11,549		158 406 (519,705)		1068.4		5,449	2
381 mm (15 in.)	9 688 896 (10,680,000)	4264 (4700)	2,272	13.7 (45)	31 163 (102,240)	173.7 (570)	179.4	5.1	915	
	11 303 712 (12,460,000)	3556 (3920)	3,178	13.7 (45)	43 589 (143,010)	161.5 (530)	269.8	5.1	1,376	
	11 303 712 (12,460,000)	2 874 (3168)	3,933	13.7 (45)	53 945 (176,985)	140.2 (460)	384.8	5.1	1,962	
Total	32 296 320 (35,600,000)		9,383		128 697 (422,235)		834.0		4,253	1

Example, 9-in. fleet: $\text{Min. drill fleet} = \frac{16{,}046 \text{ hr}}{8544 \times 0.83 \times 0.64} = 4$

Table 8. Drill Operating Costs Without Depreciation

	$ per operating hr			
	229 mm (9 in.)	270 mm (10⅝ in.)	311 mm (12¼ in.)	381 mm (15 in.)
Labor costs*	34.50	34.50	34.50	34.50
Power of diesel costs	4.05	4.80	5.60	6.90
Repair costs	30.00	34.50	39.75	45.75
Related equipment costs	3.90	5.15	5.55	8.25
Bit costs†				
(soft form.)	7.06	11.47	16.18	19.00
(med. form.)	10.25	14.90	22.00	25.98
(hard form.)	11.47	17.33	25.88	30.67

$$\text{* Labor costs} = \frac{\text{driller, helper, \& indirect labor hourly rate} + 50\%\ \text{hrly rate (fringe)}}{\text{ratio } \dfrac{\text{net drilling time}}{\text{8 hr per shift}}}$$

$$\text{Example: } \frac{\$14.66 + 0.5 \times \$14.66}{\dfrac{5.1}{8}} = \$34.50$$

† Bit costs in \$/m (\$ per ft):

	229 mm (9 in.)	270 mm (10⅝ in.)	311 mm (12¼ in.)	381 mm (15 in.)
Bit cost:				
soft form.	0.39 (0.12)	0.43 (0.13)	0.49 (0.15)	0.56 (0.17)
med. form.	0.62 (0.19)	0.62 (0.19)	0.72 (0.22)	0.82 (0.25)
hard form.	0.85 (0.26)	0.85 (0.26)	0.98 (0.30)	1.12 (0.34)

Example: 9-in. drill fleet and soft formation

$$\text{Bit cost/opr. hr} = \frac{\text{m/shift} \times \text{bit cost/m} \quad (\text{ft/shift}) \times (\text{bit cost/ft})}{\text{5.1 opr. hr/shift}}$$

$$= \frac{91.4\ \text{m/shift} \times \$0.39/\text{m} \quad (300\ \text{ft/shift}) \times (\$0.12/\text{ft})}{\text{5.1 opr. hr/shift}} = \$7.06/\text{hr}$$

the operation. It is highly recommended that electric drills be utilized when the drill is required to cover relatively short distances between drill-hole patterns because of the elimination of the diesel-electric generator and subsequent repair costs, and the current costs of petroleum fuel. If, however, the drill is to be moved over long distances between drill patterns, the savings generated with the electric drill might be offset by the time required to change the electric feed cable or provide an auxiliary moving unit.

The repair and maintenance costs include repairs due to breakdowns as well as those for preventive maintenance.

The related equipment costs is a catchall group which includes such things as warehousing and storage costs of spare parts and new and repaired stabilizer and drill steel costs.

Drill bits are the primary consumable material costs in drilling. Bit costs vary with the type of formation to be drilled, the pulldown pressure used, operator skill, the type of bit used, and the rotary speed of the drill. Bit costs per meter (foot) are estimated from a typical operating open pit mine with similar rock types. The bit cost per meter (foot) is then reduced to bit cost per operating hour by multiplying the bit cost per meter (foot) by the drill's production ratio [meters (feet) per operating hour] in the varying rock formations.

The total operating and maintenance costs for the year can be calculated by using the following formula:

Operating and maintenance costs= operating and maintenance cost per hr × drill operating hr.

Total operating and maintenance costs for the four fleets are evaluated in Table 9.

Economic Drill Evaluation

Table 10 develops the drill evaluation economic analysis from Tables 7, 8, and 9. The costs are evaluated on an after-tax cash flow basis, assuming a 10-year economic life. The economic equations used are outlined in Table 10. The analysis presented shows that the most economical drill fleet would be one drill capable of drilling a 381-mm (15-in.) diam hole. The initial economic evaluation demonstrates the value of equipment size.

The economic evaluation presented was based on the basic assumption that all the drill fleets would have the same availability and utilization figures. For the 270-mm (10⅝-in.), 311-mm (12¼-in.), and 381-mm (15-in.) fleets, this is not practical because each drill would have to routinely service more than one shovel operating on more than one level, thus increasing the amount of long drill move time required. Table 11 lists the changes that would be necessary in the drill move time to permit drilling for more than one shovel per drill on multiple levels. The additional drill move

Table 9. Annual Drill Operating and Maintenance Costs

Drill fleet size	Form. type	Form. tonnage, tpy (stpy)	Drill operating hr	Drill opr. & maint. costs, $/hr	Annual drill opr. & maint. costs, $
229 mm (9 in.)	Soft	9 688 896 (10,680,000)	3,152	79.51	250,616
	Medium	11 303 712 (12,460,000)	5,014	82.70	414,658
	Hard	11 303 712 (12,460,000)	7,880	83.92	661,290
		32 296 320 (35,600,000)	16,046		1,326,564
270 mm (10⅝ in.)	Soft	9 688 896 (10,680,000)	1,719	90.42	155,432
	Medium	11 303 712 (12,460,000)	2,758	93.85	258,838
	Hard	11 303 712 (12,460,000)	4,056	96.28	390,512
		32 296 320 (35,600,000)	8,533		804,782
311 mm (12¼ in.)	Soft	9 688 896 (10,680,000)	1,172	101.58	119,052
	Medium	11 303 712 (12,460,000)	1,770	107.40	190,098
	Hard	11 303 712 (12,460,000)	2,507	111.28	278,980
		32 296 320 (35,600,000)	5,449		588,130
381 mm (15 in.)	Soft	9 688 896 (10,680,000)	915	114.40	104,676
	Medium	11 303 712 (12,460,000)	1,376	121.38	167,019
	Hard	11 303 712 (12,460,000)	1,962	126.07	247,349
		32 296 320 (35,600,000)	4,253		519,044

Table 10. Minimum Drill Evaluation, Economic Analysis

Drill fleet size	No. of drills	Capital cost per drill, $	Total capital cost, $ (CC) (× 000)	Depr. (D), straight line for 10 yr, $ (× 000)	Ann. opr. & maint. cost, $ (AOMC) (× 000)	Ann. cash flow cost,* $ (ACFC) (× 000)	Present value net cash flow cost,† $ (PVNCFC) @ 12% 10-yr life (× 000)
229 mm (9 in.)	4	350,000	1400	140	1 326	622	4914
270 mm (10⅝ in.)	2	500,000	1000	100	805	351	3096
311 mm (12¼ in.)	2	715,000	1430	143	588	237	2769
381 mm (15 in.)	1	795,000	795	80	519	231	2100

* ACFC = (AOMC + D) (1 − tax rate (0.48)) − D.
† PVNCFC = CC + ACFC × $PVF^{12\%, 10\text{-yr}}$

$$PVF^{12,10} = \frac{(1 + i)_n - 1}{i(1 + i)_n} = \frac{(1 + 0.12)_{10} - 1}{0.12(1 + 0.12)_{10}} = 5.65.$$

time needed is based on 2 hr of additional moving time per drill per day per shovel serviced over one shovel. For example, for the 270-mm (10⅝-in.) fleet, a minimum of two drills would be employed which would have to drill for four shovels; therefore, each drill would have an additional 2 hr of moving time per day. Actual long drill move time for different operations could vary widely due to the geometric configuration of the operation. Long drill move time and the geometric configuration will determine the level of drilled muck inventories needed to balance drilling, blasting, and loading schedules.

Table 11 concludes by calculating the actual number of drills required with varying availability and utilization figures. In this example, the 270-mm (10⅝-in.) and the 381-mm (15-in.) fleets increased by one drill per fleet while the other two fleets remained the same. If the calculations show an increase of more than one drill per fleet, then a second iteration based on the new fleet size should be run to verify the additional drills.

Table 12 lists the economic evaluation for the four drill fleets that were evaluated for the variable availability and utilization figures. This evaluation shows the 380-mm (15-in.) fleet size to be the most economical with the minimum cash flow.

The evaluation thus far has not talked about the

Table 11. Actual Number of Drills Needed Calculations

Drill fleet size	Min. No. of drills needed	Additional long drill move time needed/ avail. op. day	Ttl. increase in long drill moves,* hr	Increase in long drill moves/drill	Adjusted net drilling time/drill†	Ttl. drilling hr needed, from Tbl. 7	No. of drills needed
229 mm (9 in.)	4	0	-0-	-0-	4536	16,046	4
270 mm (10⅝ in.)	2	4	1184	592	3944	8533	3
311 mm (12¼ in.)	2	4	1184	592	3944	5449	2
381 mm (15 in.)	1	6	1776	1776	2760	4253	2

* Total increase in long drill moves = add. long drill move time/avail. opr. day × avail. opr. day for 270 mm (10⅝ in.) fleet total increase in long drill moves = 4 hr/day × 296 days = 1184 hr.
† Adjusted net drilling time/drill = net drill hr from Table 6 − increase in long drill move hr/drill for 270 mm (10⅝ in.) fleet adj. net drill hr = $4563^{hr} - 592^{hr}$ = 3944 hr.

Table 12. Actual Drill Evaluation, Economic Analysis

Drill fleet size	No. of drills	Capital cost per drill, $	Total capital cost, $ (CC) (× 000)	Depr. (D) straight line for 10 yr, $ (× 000)	Ann. opr. & maint. cost, $ (AOMC) (× 000)	Ann. cash flow cost, $ (ACFC) (× 000)	Present value net cash flow cost, $ (PVNCFC) @ 12% 10-yr life (× 000)
229 mm (9 in.)	4	350,000	1400	140	1326	622	4914
270 mm (10⅝ in.)	3	500,000	1500	150	805	347	3460
311 mm (12¼ in.)	2	715,000	1430	143	588	237	2769
381 mm (15 in.)	2	795,000	1590	159	519	194	2686

benefits of using a smaller drill in the ore zone to permit greater quality control of the ore because of the close drill pattern requirements. The importance of this variable will depend on the variability of the ore with respect to ore grade and possible blending requirements of the material. This variable will not be considered in depth but is mentioned to remind the reader of the importance of this factor to some operations which could require the use of the smaller more expensive drill.

Summary

To summarize, this discussion dealt with a quantitative approach for determining drill economic selection of a new operation by using primarily operating and cost data from existing open pit operations. Variables affecting each one of the operating parameters were provided to aid in the evaluation if actual operating data were not available. The procedure used in this evaluation can be adapted easily for an evaluation of the replacement or expansion of existing equipment.

References

Chitwood, B., and Norman, N. E., 1977, "Blasthole Drilling Economics: a Look at the Costs Behind the Costs," *Engineering and Mining Journal,* June, p. 168.

Hoppe, R., 1976, "Open-Pit Mining—Rotary Drills Dominate in Open Pits," *Engineering and Mining Journal,* June, p. 191.

Williamson, T. N., 1968, "Rotary Drilling," *Surface Mining,* E. P. Pfleider, ed., AIME, New York, p. 300.

19 Shovel and Haulage Truck Evaluation

John T. Crawford III
Kennecott Minerals Co.

John T. Crawford is manager of operations control —mining for Kennecott Minerals Co. Prior to this, he held numerous positions at operating properties, central staff groups, and corporate headquarters. He received a bachelor of mining engineering degree from the University of Minnesota where he graduated with honors and was a member of Tau Beta Pi. He has an MBA degree from Stanford University.

Mr. Crawford has been active in AIME as chairman of the Eastern Nevada Section, cochairman of the Open Pit Mine Planning and Design Workshop, and has presented technical papers on mine planning and equipment selection and evaluation.

INTRODUCTION

Loading and hauling systems are evaluated in open pit mining for new properties, expanding existing operations, changing systems or equipment, and equipment replacement analysis. The objective of such evaluations is to determine the equipment needed to handle planned tonnages at the lowest cash flow cost. Cash flow costs are used to combine capital outlays with operating and maintenance costs. Leasing costs could be substituted for capital costs with appropriate modifications to calculation procedures. These modifications will not be covered.

This chapter will describe methods of evaluating loading and hauling systems, concentrating on some of the more often raised questions concerning quantitative data. While the methods described are not the only ones in current use, they have a record of proven reliability. The shovel and haulage truck system is used in the examples because of its common use in open pit mines. In medium to large open pit mines, electric shovels and diesel-electric drive haulage trucks are normally used. The basic ideas presented in this chapter apply to all loading and hauling systems, with appropriate modifications. The process of performing a shovel and haulage truck evaluation can be time-consuming and tedious because of the quantity of data involved, but it is not particularly difficult.

The examples used to illustrate shovel and haulage truck evaluation are based on a hypothetical open pit mine operating 350 days per year producing 31 750-000 t (35,000,000 st, 100,000 stpd) annually, of which 11 110 000 t (12,250,000 st, 35,000 stpd) are ore. Selection of the optimum shovel and haulage truck fleets will be shown. The principal sources of data for the examples or any other equipment evaluation are equipment manufacturers and actual operations. Whenever possible, actual operating experience should be used to verify estimates. The quantitative data presented herein are unrelated to a specific existing mining operation.

There are three basic elements which must be considered in evaluating a shovel and haulage truck system: (1) ore and waste production schedules, haulage routes, and operating conditions; (2) equipment productive capacities which include availability and utilization in percent, productivity, typically, tons per hour, and effects of interaction between shovels and haulage trucks on productivity; and (3) capital ($) and operating costs ($ per hour).

It is generally preferred that an evaluation involving the selection of both shovels and trucks begin with the shovels. This is because truck performance is influenced more by the shovel choice than vice versa.

SHOVEL EVALUATION

Shovel Utilization and Productivity

The general range of rock types, working room, and other operating conditions in most open pit mines, permit a wide range of shovel sizes, 5 to 19 m^3 (6 to 25 cu yd), to be operated efficiently. Cramped working room and low bank heights will reduce the efficiency of the larger units. Shovel point sheaves should be 1.5 m (5 ft) or less above the bank crest with a 45 to 50° boom angle. This defines the minimum bank height for efficient shovel operation. Bank heights considerably in excess of the point sheave height can be used within safe limits, depending on the caving nature of the blasted bank. They are generally in the range of 9 to 15 m (30 to 50 ft); 18 m (60 ft) and more can be used safely with the largest shovels. Excessive bank height may prevent double spotting of haulage trucks for loading, thereby decreasing shovel productivity.

The minimum working width at the shovel to provide double spot loading should be 30 to 46 m (100 to 150 ft) depending on the shovel-haulage truck combination. Roads should generally be 24 to 30 m (80 to 100 ft) in width, with maximum ramp grades of 8 to 10%.

Ninety-one-ton (100-st) haulage trucks with 11-m^3 (15-cu yd) shovels operate effectively with 37 to 40 m (120 to 130 ft) of working surface, while 136 to 154-t (150 to 170-st) trucks and 19-m^3 (25-cu yd) shovels need 43 to 46 m (140 to 150 ft) of room.

In determining shovel productive capacity, two things must be considered: utilization and productivity. Electric shovels can normally sustain 75 to 80% availability and 80 to 90% utilization of availability for a substantial portion of their nominal 20-year life. This amounts to 640 to 770 operating shifts annually, or 4200 to 5200 operating hours for a 21 shifts per week operation. Allowing for reasonable delays, the operating hours per shift range from 6.50 to 6.75. Specific availability and utilization data for both shovels and haulage trucks depend on calculation methods and operating and maintenance practices used at a particular operation.

The formula for annual operating hours is:

$$\text{Operating hr} = \text{possible annual shifts} \times \%\ \text{availability} \times \%\ \text{utilization of available} \times \text{hr per shift.}$$

A minimum of three shovels is required to achieve a fleet utilization of 65% with two units operating. Additional units may be needed for reasons other than productive capacity such as to achieve proper ore blends or sustain production continuity. For instance, if experience shows that two units are often down concurrently, and two units are needed to achieve production goals, then four units are needed. A consistent need for three operating units usually requires a five-unit fleet. The shovel fleet should have somewhat greater capacity than the associated haulage truck fleet to avoid bottlenecks.

Shovel productivity, tons per hour, is determined by dipper size, swing time, truck box size, and truck spotting conditions, as shown in the following equation:

$$\text{t (tons)/hr} = \frac{60\text{ min}}{\left[\dfrac{\text{truck box m}^3\text{ (cu yd)}}{\text{shovel dipper m}^3\text{ (cu yd)}} \times \begin{matrix}\text{shovel}\\\text{swing}\\\text{time}\end{matrix}\right] + \begin{matrix}\text{truck}\\\text{spot}\\\text{time}\end{matrix}} \times \begin{matrix}\text{truck box}\\\text{t (tons)}\end{matrix}$$

Shovel dipper m^3 (cu yd) (effective) = dipper struck capacity × dipper fill factor.

Truck box m^3 (cu yd) (effective) = truck struck capacity × box fill factor.

The shovel swing time is determined by the machine operating characteristics, material type, and swing arc to the truck. The effective dipper and truck box sizes are functions of equipment design, material type, and loading practices. The spot time factor normally varies between truck-shovel configuration and single or double spot loading conditions. In determining the number of swings to fill a truck [truck box m^3 (cu yd) divided by shovel dipper m^3 (cu yd)], it is advisable to round up fractional swings greater than 0.10 to a full swing.

In well-blasted, free-flowing material, the minimum shovel swing time will be 30 to 35 sec and the dipper fill factor will approach 100%. Allowing for minimal delays, an attainable target shovel productivity approaches 1185 t/dipper m^3 (1000 st per cu yd) per shift. This rule of thumb does not apply to coal or other lightweight materials. Even with good estimating methods, achieving planned shovel productivity in practice is heavily dependent on having adequate truck coverage to minimize shovel delays.

The methodology of shovel evaluation is illustrated by an example developed in Tables 1 to 3. It can be applied to front-end loaders, backhoes, and draglines. Table 1 presents representative cost and productivity data for a wide range of electric shovel sizes. These data exclude maintenance facilities and support equipment and may vary between companies depending on quoted prices, operating experience, and accounting practices.

The productivity data from Table 1 are used in conjunction with assumed availability and utilization in Table 2 to determine the number of units for the indicated shovel sizes to handle 31 750 000 t/yr (35,000,000 stpy) in 350 operating days. Table 2 shows the formula used to calculate the required units. The fractional units required are rounded up if greater than 0.10 units. In actual practice, estimating the availability, utilization, and productivity will depend on the specific nature of the operation being evaluated.

Table 1. Electric Shovel Data

Shovel size, m^3 (cu yd)	Capital cost, $ *	Oper. & maint. cost, $/hr †	Metric tons (short tons) per hr
6 (8)	1,000,000	40- 80	725-1100 (800-1200)
9 (12)	1,300,000	40-110	900-1600 (1000-1800)
11 (15)	1,500,000	40-140	900-2040 (1000-2250)
15 (20)	2,100,000	40-140	1100-2720 (1200-3000)
19 (25)	3,000,000	40-140	1100-3400 (1200-3750)

* Includes freight, sales tax, and erection charges.

† Includes direct labor, operating supplies, lubrication, electric power, teeth and adaptors, hoist cables, and maintenance costs.

Table 2. Shovel Evaluation—Fleet Units Required

31 750 000 t/yr (35,000,000 stpy)

Annual operating hr (76% availability, 85% utilization of available units)

Annual possible shifts × **%A** × **%U/A** × hr/shift = operating hr

350 days × 3 shifts/day × 0.76 × 0.85 × 6.65 hr/shift = 4500 hr

Shovel size m^3 (cu yd)	Metric tons (short tons) per hr	Fleet units required
9 (12)	1180 (1300)	6
11 (15)	1450 (1600)	5
15 (20)	1815 (2000)	4
19 (25)	2175 (2400)	4

$$\text{Fleet units required} = \frac{\text{annual metric tons (short tons)}}{\text{annual operating hr} \times \text{t/h (stph)/unit}}$$

$$\text{For 9 m}^3\text{ (12 cu yd): } \frac{31\,750\,000\ (35{,}000{,}000)}{4500 \times 1180\ (1300)} = 5.98 \text{ or } 6$$

Table 3. Shovel Evaluation—Economic Analysis

Shovel size, m^3 (cu yd)	No. of units	Capital (CC) cost, \$ (000)	Depr. (D) St. line 20 yr, \$ (000)	t/hr (stph)	Ann. oper. hr. 31.75 MM t (35 MM st)	Op. & maint. cost, \$/hr	Ann. op. & maint. cost (AOMC), \$ (000)	Ann. cash flow cost* (ACFC), \$(000)	Present value net cash flow for cost @ 12% 20 yr. (PVNCFC)† \$ (000)
9 (12)	6	7,800	390	1180 (1,300)	26 923	75	2,019	863	14,245
11 (15)	5	7,000	350	1450 (1,600)	21 875	80	1,750	742	12,542
15 (20)	4	8,400	420	1815 (2,000)	17 500	85	1,488	572	12,672
19 (25)	4	12,000	600	2175 (2,400)	14 583	90	1,312	394	14,947

* ACFC = (AOMC + D) (1 − tax rate (0.48)) − D
† PVNCFC = CC + ACFC × PVF 12%, 20 years

Shovel Economic Analysis

Table 3 develops the shovel evaluation economic analysis from Tables 1 and 2. The costs are evaluated on an aftertax cash flow basis, assuming a 20-year economic life. The pertinent calculation formulas are shown in the table. The resulting data show that 11-m^3 (15-cu yd) and 15-m^3 (20 cu yd) shovels are closely competitive. A firm choice has to await the haulage truck evaluation because of the effect loading time has on hauling costs. Using the 15-m^3 (20-cu yd) unit will yield lower hauling costs compared to the 11-m^3 (15-cu yd) unit.

For simplification of the example, the effects of electric power sources, maintenance facilities, and other support items have been deleted. They would show up in capital, operating, and maintenance costs.

HAULAGE TRUCK EVALUATION

Haulage Truck Utilization and Productivity

Having narrowed the choice of candidates for the shovel fleet, the selection of haulage trucks can proceed. As with shovels, truck evaluation starts with utilization and productivity.

The typical availability pattern for a haulage truck is shown in Fig. 1. During the first two years of life, availability will generally average 75% and may range up to 85 to 90%. It can be expected to decrease at a rate of up to two percentage points per year over the remainder of a nominal ten-year life. In general, maximum utilization is 85 to 95% of available and operating hours per shift approach seven. At 80% availability, 90% utilization of available, and 7 hr per shift, 5300 operating hr can be achieved annually.

Haulage truck productivity is generally assessed in terms of tons per hour. Tons per hour are payload carried per trip times the trips per hour. Because of the variables involved, estimating haulage truck productivity is probably the most complex portion of shovel and haulage truck evaluation.

The truck factor or payload is based on the box size, material type, and loading conditions. If the truck factor is stated in dry tons (short tons), then the moisture content must be known because the water weight being carried will affect the truck's travel time performance; the heavier the gross payload, the slower the speed. Because the truck payload factor has a critical influence on production reporting and mining plan monitoring in actual operations, it should be checked on a continuing basis and raised or lowered as needed. Two methods which have proven reliable are weighing and mill tonnage vs. mine tonnage balance.

Trips per hour are a function of the round trip cycle time. Over the years, many methods have been used to arrive at a satisfactory cycle time estimate. Computer methods have proven to be popular and efficient. Both probabilistic and deterministic approaches are used. The author's preference is for the deterministic approach because of greater adaptability in handling different shovel-truck configurations, cost effectiveness, and easier application to operations performance monitoring. If an in-house profile simulator is unavailable, the haulage truck suppliers being considered will often analyze profiles as a customer service. For flexibility, convenience, and operations performance monitoring, an in-house computer profile simulator is advisable.

Whether manual or computer methods are used, estimating the cycle time involves the following time elements: (1) loading, (2) traveling loaded to the dump, (3) dumping, (4) returning empty to the shovel, (5)

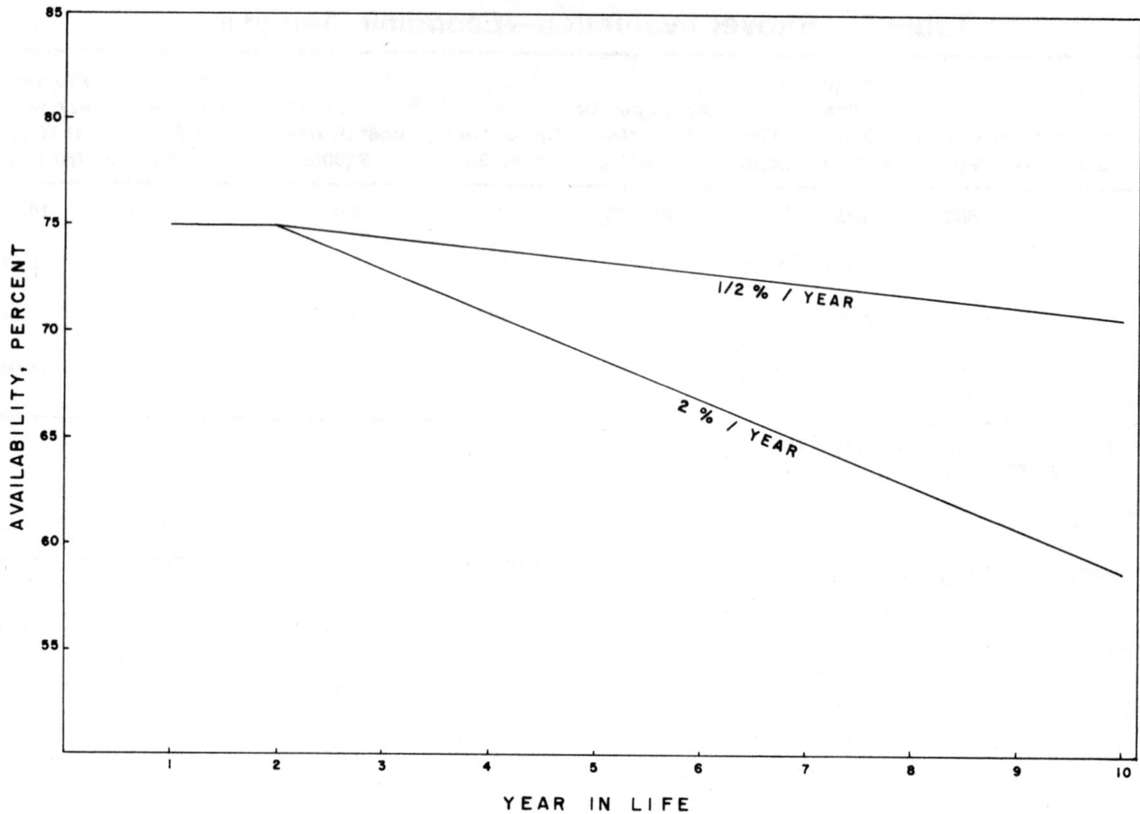

Fig. 1. Haulage truck availability.

spotting at the shovel, and (6) delay time. The cycle time is the sum of these elements.

The nontravel elements are constants for a given shovel-truck combination. The total dumping, spotting, and delay time will generally be in the range of 4 to 6 min per cycle. The loading time uses the following formula:

$$\text{load time} = \frac{\begin{array}{c}\text{truck box m}^3\\ \text{(cu yd)}\end{array}}{\begin{array}{c}\text{shovel dipper m}^3\\ \text{(cu yd)}\end{array}} \times \text{shovel swing time.}$$

The travel times are based on calculated interactions between the length and grade of haulage profile segments, the haulage truck performance, and dynamic braking charts supplied by the manufacturer, and speed limits dictated by operating conditions. The principal advantage of the computer over manual methods, aside from time-saving, is its ability to handle many different profiles, performance, and dynamic braking curves, and to use these data to account for the effects of acceleration and deceleration on travel time.

Manual methods of calculating travel times use the manufacturer nomograph shown in Figs. 2 and 3 to estimate travel speeds on various profile segments. Entering in a truck's gross weight and the grade and rolling resistance of a profile segment into the nomograph will yield the maximum steady state truck speed for the segment, as shown in the figures.

For example, the speed for a truck having a gross weight of 181 000 kg (400,000 lb) on a +8% ramp with a 2% rolling resistance would be 13.2 km/hr (8.2 mph). (See Fig. 2). With the same configuration, but traveling downhill loaded, the speed would be 2.9 or 6.6 km/h (1.8 or 4.1 mph), depending on the dynamic braking option (Fig. 3). On downhill hauls or for trucks returning empty, speed limits may supersede higher performance curve speeds.

When using a computerized simulator, the basic data for the various shovels and trucks to be simulated are stored in a standards file. This file contains the dipper size and swing time for various shovels. Performance and dynamic braking curves, dump, spot and delay times, truck box capacity, and empty and loaded weights are provided for haulage trucks. Also, speed limits and rolling resistance are supplied.

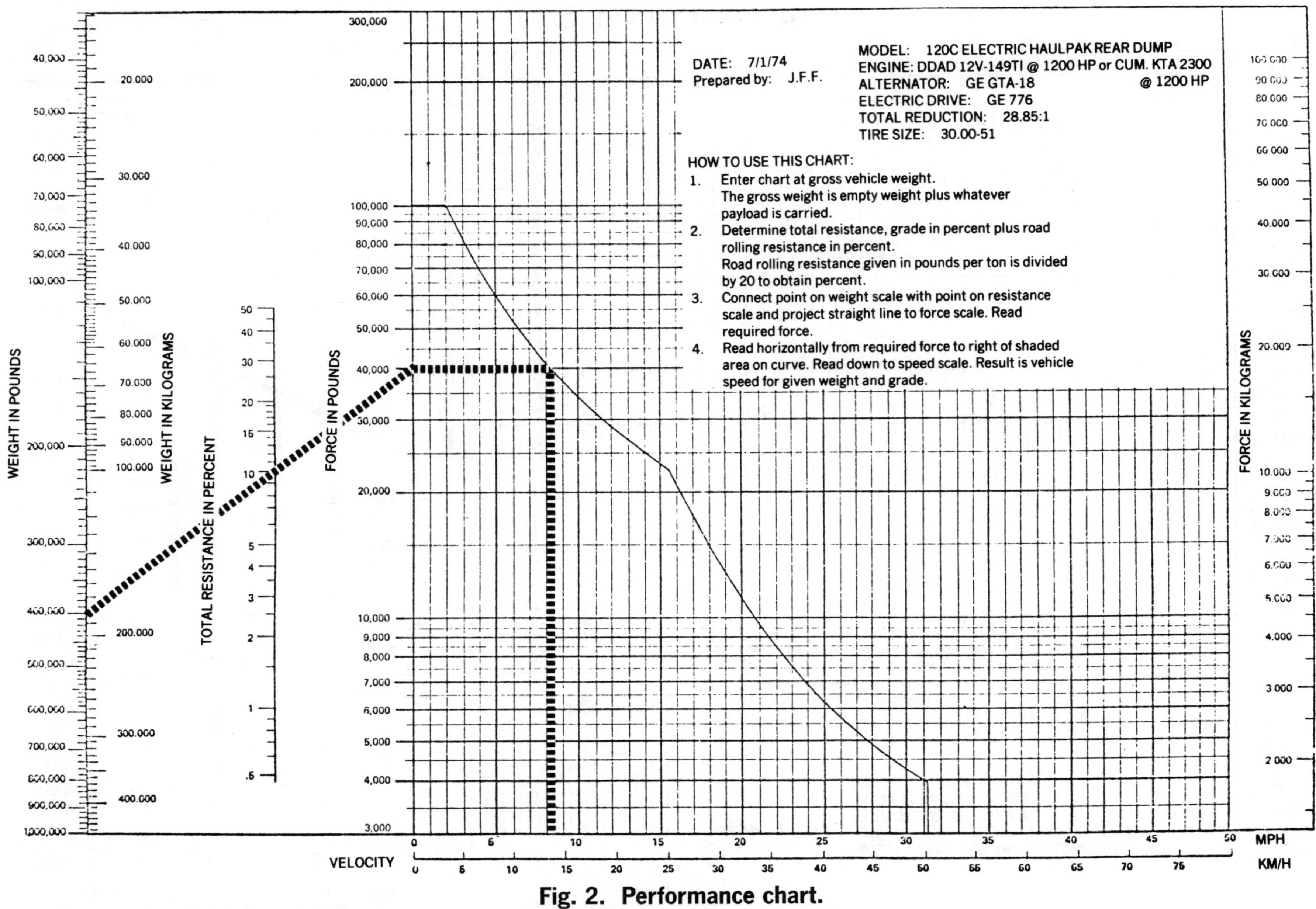

Fig. 2. Performance chart.

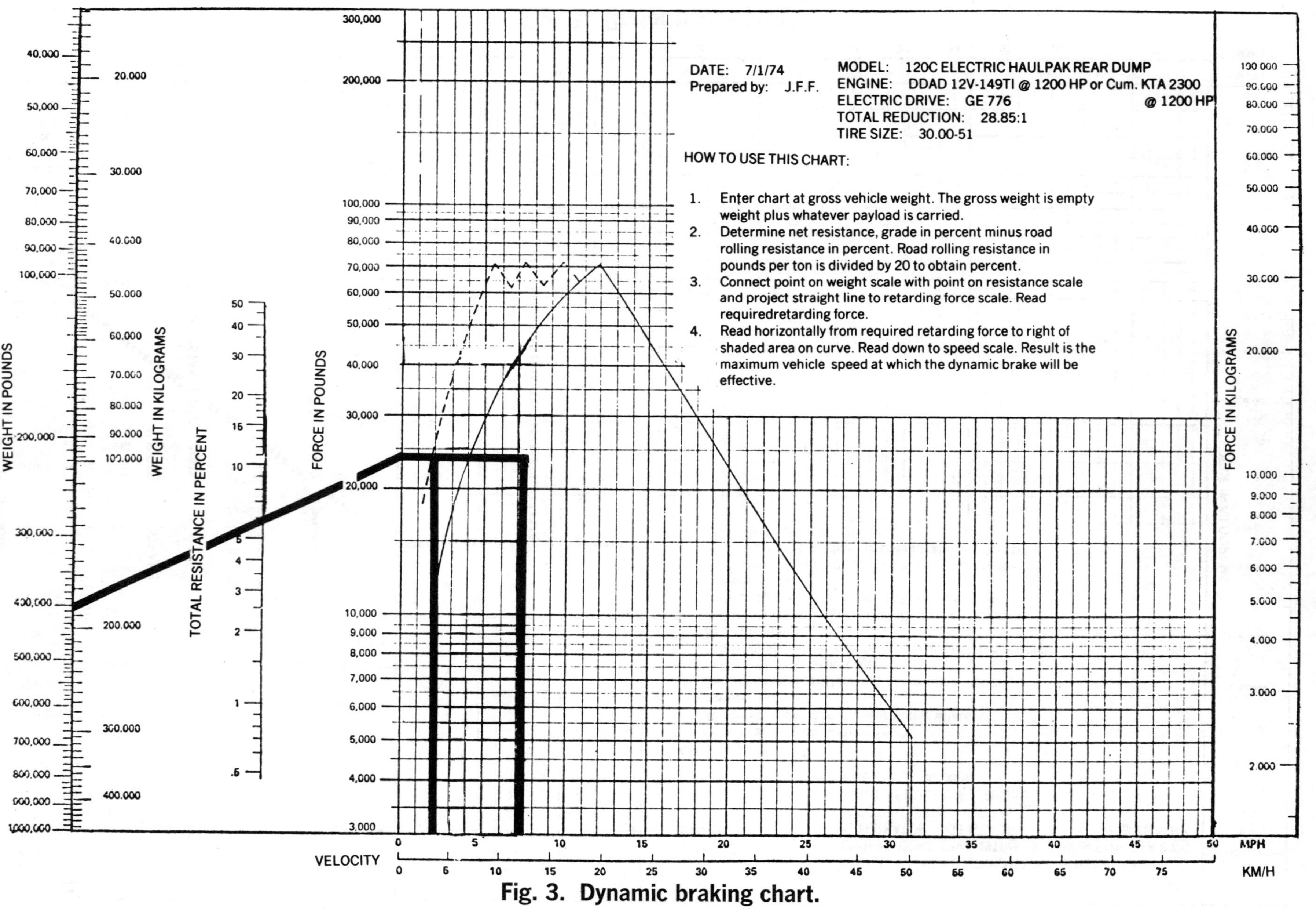

Fig. 3. Dynamic braking chart.

The shovel-profile descriptions are input to the simulator. Drawing on the standards file, the computer calculates and provides output to each shovel-profile combination. Table 4 shows a typical output. The key data are the simulated cycle time and the tons (short tons) per hour. As can be seen, many different types of trucks can be simulated simultaneously and the data presented in a compact format. The potential effect of truck size on shovel productivity is readily discernible. The estimated truck coverage for the shovel is a helpful operating assist for production scheduling.

It is necessary to compare the simulation results with actual operating data when possible. Significant portions of the simulations are based on theoretical data, i.e., performance and dynamic braking curves and unpredictable variations in operating conditions, such as adverse weather, go unsimulated. This means deriving

Table 4. Haulage Truck and Shovel Effectiveness Report Shovel—Truck Assignment Sheet*

Profile no. 4-85, 3-24-77	Bank cond	Op. hr avail shovels	Time trucks	Spot cond.	Truck fleet	Load cap.	Cycle time	per truck Trips/ hr	per truck t/h	per shovel Load time	per shovel t/h	Trucks to assign
	M	60.0	60.0	2	1	65	27.48	2.2	141	2.24	1566	11.0
	M	60.0	60.0	2	2	67	27.94	2.1	143	2.24	1614	11.2
	M	60.0	60.0	2	3	67	29.17	2.1	137	2.24	1614	11.7
	M	60.0	60.0	2	4	100	29.94	2.0	200	3.36	1662	8.3
	M	60.0	60.0	2	5	100	27.75	2.2	216	3.36	1662	7.7
	M	60.0	60.0	2	7	100	30.24	2.0	198	3.36	1662	8.4
	M	60.0	60.0	2	8	100	31.15	1.9	192	3.36	1662	8.6
	M	60.0	60.0	2	9	100	28.40	2.1	211	3.36	1662	7.9
	M	60.0	60.0	2	10	100	28.11	2.1	213	3.36	1662	7.8
	M	60.0	60.0	2	18	120	30.40	2.0	236	3.92	1726	7.3
	M	60.0	60.0	2	19	150	30.50	2.0	295	5.04	1701	5.8
	M	60.0	60.0	2	20	120	30.57	2.0	235	3.92	1726	7.3
	M	60.0	60.0	2	21	150	29.94	2.0	300	5.04	1701	5.7
	M	60.0	60.0	2	22	150	29.83	2.0	301	5.04	1701	5.6
	M	60.0	60.0	2	23	113	29.88	2.0	226	3.92	1625	7.2
	M	60.0	60.0	2	24	113	29.98	2.0	226	3.92	1625	7.2
	M	60.0	60.0	2	25	100	29.15	2.1	205	3.36	1662	8.1
	M	60.0	60.0	2	26	100	26.88	2.2	223	3.36	1662	7.4

Shovel 4 is in Shovel Fleet 2

Profile no	Date	Leg	Horiz. distance	Slope distance	Grade	Rolling resistance	Going Speed limit	Going End speed	Return Speed limit	Return End speed
4-85	3-24-77	1	750	750	0.0	3.50	25	25	25	0
		2	2250	2258	8.89	3.50	25	25	15	15
		3	2020	2026	7.92	3.50	25	25	18	18
		4	470	471	8.51	3.50	25	25	15	15
		5	850	853	9.41	3.50	25	25	15	15
		6	275	275	0.0	3.50	25	25	25	15
		7	600	601	7.50	3.50	25	25	18	18
		8	1250	1250	−0.64	3.50	22	22	25	25
		9	850	850	−2.00	3.50	22	22	25	25
		10	1200	1200	3.33	3.50	25	25	25	25
		11	650	650	2.31	3.50	25	0	25	25

Type 1 Shovel elev. = 6385; dump elev = 6940; slope dist = 11 188; pos. foot lift = 580; neg. foot lift = 25.
Total horizontal distance of legs with positive grade = 8040.
Total horizontal distance of legs with negative grade = 2100.
Distances and lifts are in feet and for loaded truck only.

Table 5. Haulage Truck Operating Performance Report Period Beginning 12/01/76 Ending 12/31/76

Fleet No	Truck No	MTL	Profile Sh	Profile Dn	Total trips	Total tons*	Total oper hr	Total miles	Tph	Oper hr/shft	Truck % effective	Average Slope dist	Average Postv. lift	Average Postv. grade	Average Negtv. lift	Average Negtv. grade	Ton-miles per hr	Work Op. mile-ton /opr hr
07	463	O	04	06	8	560	3.5	16	160	6.57	84.86	10,728	259	7.55	266	4.00	320	27
07	463	O			8	560	3.5	16	160	6.57	84.86	10,728	259	7.55	266	4.00	320	27
07	463				14	980	5.3	24	185	6.52	89.81	9,026	275	7.44	161	3.74	317	26
07	464	L	04	59	5	350	4.0	9	88	7.10	46.92	9,221	435	6.29	20	1.24	158	13
07	464	L	04	69	20	1,400	7.0	26	200	7.00	84.48	6,757	297	7.32	22	1.82	260	21
07	464	L			25	1,750	11.0	34	159	7.04	70.82	7,250	325	7.01	22	1.67	216	18
07	464				25	1,750	11.0	34	159	7.04	70.82	7,250	325	7.01	22	1.67	216	18
07	465	L	04	59	4	280	2.4	7	117	6.47	63.01	9,221	435	6.29	20	1.24	204	17
07	465	L	04	69	24	1,680	7.9	31	213	6.61	89.45	6,757	297	7.32	22	1.82	275	22
07	465	L			28	1,960	10.3	38	190	6.57	83.34	7,109	317	7.09	22	1.71	258	21
07	465	O	04	06	28	1,960	12.7	58	154	6.96	82.52	10,912	264	7.37	264	4.06	320	27
07	465	O			28	1,960	12.7	58	154	6.96	82.52	10,912	264	7.37	264	4.06	320	27
07	465	W	04	73	16	1,120	6.3	29	178	6.69	97.06	9,421	443	6.22	20	1.24	322	27
07	465	W			16	1,120	6.3	29	178	6.69	97.06	9,421	443	6.22	20	1.24	322	27
07	465				72	5,040	29.3	124	172	6.76	85.93	9,102	324	6.88	116	3.42	296	25
07	466	L	04	59	4	280	2.6	7	108	7.56	58.13	9,221	435	6.29	20	1.24	188	16
07	466	L	04	69	4	280	1.9	5	147	7.98	62.28	6,757	297	7.32	22	1.82	184	15
07	466	L			8	560	4.5	12	124	7.75	59.89	7,989	366	6.67	21	1.49	187	16
07	466	O	04	06	1	70	0.6	2	117	8.00	61.39	10,728	259	7.55	266	4.00	233	20
07	466	O			1	70	0.6	2	117	8.00	61.39	10,728	259	7.55	266	4.00	233	20
07	466				9	630	5.1	14	124	7.77	60.07	8,293	354	6.73	48	2.42	192	16
07		L	04	59	32	2,240	15.5	56	145	6.94	77.66	9,221	435	6.29	20	1.24	253	21
07		L	04	69	54	3,780	18.6	69	203	6.86	85.77	6,757	297	7.32	22	1.82	260	21
07		L			86	6,020	34.1	125	177	6.89	82.09	7,674	348	6.80	21	1.56	257	21
07		O	04	06	37	2,590	16.7	76	155	6.86	82.24	10,867	263	7.41	265	4.05	319	27
07		O			37	2,590	16.7	76	155	6.86	82.24	10,867	263	7.41	265	4.05	319	27
07		W	04	73	16	1,120	6.3	29	178	6.69	97.06	9,421	443	6.22	20	1.24	322	27
07		W			16	1,120	6.3	29	178	6.69	97.06	9,421	443	6.22	20	1.24	322	27
07					139	9,730	57.1	230	170	6.86	83.78	8,725	337	6.82	86	3.10	282	23
08	301	L	04	59	6	642	2.8	10	229	7.47	85.76	9,221	435	6.29	20	1.24	382	34
08	301	L	04	69	140	14,980	53.1	181	282	6.91	82.53	6,826	297	7.14	21	1.80	365	30
08	301	L	05	59	83	8,881	37.4	147	237	7.05	84.74	9,348	437	6.12	30	1.97	421	36

*Metric equivalents: 1 st × 0.907 184 7 = t; 1 mile × 1.609 344 = km.

cycle time adjustment factors to adjust simulation data to approximate actual average operating conditions more closely.

When operating data are available, an operations performance monitoring system can be used to derive the cycle time adjustment factors. Actual operating profiles are simulated and the resulting data are compared to actual operating results. The type of report shown in Table 5 provides the adjusting factors (truck percent effective) and serves as a valuable operating management tool. Engineering judgment must be used to derive the adjustment factor if such actual operating data are unavailable.

The truck percent effectiveness is the cycle time adjustment factor as shown in the following formulas:

$$\text{truck effectiveness (\%)} = \frac{\text{actual trips} \times \text{sim. cycle time}}{\text{actual operating hr}} \times 100$$

$$\text{adjusted simulated cycle time} = \frac{\text{simulated cycle time}}{\text{truck effectiveness}}.$$

An ongoing operating performance monitoring system provides continual monitoring for operating management. In initiating such a system, it is desirable to set the simulation standards so that approximately 100% effectiveness will be achieved under average operating conditions. The standards file should remain undisturbed except for known physical changes in the equipment, such as payload, engine size, etc., or in operating conditions, i.e., speed limits. Fluctuations in effectiveness will then account for random performance variations caused by weather, road conditions, deteriorating truck performance, etc. The standards file should be reviewed every two to three years.

Table 6 contains haulage truck data to be used in the example illustrating haulage truck evaluation. This evaluation will be developed in Tables 7 to 9. The qualifying comments regarding similar shovel data also apply to these data.

Table 7 develops the estimated productivity, tons per hour, for 91 to 227-t (100 to 250-st) haulage trucks working with 9 to 19-m³ (12 to 25-cu yd) shovels. The compositing of ore and waste profile cycle times and the use of the truck effectiveness factor are illustrated; appropriate formulas for calculations are shown. If necessary, there could be several ore and waste profiles simulated by level and/or dumps and composited by the weighting method shown.

The resulting tons per hour account for the effect of truck size and shovel size. As each truck size is loaded by a progressively larger shovel, the cycle time is reduced, accompanied by a corresponding improvement in productivity.

Table 6. Diesel-Electric Drive Haulage Truck Data

Truck size, t (st)	Capital cost, $*	Oper. & maint. cost, $/hr†
77(85)	365,000	25 - 50
91(100)	490,000	25 - 60
109(120)	530,000	25 - 60
136(150)	600,000	25 - 75
154(170)	605,000	25 - 75
181(200)	1,100,000	35 - 90
227(250)	1,125,000	35 - 90

* Includes freight, sales tax, and erection charges.
† Includes direct labor, operating supplies, fuel, lubrication, tires, and maintenance costs.

Table 8 shows how the number of fleet units required is derived from Table 7 and the estimated availability and utilization. Fractional units required are rounded up if greater than 0.10 units. As a rule of thumb, the number of units in the fleet should not drop below ten without sound justification because excess productive capacity may be tied up in one unit. With too few units in the fleet, it may be difficult to make up production caused by erratic availability.

Haulage Truck Economic Analysis

Table 9 shows the haulage truck economic analysis derived from Tables 6 to 8. The present value calculation is extended to 20 years to be compatible with the economic life of the shovels. The figures reflect replacement of the haulage trucks in the tenth year. The costs exclude consideration of maintenance facilities and support equipment for simplicity. The data in Table 9 clearly favor the 154-t (170-ton) truck regardless of shovel size.

COMBINED EVALUATION

Table 10 combines the present value net cash flow cost data for the shovels and trucks to determine the optimum shovel-truck combination to meet the required production and haulage profiles. The 15-m³ (20-cu yd) shovels and the 154-t (170-st) haulage trucks are preferred. The cost benefits to the haulage trucks favor the 15-m³ (20-cu yd) shovel over the 11-m³ (15-cu yd) unit. In this combination, four 15-m³

Table 7. Haulage Truck Evaluation - Productivity

Simulated cycle time with 9 m³ (12 cu yd) shovel (min)	Truck size					
	91 t (100 st)	109 t (120 st)	136 t (150 st)	154 t (170 st)	181 t (200 st)	227 t (250 st)
Ore - 11 110 000 t (12, 750, 000 st) annually; haul distance: 2259 m (7412 ft); positive lift: 102 m (335 ft); negative lift: 0 m (0 ft).						
	20.34	21.84	22.21	23.01	24.19	25.94
Waste - 20 640 000 t (22, 750, 000 st) annually; haul distance: 3191 m (10,470 ft); positive lift: 112 m (366 ft); negative lift: 2 m (6 ft).						
	23.43	25.11	25.30	26.10	27.28	29.03
Composite	22.35	23.97	24.22	25.02	26.20	27.95
Composite Cycle Time $= \frac{\Sigma \text{ cycle time} \times \text{t (st)}}{\Sigma \text{ t (st)}}$						
Productivity						
9 m³ (12 cu yd):						
Cycle time (min.)	22.35	23.97	24.22	25.02	26.20	27.95
t (st)/hr @ 90% Efficiency	220(242)	245(270)	303(334)	333(367)	374(412)	438(483)
11 m³ (15 cu yd):						
Cycle time	21.80	23.42	23.12	23.92	25.10	26.30
t (st)/hr	225(248)	251(277)	318(350)	348(384)	390(430)	465(513)
15 m³ (20 cu yd):						
Cycle time	21.25	22.88	22.57	22.82	23.45	24.65
t (st)/hr	230(254)	257(283)	326(259)	365(402)	418(461)	497(548)
19 m³ (25 cu yd):						
Cycle time	20.70	22.32	22.02	22.27	22.90	23.55
t (st)/hr	237(261)	263(290)	334(368)	374(412)	428(472)	520(573)

$$\text{t (st)/hr} = \frac{60 \text{ min}}{\text{cycle time/eff.}} \times \text{truck payload factor}$$

For 91-t (100-st) trucks with 9-m³ (12-cu yd) shovel: $\frac{60 \text{ min}}{22.35 \text{ min}/0.90} \times 91 \text{ t (100 st)}$
$= 220$ t (242 st)/hr

(20-cu yd) shovels and eighteen 154-t (170-st) haulage trucks are needed according to the data used.

SUMMARY

To summarize, this chapter has presented reliable, comprehensive methods for evaluating shovel and haulage truck selection on a combined basis. The strength of the methods lies in the cash flow analysis basis. It should be noted, however, that in a specific situation, some details may differ. Maintenance facilities and support equipment must be considered. It may be appropriate to include indirect labor costs. While the cash flow analysis endeavors to deal with true economic costs, varying accounting practices followed by different companies may make it difficult to ascertain these costs and therefore alter the results. While the examples have dealt with shovel and haulage truck fleets of uni-

Table 8. Haulage Truck Evaluation—Fleet Units Required

31 750 000 t/yr (35, 000, 000 stpy)

Annual operating hr (72% availability, 95% utilization/available unit) = 5000 hr.

Truck size, t (st)	Shovel size: 9 m³ (12 cu yd)	11 m³ (15 cu yd)	15 m³ (20 cu yd)	19 m³ (25 cu yd)
91(100)	29	29	28	27
109(120)	26	26	25	25
136(150)	21	20	20	19
154(170)	19	19	18	17
181(200)	17	17	16	15
227(250)	15	14	13	13

$$\text{Fleet units required} = \frac{\text{annual t (st)}}{\text{annual operating hr} \times \text{t (st)/hr/unit}}$$

For 91-t (100-st) truck with 9-m³ (12-cu yd) shovel:

$$\frac{31\,750\,000 \text{ t } (35,000,000 \text{ st})}{5000 \text{ hr} \times 220 \text{ t } (242 \text{ st})/\text{hr}} = 28.92 \text{ or } 29$$

Table 9. Haulage Truck Evaluation—Economic Analysis

Truck size t (st)	No. of units	Capital (CC) cost, $(000)	Depr. (D) st. line, 10 yr, $(000)	t(st)/hr	Ann. oper. hr 31.75MM t (35MM st)	Oper. & maint. cost, $/hr	Ann. oper. and maint. cost (AOMC), $(000)	Ann. cash flow cost (ACFC), $(000)	Present value net cash flow cost @ 12%, 10 yr (PVNCFC), $(000)	PVNCFC @ 12%, 20 yr, $(000)
109 (120)										
9 m³ (12 cu yd)	26	13,780	1,378	245(270)	129,630	40	5,185	2,035	25,277	33,416
11 m³ (15 cu yd)	26	13,780	1,378	251(277)	126,354	40	5,054	1,967	24,892	32,907
15 m³ (20 cu yd)	25	13,250	1,325	257(283)	123,675	40	4,947	1,936	24,191	31,980
19 m³ (25 cu yd)	25	13,250	1,325	263(290)	120,690	40	4,828	1,874	23,841	31,517
136 (150)										
9 m³ (12 cu yd)	21	12,600	1,260	303(334)	104,690	50	5,235	2,117	24,562	32,471
11 m³ (15 cu yd)	20	12,000	1,200	318(350)	100,000	50	5,000	2,024	23,436	30,982
15 m³ (20 cu yd)	20	12,000	1,200	326(359)	97,493	50	4,875	1,959	23,068	30,495
19 m³ (25 cu yd)	19	11,400	1,140	334(368)	95,109	50	4,755	1,926	22,280	29,454
154 (170)										
9 m³ (12 cu yd)	19	11,495	1,150	333(367)	95,368	55	5,245	2,176	23,789	31,448
11 m³ (15 cu yd)	19	11,495	1,150	348(384)	91,146	55	5,013	2,055	23,106	30,546
15 m³ (20 cu yd)	18	10,890	1,089	365(402)	87,065	55	4,789	1,967	22,006	29,091
19 m³ (25 cu yd)	17	10,285	1,029	374(412)	84,951	55	4,672	1,936	21,223	28,057
181 (200)										
9 m³ (12 cu yd)	17	18,700	1,870	374(412)	84,951	65	5,522	1,974	29,852	39,464
11 m³ (15 cu yd)	17	18,700	1,870	390(430)	81,395	65	5,291	1,854	29,173	38,566
15 m³ (20 cu yd)	16	17,600	1,760	418(461)	75,922	65	4,935	1,721	27,326	36,124
19 m³ (25 cu yd)	15	16,500	1,650	428(472)	74,153	65	4,820	1,714	26,187	34,618

Table 10. Combined Shovel-Haulage Truck Evaluation Summary

Present value net cash flow @ 12%, 20 yr	Truck size					
	91 t (100 st)	109 t (120 st)	136 t (150 st)	154 t (170 st)	181 t (200 st)	227 t (250 st)
Shovel size:						
9 m³ (12 cu yd)	$47,596,389	$47,660,707	$46,715,395	$45,692,450	$53,708,283	$50,204,740
11 m³ (15 cu yd)	45,418,433	45,449,445	43,524,160	43,088,236	51,108,221	46,266,317
15 m³ (20 cu yd)	44,622,557	44,652,050	43,166,734	*41,762,991*	48,796,135	44,127,286
19 m³ (25 cu yd)	45,922,960	46,463,088	44,400,486	43,003,309	49,564,532	45,644,334

$PVNCFC_{Comb.} = PVNCFC_{Shovel} + PVNCFC_{Truck}$

form unit size, the methods described can be applied equally well to evaluate situations requiring fleets of mixed unit size.

Many operating and cost parameters normally vary between years. Production requirements, hauling routes, operating conditions, and equipment aging will affect equipment productive capacity and fleet unit needs. Operating conditions, equipment aging, and inflation may cause progressively increasing hourly operating and maintenance costs.

20 Maintenance and Ancillary Facilities

Donald C. Myntti
AMAX Coal Co.

Donald C. Myntti is the director of maintenance engineering for AMAX Coal Co. He has had over 30 years experience in heavy open pit mining equipment maintenance. He has been associated with the Oliver Iron Mining Div., Western Contracting Corp., Hawthorne Machinery Co., and Kennecott Copper Corp.

Mr. Myntti received his BS degree in mechanical engineering from the University of Minnesota. He is a member of SME-AIME and is a registered professional engineer in Minnesota and Utah.

INTRODUCTION

A major segment in a successful heavy equipment maintenance and repair program is the provision of well-laid out and well-equipped shop and service facilities. The facilities described here will be those required for a large, high tonnage hard rock open pit mine utilizing rotary drills, shovels, haulage trucks, support equipment, and service trucks. The term support equipment covers the equipment required to support the drill, shovel, and truck operations; these are crawler tractors, road graders, front-end loaders, water trucks, rollers, rubber-tired tractors, etc. Service trucks are the automotive type consisting of rubber-tired over-the-road equipment.

The primary structures generally involved in the total maintenance effort are as follows: (1) the main shop with an attached warehouse (the structure could also house the tire and lube shops plus the equipment cleaning stalls), (2) a service station with bulk petroleum product storage tanks, (3) a heating plant, (4) a security station (if this is a corporate requirement), (5) operating department and employee assembly buildings, and (6) an electrical substation.

Under certain conditions, many of these facilities with the exception of the substation could be combined in one or several structures. This arrangement could have a great advantage especially in areas where the heating of the structures is extremely critical.

THE MAIN SHOP—WAREHOUSE STRUCTURE

The general arrangement and size of a long-term maintenance shop-warehouse structure must reflect basically the size and quantity of the mobile equipment required to safely move the target quantities of material and the scope of maintenance activity. The shop is usually divided along equipment lines, each with its own features. The primary areas are: haulage truck, support equipment, and service truck, along with the tire, lube, and welding shops, the cleaning area, tool room, and possibly a component rebuild room.

The number of stalls in each area depends largely on the number of units in each of the three fleets and the maintenance and repair demands of the fleets. The ratios of the repair stalls to units of equipment currently being strived for at a large open pit copper mine based on a 21 shift per week work schedule are as follows: (1) 136-t (150-st) haulage trucks: one repair stall for four units, (2) support equipment: one repair stall for 12 units, (3) service trucks: one repair stall for 20 units, (4) tire shop: one stall for 30 haulage trucks, and (5) lube shop: one stall for haulage trucks, one for support equipment. The ratios differ between mines and they undergo continual change as various conditions improve or deteriorate.

The size of the repair stall and therefore the structural skeleton of the building is determined by the length, width, and height of the largest unit of equipment. A 136-t (150-st) haulage truck of the rear-end dump type, for example, requires a stall 13.7 m (45 ft) wide, 19.8 m (65 ft) long. The elevation of the crane rail of the overhead crane must be such that the overhead crane can pass over the elevated bed of a rear dump type haulage truck.

The primary factors for the determination of the location of the shop-warehouse structure are that it be located beyond the ultimate stripping limits of the open pit mine and adjacent to a major haul road.

The location of the main structure is extremely critical because so many designated areas are required in its immediate vicinity. There should be much available space on all four sides of the structure. For each of the equipment fleets, haulage trucks, support equipment, and service trucks, there should be an equipment ready line and an equipment bad order line for units scheduled to be shopped. An area for the exterior storage of certain warehouse stock items would be beneficial. The space requirements for the orderly handling of tires are extensive. Equipment traffic patterns around the shop require considerable space as do the roads to the warehouse receiving dock. Elevated storage docks also require substantial space. An area for the parking of idle or surplus equipment is also needed. The storage of road salt may be required at some northern mines. The employee's parking lot will be discussed later. An equipment tie area is needed nearby for temporary parking of operational equipment during shift change. And one must always keep in mind the provision for the orderly expansion of the facilities should future business conditions warrant such action.

The design of the structures is also affected by many factors such as the range of ambient temperatures, rain and snow conditions, and prevailing winds.

Solar orientation of the main shop structure should be mentioned. It could be of significant importance in northern areas, primarily in the prevention of ice build-up near the overhead and man doors. If the large overhead doors are located on one side of the building, that side should face south so that advantage can be taken of solar radiant heat. Should the large overhead doors be located on both sides of the building, the long axis of the structure then should lie on the north and south line.

Overhead cranes of adequate capacity should be

provided over all shop repair areas. The capacity of each crane is determined by the weight of the heaviest component which is frequently lifted. Radio control of cranes is fairly new and should be seriously considered for new overhead crane installations.

A high percentage of the buildings constructed in recent years has been of the steel prefab insulated type. This type of construction generally offers the lowest possible unit cost because of the standardized modular concept, and it lends itself to easy future expansion.

The selection of the type of overhead door is important more so in areas where high winds are encountered. Insulated, vertical lift, exterior mounted overhead doors equipped with windows perform well. In order to prevent damage to the door frames by equipment moving through the door openings, substantial concrete or pipe barricades should be provided outside of each opening.

One extremely important point to remember in the layout of the facilities is that water flows downhill and that drainage should be away from the buildings and not into them. It seems as if this basic physical law is frequently overlooked. In fact, good positive drainage of surface water from the entire maintenance and warehouse area is required.

It is recommended that exterior concrete aprons be installed along the perimeter of the shop in order to improve surface conditions. The aprons improve housekeeping near the shop and facilitate the completion of running repairs which can be undertaken out of doors.

If the mine and maintenance shops are located at great distances from vendor facilities, serious consideration should be given to the provision of certain machine tools in the shop building in order to have limited manufacturing capabilities and certain welding equipment and apparatus to undertake welding repairs and fabrication. An electrical equipment repair shop which includes power cable repairs may be worthy of investigation. The establishment of a rebuild room for equipment components may be required under these circumstances. This approach would increase the size of the shop, warehouse, and the maintenance crew. The provision of a dynamometer for the run in of overhauled diesel engines should be investigated.

In the day-to-day operation of a mobile equipment maintenance shop, many auxiliary items are required, some of which are: portable hydraulic jacks, welding machines, mobile sweepers, fork lift trucks, hydraulic press, welding rod ovens, hose making machines, coolant tanks, waste oil tanks, scrap and trash containers, glass bead cleaning machines, oxygen-acetylene cutting outfits, containers for oil absorbent, painting equipment, parts washing sinks, bearing heaters, and many special lifting slings and devices. Convenient areas for the proper storage of these units, when not in use, must be provided to prevent undue congestion in the working and passage areas.

Employees' tool boxes require suitable storage space when not in use. The total area required for the storage of tool boxes can be sizable.

The shop-warehouse structure should also include the following facilities: supervisory offices, clerical offices, conference room, training room, copier room, telephone equipment room, computer room, first aid room, and separate lunchrooms, locker rooms, and toilets for the supervisory and hourly groups.

EQUIPMENT CLEANING FACILITY

The equipment cleaning area is an extremely important facility which generally has been overlooked or underdesigned in the past. Cleaning of equipment prior to shopping is required as a very important part of the inspection and repair program. The facility should be provided with high pressure cold water, a steam cleaner, and a proven system for the positive capture of large quantities of mud and water along with a pumping or carrying capability for the disposal of the mud and water. The structure should provide the drive through feature for easy movement of equipment. The cleaning area could be part of the main shop or be in a separate building located convenient to the main shop. In warm areas, open air type cleaning slabs work well. One or more stalls may be required for the cleaning of mobile equipment prior to shopping.

TIRE SHOP

An efficient tire shop operation is dependent on the facilities available to the tire crews. One or more tire handlers consisting of fork lift trucks equipped with special tire handling attachments should be provided to simplify tire work. The size of the tire handler must be determined by the weight of the heaviest inflated, mounted tire assembly and by the outside diameter of the largest tire in use.

Additional facilities required in a tire shop would be: (1) a compressed air system which provides dry air at a minimum pressure of 1033 kPa (150 psi) and which incorporates the use of a super large bore tire valve system, (2) a tire press which safely mounts and demounts tires from rims, (3) a safety inflation cage to be used when the mounted tire is being inflated with compressed air, and (4) mounted tire storage racks.

The size of the tire shop stalls should be determined by the largest unit to be driven in for tire service. Adequate space should be provided for the maneuvering of the tire handlers to each of the tire positions of a rubber-tired unit.

Inasmuch as the tire shop operation usually involves large quantities of tires of different sizes and manufacturers, the space requirements around a tire shop are large. At least seven distinct outside areas are required for the proper flow and the orderly handling of tires: (1) mounted bad order tires which have been removed from equipment, (2) unmounted bad order tires which are to be inspected and categorized, (3) bad order tires which are to be sent to the vendor or recapper, (4) scrap tires awaiting transport to the disposal area, (5) repaired or recapped tires (unmounted), (6) new tires (unmounted), and (7) new, repaired, or recapped tires (mounted).

These remarks on the tire operation are based on the mining company providing the facilities and personnel to undertake the work. If the tire work has been contracted to a tire company, the need to provide company facilities and personnel is obviously eliminated. However, adequate space must still be provided for the tire contractor so that his crews can efficiently perform all of the required tire work.

LUBE SHOP

The lube facilities are a very essential part of a maintenance shop inasmuch as an active lubrication program is the major segment of an equipment preventive maintenance (PM) program. In this area, facilities, such as storage tanks, pumps, piping, meters, and hose reels should be provided to add the following lubes and fluids in accordance with manufacturers' recommendations regarding mobile equipment which is driven into the lube stall: engine oil, hydraulic oil, transmission fluid, gear lube, multipurpose grease, coolant, and compressed air. The use of bulk delivery and storage systems should be encouraged.

The lube stalls can be a part of the main shop or be included in a separate tire/lube building located near the main shop. The size of the lube shop stalls is determined by the size of the largest unit driven in for lube service. An overhead crane is not required in this application.

Diesel engine lube oil samples are to be obtained at specific intervals and at oil change time at the lube shop for analysis purposes. Waste oil evacuation and storage systems are becoming standard equipment at many mines.

It should be pointed out that support equipment such as crawler tractors which work in isolated areas for extended periods of time can be serviced in the field from lube trucks dispatched from the lube shop.

SERVICE TRUCK SHOP

Each mine generally has a large fleet of service truck units and for those units which are to be serviced and repaired at the mine, certain minimum facilities are required. Two and three postfloor hoists are required for the proper servicing and component replacement activity of the usual fleet of two or three axle units.

FIELD SERVICE STATION

The field service station, in order to be most effective, must be located adjacent to the major haul road and be positioned so that the approach and exit can readily accommodate the flow of traffic. In larger mines, it may be necessary to provide more than one service station if a number of haul roads are in active use.

Petroleum products, such as fuel oil, engine oil, hydraulic oil, and coolant, which are dispensed at the service station, should be handled in bulk if the annual usage quantities are adequate to effect economies. This arrangement necessitates delivery of the products generally by over-the-road tank trucks on access roads which must always be passable.

The storage tanks, if the bulk delivery system is used, should have the capacity to provide about one month's usage for each product. The capacity of each storage tank must also reflect the size of the delivery unit.

If the natural terrain permits gravity flow from the bulk storage tanks to the service station, the need for pumps is eliminated. In northern areas, the use of heat tape to warm the distribution pipes greatly assists the flow of the oils in low-ambient temperatures.

The service station crew and facilities can also be utilized to perform periodic lube jobs on rubber-tired mobile equipment which can be scheduled into the station. One additional task that could be undertaken at the service station is the making of daily hot inflation tire checks on rubber-tired mobile equipment. A complete compressed air system is required at the service station to do the lube jobs and hot air checks.

HEATING PLANT

As natural gas and fuel oil become unavailable or too expensive for industrial heating purposes, alternate fuels and systems must be considered.

In areas where heating systems are required, serious consideration should be given at this time to the construction of coal-fired heating plants, if not now in use, in the near future in order to assure a continuous supply of heating fuel.

The space requirements for a coal-fired plant will be substantial in that the coal hopper, coal storage area, heating plant building, ash disposal system, and access roads will require space far in excess of that needed for plants burning natural gas or fuel oil.

The use of coal-fired heating plants will generally require extensive steam and condensate distribution systems between the plants and the buildings in a mine maintenance complex. The distribution system should be properly designed in order to minimize maintenance requirements. Electric heating of maintenance facilities is in use on a limited basis and may be a viable possibility on a more extensive basis.

OPERATING DEPARTMENT BUILDINGS

Operating department buildings house the supervisory and clerical offices, a conference room, a training room, an assembly hall, locker rooms, and toilets and are generally located close to the employees' parking lot in order to facilitate the movement of employees from the parking lot to company facilities.

Another building which could be located near the main gate on the road to the mine is the security or plant protection station. A flagpole and a large attractive sign which shows the name of the company and mine should be provided. The location of the gasoline pump in the close vicinity of the security station would be a good move.

ELECTRICAL INSTALLATION

The electrical installation for the maintenance facility is critical and should be properly engineered by competent electrical engineers who are familiar with this type of facility. The electrical substation which is to serve the maintenance buildings and the nearby facilities obviously must have adequate capacity for the estimated electrical load and for limited future increases in power requirements.

The exterior and interior lighting of the facilities in this day of energy awareness should utilize lamps of the metal arc or high pressure sodium type which provide maximum lumen output per watt. The automatic control of the exterior lights can be accomplished by the use of photoelectric cells so that the lights operate only when needed. One word of caution should be mentioned. Overhead power lines must not be installed in the vicinity of the shop inasmuch as mobile cranes generally make numerous heavy lifts in the shop area.

The shop electrical system in northern areas should be sized to permit the use of electric heaters in the engine blocks of mobile equipment parked outside of the shop and at the ready/bad order lines during extended periods of scheduled downtime during freezing temperatures.

In an area where power outages occur more frequently than can be tolerated, it may be feasible to design into the substation the capability of temporarily hooking up an auxiliary diesel-powered generator set to provide sufficient power to operate lights, overhead doors, the heating plant, and perhaps the air compressor and a few welding machines. Should the buildings be electrically heated, the electrical installation will be much more extensive.

UTILITIES

Adequate utilities must be provided to the maintenance area in order to have a complete facility. The following utilities must be available in the area in order that each structure has its required utilities: electric power, telephone service, culinary water, water for fire protection and for equipment washing, heating plant fuel, sewage systems, and compressed air.

HOUSEKEEPING

Solid safety and maintenance programs are based on good housekeeping in and about all facilities. It begins with the capture of trash and scrap at the point and time of generation in and about the shops and ends with proper disposal. The trash generally can be transported to a landfill area for burial, and the scrap metal should be segregated by type at a nearby area for easy loadout by scrap metal contractors.

LOADING RAMP

Required for the proper and safe handling of large components or assembled units during the receipt or loadout of equipment is a loading ramp which would permit the side or end loading of over-the-road lowboys. If one is fortunate enough to have a railroad spur onto the property, a loading dock permitting the side or end loading onto or off railroad cars would be most advantageous.

PARKING LOT

The employee parking lot for private cars should be located as close to the maintenance buildings and the employee assembly hall as possible. At some northern mines, electrical distribution systems have been installed at the parking lots to provide 110-v current for cold weather use on engine block heaters of private cars. The lots should have easy traffic patterns which facilitate the smooth flow of traffic. The size of the lot should be determined by the sum of the cars used by the day and afternoon shift personnel.

HELICOPTER PAD

A recent refinement at several mines has been the provision of concrete pads for use by helicopters for the delivery of dignitaries or repair parts.

FIRE PROTECTION

To protect the corporate assets from damage by fire, a system of exterior hydrants, interior fireplugs, interior sprinkler systems, and portable fire extinguishers is required. The hydrants and hoses should be protected from the elements by houses and from possible equipment damage by substantial barricades, which must be visible at all times including during winter time in areas of heavy snowfall.

If a fire truck is available, proper storage must be provided, especially in areas where freezing temperatures are encountered. A battery charger should be provided to keep the fire truck battery in a fully charged state at all times.

MAINTENANCE OF FACILITIES

Routine planned maintenance of the facilities provided in the maintenance structures must be undertaken. These items are such major components as overhead cranes, air compressors, heating systems, welding machines, overhead doors, etc., and they must be serviced in accordance with the manufacturer's recommendations. This program requires that the maintenance manuals, parts books, and lube charts for each unit be in the files for ready reference by maintenance personnel. Serious consideration must be given to the establishment of a supervisory position of facility engineer who would engineer and implement repair, modification, and expansion plans for the maintenance facilities after a certain size is reached.

The maintenance facilities should be sturdy and durable in order to withstand the severe service to which they will be subjected. The facilities should require minimum maintenance so that the efforts of the maintenance crews working on mobile equipment will not be diluted by diversion to facility repairs.

DESIGN INPUT

When the design time for maintenance facilities approaches, it is essential that input be obtained from many sources outside of the company in order to assure that adequate services from the outside are available and that adequate accommodations within the buildings are provided.

These people would be representatives from the electrical utility company, telephone company, computer company, the fire insurance carrier, and possibly others. It would be advisable to consult with various governmental agencies such as zoning boards, EPA, OSHA, and perhaps other agencies to determine the minimum legal requirements which must be met.

It would be worth the time and money to send several key people from the mine or division maintenance and engineering departments to inspect existing installations similar to those proposed in order to obtain current, firsthand information from appropriate counterparts on what should and should not be done in the design and construction of maintenance facilities.

Input for layout details from various company departments in addition to maintenance is required to assure that important features which would assist them in carrying out their functions are not overlooked. Some of the functions of the departments are: (1) warehouse supervisors should determine the sizes of the heated and unheated warehouses, the type and amount of shelving and storage bins, size of the exterior storage yard which includes pallet racks, quantity and size of forklift trucks, etc.; (2) representatives from the safety department should give recommendations as to the required fire fighting equipment for the elimination of potential hazards; and (3) the security department is concerned with the orderly control of personnel and vehicles into and from the mine and maintenance areas.

The parameters discussed in this section should apply whether the layout of the maintenance facilities is prepared by an outside consulting firm or by an in-house engineering department.

GENERAL APPEARANCE

For many years I have harbored the impression that many mining companies have deliberately constructed ugly maintenance buildings and, I might add, undersized structures. Most of the shops I have visited over the years have been extremely congested as well as unsightly.

Future mine maintenance structures should be functional, neat, and attractive in appearance. The morale of all employees, both supervisory and hourly, will be higher due to the fact that they will be working in more pleasant surroundings. Several companies in recent years have made great strides in providing attractive maintenance buildings and they are to be commended for their enlightened approach. And there is nothing wrong with having some lawn and shrubs near mine buildings!

LONG-RANGE MAINTENANCE FACILITY PLANS

All of the effort that goes into the layout and design of the original maintenance facilities at a mine is based on providing tailor-made facilities that are to be constructed at the lowest possible capital cost and are adequate to provide the maintenance function for the equipment fleet assigned to the mine. As time passes, changes inevitably take place, and the facilities must be altered and expanded to keep pace with the numerous changes.

Long-range master plans for maintenance facilities are required in the first step to assure that adequate facilities are on hand as equipment fleets change or expand in size or quantities. A master plan, which should be updated annually, covering the next five-year period is an excellent approach. The plan should also specify action to be taken to correct all of the recognized deficiencies in and around the maintenance facilities.

CONCLUSION

The initial maintenance facility arrangement at a new mine will, of course, be a compromise. The problems associated with the requirements of the interested and concerned parties involved in the design, construction, operation, maintenance, and protection of the facilities must be resolved. Should the mobilization of the equipment fleet be scheduled to take place over a period of several years in accordance with an orderly mine development plan, the initial maintenance facility must have the required basic components, i.e., main shop, warehouse, cleaning building, tire shop, and lube shop. As additional equipment is received, the maintenance facilities should be expanded on a timetable which has the facilities in place prior to the receipt of the additional equipment.

Contents

W. A. Hustrulid, editor

INTRODUCTION TO SECTION 4

In the previous three sections, the various elements which enter into mine planning and design have been described in some detail. Overall project planning requires the integration of all of these elements such that the ore body, planned extraction rate, pit geometry, equipment selection, etc., are compatible with one another and as a whole yield the desired return on investment. The engineering, operational, and financial constraints must be thoroughly understood for each element and an overall project optimization obtained by utilizing various feedback or iteration loops.

A demonstration of how this optimization process is applied to a uranium ore body is described in this section.

21 Circular Analysis—Open Pit Optimization

Gerald C. Dohm, Jr.
Minerals Exploration Co.
Union Oil of California

Gerald C. Dohm, Jr. graduated with a B.A. in geology from the University of North Dakota. He went to work for Anaconda in Butte, MT, in various engineering assignments including underground, open pit, and field projects. Transferred to the New Mexico uranium operations in late 1967, his experience there included development drilling and operations supervision, mine design and scheduling, forecasting and financial analyses, and computer applications. Dohm was promoted to chief mine engineer in late 1974 simultaneous with his joining Minerals Exploration Co. as project engineer in charge of the design, development, and licensing of a major uranium property in Wyoming. During this period he established the Mine Development Group and was promoted to development manager in late 1976. He transferred to Tucson, AZ in early 1977. His current responsibilities include property appraisals, submittal evaluations, environmental base-line studies, feasibility studies, and license applications.

INTRODUCTION

After a mining company has discovered a mineral deposit, the problem is then how to mine and process that deposit the best way. The principal problem facing managers or engineers who must decide on mine plant size, equipment selection, and long-range scheduling is how one can optimize a property not only in terms of efficiency but also as to project duration. Mine optimization is dependent upon the interaction of contributing parameters which leads to maximizing the net present value (NPV). Realistic mine planning is the key to the analysis. The process is to be used when evaluating undeveloped properties or when a complete revamping of existing facilities is scheduled. The process is not intended for detailed mine design but rather for property evaluation and feasibility studies.

The derivation of the various input parameters will only be discussed to the point where a meaningful correlation can be made with the overall analysis. The interrelationship of the parameters will be the central point because most company guidelines and financial analysts have their own standards when deriving the various inputs.

Assume that the data is totally acceptable and provides at least the following categories of information and guidelines: (1) mineral and waste inventory; (2) geologic, hydrologic, and geotechnical criteria; (3) topographic layout including property boundaries; (4) metallurgical flowsheet, recovery and design criteria; (5) access, water, power, and environmental base-line study; and (6) financial constraints (minimum rate of return, payback period, etc.).

Any optimization of an open pit-mill complex involves bringing all the parameters together, all of which are interrelated and dependent upon one another, i.e., without knowing the cutoff grade, the ultimate ore reserves cannot be calculated; without the ore reserves, the final pit limits cannot be established; without knowing the overall tonnages, the production schedule cannot be selected; and without required capacity generated capital and operating costs, the cutoff grade and total ore reserves cannot be derived. The optimization process is, in essence, a circular analysis.

The process is one of closing in on the optimum target by going through the circular analysis (Fig. 1) a sufficient number of times whereby no appreciable change is noted after further modification of the contributing parameters. Industry experience is invaluable in shortening the process but even without such a background, the process can be completed if sufficient attention is paid to the correlation of the detail.

Where does one enter the analysis? Start with a rather cursory treatment of equipment, facilities, and production scheduling to establish the basis for computing rough ore reserves, mining techniques, and process design criteria.

Again, what we are trying to do is maximize the net present value (NPV) of a given ore body. All evaluations utilize a discount rate which is equivalent to the cost of capital to a company. The resulting NPV of a mineral property is an extremely functional yardstick by which to compare various investment alternatives. A corroborating conclusion can be arrived at by utilizing the same process where the actual rate of return (ROR) is computed by utilizing the same parameters and cash flows and setting the net present value equal to zero. The comparison is useful when evaluating similar size projects where the actual ROR will probably be the deciding factor.

Maximizing the net present value is predominantly a function of production scheduling and ore cutoff grades. As a general rule of thumb, the sooner the capital write-off occurs, the better the optimization level achieved. Therefore, if the opportunity exists, some initial high-grading of an ore body could greatly enhance the cash flow if the remaining average ore grade is acceptable and/or if one seeks initial low waste to ore ratio areas.

All capital and operating costs can be expressed in the current year's dollars, as financial projections are subject to many uncertainties. It should also be assumed, for the initial evaluation, that the ratio of income to production costs will remain constant throughout the life of the property. Periodic checks should be made throughout the operating life, as modifications of the ore reserves and operational plans may result if significant changes occur in the actual operating costs, commodity value, improved technologies, etc.

For the purpose of developing the sequencing of the evaluation cycle, a hypothetical case example will be presented and evaluated step by step following the general requirements until the optimum level is approached.

Metric Equivalents

Length:
1 in.×25.4=mm
1 in.×2.54=cm
1 ft×0.304 8=m

Weight and Mass:
1 lb×0.453 592 4=kg
1 st×0.907 184 7=t

Volume and Capacity:
1 cu ft×0.028 316 8=m^3
1 cu yd×0.764 554 9=m^3

Area:
1 sq ft×0.092 903 0=m^2

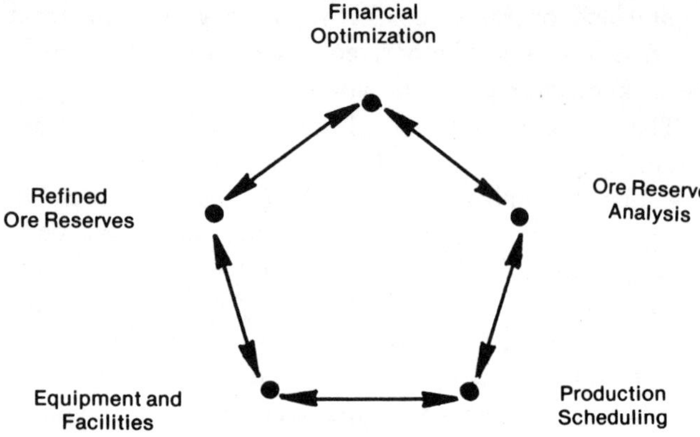

Fig. 1. Circular analysis.

A mining company has discovered and drilled out a large, low-grade, typical sandstone uranium deposit at depths that are sufficiently shallow so that the property is judged to be amenable to surface mining and conventional milling techniques. The area is situated in a rather remote region, and while there are dirt roads into the property, no power, water, or communication lines exist.

The drilling program has reduced the grid to a point, 200-ft centers in this example, sufficient to outline the mineral zone and associated waste and stripping given in Table 1.

A geotechnical study has been completed and has indicated the back slopes in Table 2 as acceptable.

The specific gravity of ore samples has been tested and has resulted in an average of 2.00.

$$\text{Tonnage factor} = \frac{2000 \text{ lb per ton}}{SPG \times 62.4 \text{ lb per cu ft}} = \text{cu ft per ton}$$

$$\frac{2000}{2 \times 62.4} = 16 \text{ cu ft per ton.}$$

The initial metallurgical flowsheets and design criteria have been developed. The process is to be a conventional acid leach-solvent extraction circuit with an average recovery of 90%. The selling price of the recovered uranium oxide is set at $40 per lb. The criteria are given in Table 3.

Table 2. Back Slope vs. Height (Horizontal:Vertical)

Vertical height, ft	Back slope
100	0.75:1
300	1.25:1
500	1.50:1

The access road, water production, power line construction, and environmental base-line studies are considered as preproduction capital items. For the purpose of calculating the cash flow, the total sum of $4,000,000 will be used in addition to the normal ancillary capital costs.

The minimum acceptable rate of return will be set at 12% for this example. We are going to establish the conditions that will make the net present value at this discount rate as high as possible.

ORE RESERVES—CALCULATIONS AND ADJUSTMENTS (Fig. 2)

Drill-Hole Evaluation

As the basic premise in establishing a starting point for the calculations, use regional operating and capital costs for both mining and milling based on operations having similar size, geology, and equipment appropriate for the setting. Estimate mining recovery and ore dilution. Utilize amenability studies on which to base mill recovery and use current commodity selling price.

Table 1. Ore and Waste Inventory

Cutoff grade*, % U308	Tonnages†, millions	Average Grade % U308	Material classification	Ratios ton per ton
0.050	8	0.120		
0.040	10	0.104	ore	1
0.030	14	0.083		
0.020	18	0.067	low-grade	
0.010	26	0.051		2
<0.010	16	not calculated	associated waste	
-0-	364	-0-	primary stripping	26

*Calculated in the first step of the cycle: ore reserves, calculations and adjustments.
†Cumulative through low-grade only.

Table 3. Mill Capital and Operating Costs

Mill feed, tpd*	Operating $ per ton	Directs and indirects, $ per ton	Capital, $ millions
1000	5.50	4.90	24
3000	5.00	4.50	30
5000	4.50	4.10	40
7000	4.00	3.70	50

*Annual production based on 365 days per year. Actual daily production is tons per day × 52 wks per yr/50 working wks per yr = tons per working day.

Area of Influence (Polygons)

The example shows drill hole No. 1. The drilling grid is 200 ft, the area of influence 40,000 sq ft, and the tonnage factor $\frac{40{,}000 \text{ sq ft}}{16 \text{ cu ft per ton}} = 2500$ tons per vertical ft. This deposit is assumed to be dry. The weight of water, when present, must be added for the determination of operational costs and capital requirements.

Regional Operating Costs

The operating costs ($ per ton) include stripping (0.35); associated waste (AW) and ore (1.00); milling (5.00); indirect costs such as taxes, interest, depreciation, and corporate overhead (2.00); and direct costs such as salaries, power, etc. (2.50).

Break-Even Analysis

Cutoff grade calculation is based on production costs utilizing the overall ratio of tons of waste to tons of ore (Table 4).

$$\frac{\text{production cost}}{\text{selling price} \times \%\ \text{recovery}} = \frac{\$21.60}{\$40 \times 0.90} = \frac{0.60\ \text{lb}}{20} = 0.030\%.$$

Normally, the initial cutoff grade and production costs are used to derive the preliminary ore and waste inventory. It is obvious that some assumptions must be made because the overall stripping ratio cannot be established until a preliminary cutoff grade is set, as the cutoff grade is a direct function of the amount of overburden to be removed. All future reserve calculations are based on actual data derived for the particular deposit.

Drill-Hole Analysis (Fig. 3)

Evaluate all drill holes from top to bottom. The economic bottom of mining is the depth where the maximum net profit is realized.

The unit evaluation is repeated for all the drill holes, even those that obviously carry a negative value. In the expansion of the total reserves both laterally and

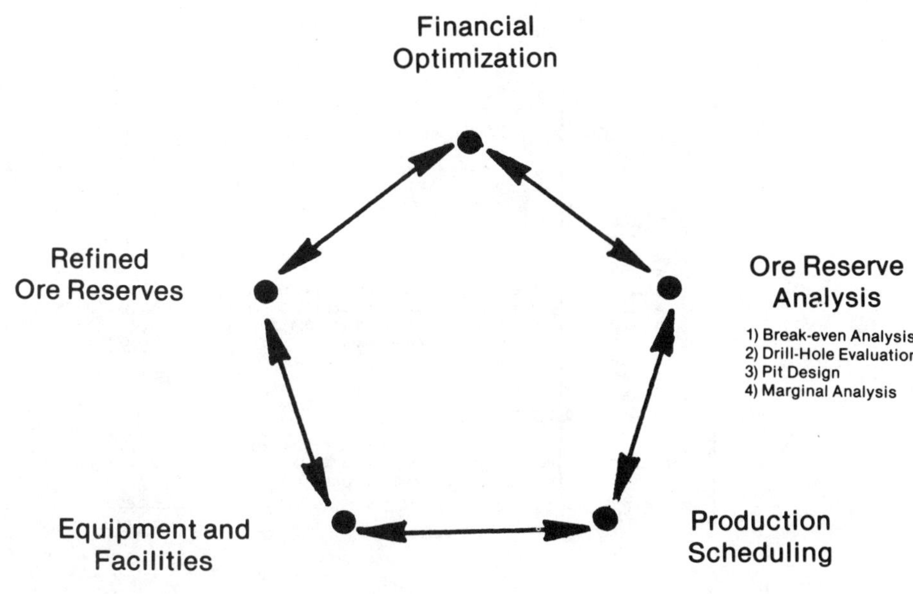

Fig. 2. Ore reserves, calculations and adjustments.

Table 4. Production Costs

Category	Ratio	$ per ton	Total cost
Stripping	26	0.35	$ 9.10
Assoc. waste	2	1.00	2.00
Ore mining	1	1.00	1.00
Milling	1	5.00	5.00
Indirects	1	2.00	2.00
Directs	1	2.50	2.50
		Total	$21.60

vertically, many instances will be found where a few good holes will more than offset some negative values, although the extent of each must be known before a final determination can be made. The complete analysis is extremely essential in establishing the ultimate pit outline as will be discussed next.

Open Pit Design

The final limits of the pit design usually do not exactly correspond to the economic limits as established in the unit evaluation. The ultimate pit is a direct function of various working, engineering, financial, and safety constraints.

Operational Efficiency: The operational efficiency increases advantageously as the horizontal section is smoothed by eliminating projections and generally irregular configurations. The unit evaluation will allow the minimizing of costs associated with the smoothing of the pit walls and inclusion of waste, where it is obvious that the positive block more than offsets the negative blocks that must be mined in order to optimize the total value and provide adequate access and working room to extract the block in question (Fig. 4).

Fig. 4. Ore block smoothing.

Back Slope: Use the steepest possible back slope with regards to the safety factor deemed acceptable for the particular pit.

Back slopes are commonly not only a function of the actual rock conditions encountered and the total vertical lift (geotechnical analysis) but also of the duration of exposure and the amount of water indigenous to or added to the formation. The latter two can be alleviated by timely back filling and by either or both a mine dewatering program and a surface runoff diversion system. Where variable heights affect the ultimate design, a graph can be utilized relating height to back slope at a safety factor where the slope is close to failure but does meet risk standards set by company guidelines.

In the example utilized herein, all back slopes are a direct function of height (Fig. 5). It must be remem-

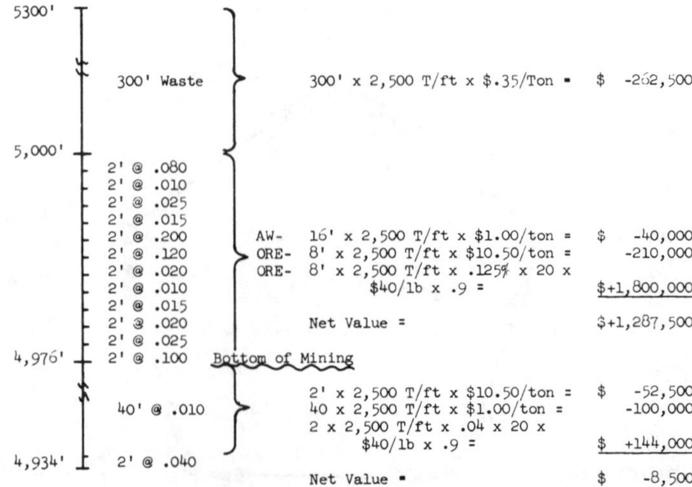

Fig. 3. Drill-hole profit/loss calculation. Metric equivalents: 1 ft×0.304 8=m; 1 st×0.907 184 7=t; 1 lb×0.453 592 4=kg.

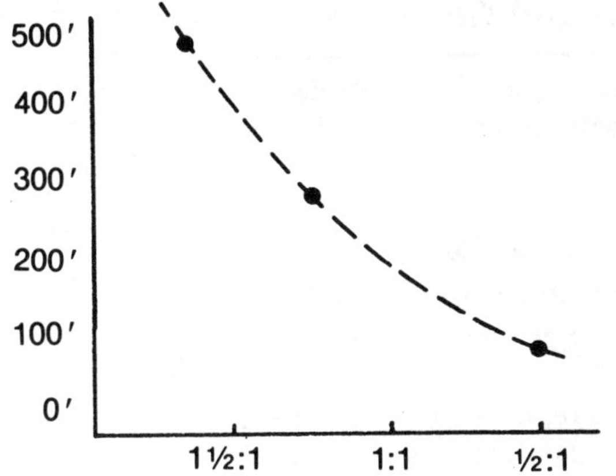

Fig. 5. Back-slope angle determination. Metric equivalent: 1 ft$\times$0.304 8=m.

bered that in rough terrain it is the difference in elevation from where the back-slope angle intersects the surface from the economic bottom of the pit that ultimately establishes the final back-slope angle resulting in a few minor adjustments during design.

Design: Design must allow for adequate operating room for maneuverability and mining flexibility in ore blending requirements.

Utilize a 0.5:1 back slope between benches for design purposes and achieve the overall back-slope angle by adjusting the widths of the final benches on the individual levels.

$$\frac{\text{vert. hgt.} \times (\text{BS ratio} - \text{BS ratio between benches})}{(\text{vertical height} \div \text{bench height}) - 1} = \frac{\text{horizontal displacement}}{\text{No. benches required}} = \text{bench width.}$$

example

$$\frac{500\text{ ft} \times (1.5 - 0.5)}{(500\text{ ft} \div 50\text{ ft}) - 1} = \frac{500\text{ ft}}{9\text{ benches}} = 56\text{ ft per bench.}$$

Individual bench widths may be adjusted to allow for ramps, lithology changes, etc., as long as the aggregate remains unchanged.

Having engineered a practical and safe design, the ore reserves, associated waste, and stripping tonnages are now calculated.

Survey Control: Close attention must be paid to survey control. If the limits are set out of position, a loss of ore and an increase of waste will result.

Ore Reserves

The total ore reserves have been shown to be the direct result of drill-hole evaluation and open pit design. The calculation utilizes the concept of a break-even cutoff grade based on known or assumed operating costs rather than a break-even stripping ratio because the latter only deals with the average grade of the ore body, while in fact the average grade can be so markedly different from one hole to the next so as to require a separate ratio for each area of influence.

Once the ultimate pit has been designed and the resulting ore and waste inventory calculated, direct mining and stripping costs are accounted for in the calculations and are considered thereafter as sunk costs. Ore reserve optimization then involves the deletion of the sunk costs from the analysis of the cutoff grade. The ensuing marginal analysis not only increases the ore reserves and tons to be milled but also lowers waste to ore ratio. The modification either has an effect on the capital requirements, operating costs, or a lengthening of the project duration.

Marginal Analysis

$$\frac{\text{operating cost-sunk cost} + \text{extra material handling}^{*}}{\text{commodity value} \times \%\ \text{recovery}} = \text{marginal cutoff grade.}$$

Example:

$$\frac{\$21.60\text{ per ton-}\$11.10\text{ per ton} + \$.50\text{ per ton}}{\$40\text{ per lb} \times 0.9} = \frac{\$11\text{ per ton}}{\$36\text{ per lb}} = 0.31\text{ lb per ton} = 0.015\%.$$

The resulting reserves and ratios are listed in Table 5.

It can be seen that the ore tonnage increases from 14,000,000 tons (Table 1) to a total of 22,000,000 tons for a 57% increase, and the grade goes from an average of 0.083% to 0.058%, a 30% decrease for an overall 10% increase in pounds.

While the larger reserves appear to be the more efficient utilization of the mineral present, it is the cash flow and resulting net present value which will dictate the final cutoff grade. For the purpose of determining the best cash flow, the marginal reserve will be used as the total amount of ore to be milled with the extremes being whether or not the marginal ore should be averaged in with the ore or milled separately at the end of the mine life. In reality, some lateral inclusion of lower grade material due to changing economics may also alter the overall reserves but will not be considered herein.

* Load and haul from stockpile to crusher

Table 5. Marginal Analysis Ore Reserves

Tons, millions	Grade, % U308	Material classification	Ratio, ton per ton
22	0.058	ore	1
20	not calculated	associated waste	0.9
364	-0-	primary stripping	16.5

A recalculation of the drill-hole profit/loss calculation (Fig. 3) results in an optimized net profit (Table 6).

The marginal ore reserve calculation is shown to be sufficiently better than the break-even calculation to justify its utilization as the final cutoff grade. The break-even analysis, therefore, only establishes the ultimate pit limits.

PRODUCTION SCHEDULING GUIDELINES (Fig. 6)

The production scheduling is an important facet of mine planning. Once the mill feed grade has been established, whether it is based on the break-even calculation, marginal analysis, or some combination thereof, maximizing the NPV of a property is a direct result of scheduling. Scheduling dictates total mine life and therefore cash flows, capital requirements, and operational costs.

The initial production scheduling is based on the preliminary ore reserves utilizing a haulage study based on the general pit design and facility layout.

The following parameters provide guidelines for the scheduling of the operation:

Minimizing Preproduction Costs

Preproduction operational costs are like equipment costs in that they are incurred before production starts. The costs are not discounted because they come first, and therefore, carry a fully weighted effect on the cash flow and payback period.

The following example only illustrates the time value of money, where two cases having the same total production costs ($\$5 \times 10^6$) and generated revenue ($\$7.5 \times 10^6$) are compared on slightly different production schedules. A discount rate of 12% is applied to each alternative.

The comparison involves a two-year preproduction period versus a one year preproduction period with a total five-year mill (revenue generating) life for each case. Each example will have a balanced production rate with stripping being completed one year prior to the final production date in both cases.

Case I. Two-Year Preproduction
Total Cost:

Stripping	$\$ 3\times10^6 \div 6$ years=$ 500,000 per year
Mining & milling	$\$ 2\times10^6 \div 5$ years=$ 400,000 per year
Total revenue	$\$7.5\times10^6 \div 5$ years=$1,500,000 per year

Case II. One-Year Preproduction
Total Cost:

Stripping	$\$ 3\times10^6 \div 5$ years=$ 600,000 per year
Mining & milling	$\$ 2\times10^6 \div 5$ years=$ 400,000 per year
Total revenue (Table 7)	$\$7.5\times10^6 \div 5$ years=$1,500,000 per year

This example is oversimplified, but it does stress the importance of maximizing cash flow by lessening the preproduction commitment.

Contract stripping may also prove to be advantageous if used to the point where a positive cash flow is realized.

Table 6. Optimized Net Profit

	Marginal	Cutoff
300 ft stripping × 2500 tons per ft × $0.35 per ton =	$ −262,500	$ −262,500
4 ft associated waste × 2500 tons per ft × $1.00 per ton =	−10,000	−40,000
20 ft ore × 2500 tons per ft × $10.50 per ton =	−525,000	−210,000
20 ft × 2500 tons per ft × 0.062% × 20 × $40 per lb × 0.9 =	+2,232,000	+1,800,000
	$+1,434,500*	$+1,287,500

*Represents an increase of $147,000 or 11.4% over the evaluation utilizing the cutoff evaluation grade of 0.030% U308.

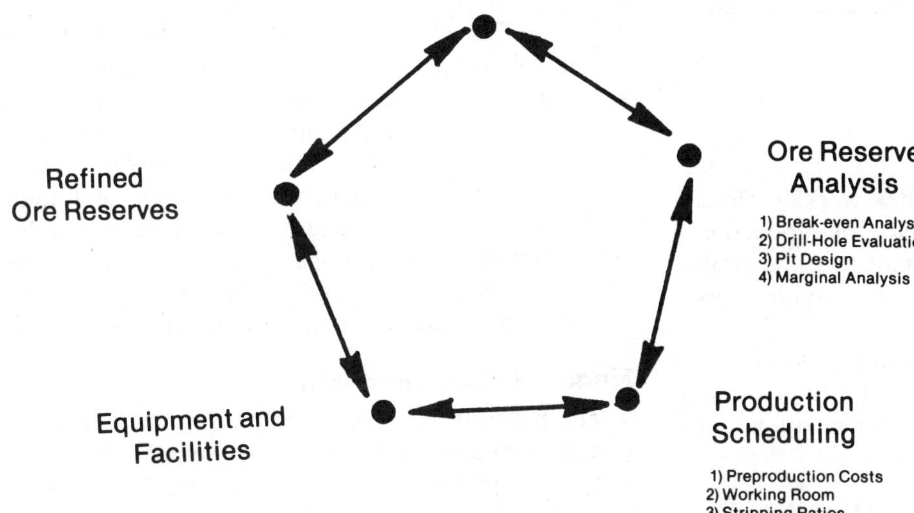

Fig. 6. Production scheduling guidelines.

Here again, a cash flow analysis should be the determining factor.

Assuring Adequate Working Room

Especially in projects where the stripping ratio varies from one area to another, flattened working slopes should be used up to the point where the final pit configuration is intercepted. As mentioned under Pit Design, the flattening is accomplished by widening the benches. The resulting uncongested working areas provide for a smoother operation and minimize safety hazards, both operational, and also the slope stability increases as the back slope is flattened. Allowing the degree of slope flattening to fluctuate with the overburden ratio allows for a constant production rate and a predictable ore availability (Fig. 7).

Table 7. Cash Flow

Discount factor @ 12% (end of period)		2-yr Preprod ACF*	2-yr Preprod DCF†	1-yr Preprod ACF	1-yr Preprod DCF
Year 1	0.89	$ −500‡	$ −445	$ −600	$ −534
2	0.80	−500	−400	+500	+400
3	0.71	+600	+426	+500	+355
4	0.64	+600	+384	+500	+320
5	0.57	+600	+342	+500	+285
6	0.51	+600	+306	+1,100	+561
7	0.45	+1,100	+495		
Net present value			$+1,108		$+1,387
Total undiscounted value		$ 2,500		$ 2,500	

*Actual cash flow.
†Discounted cash flow.
‡Dollars in thousands.

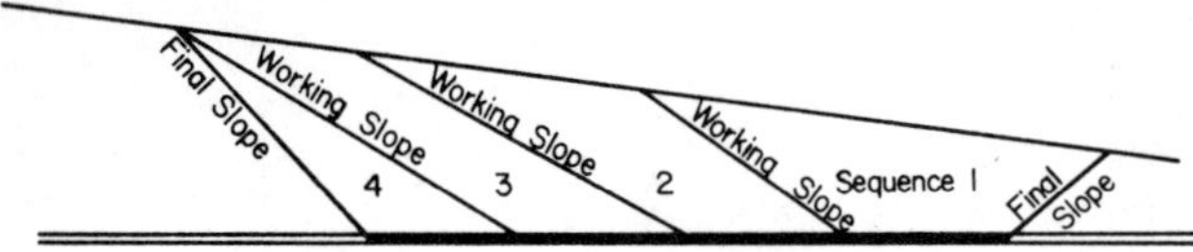

Fig. 7. Working slopes.

Smoothing of the Stripping Ratios

When scheduling the operations, it is the average stripping ratio (tonnage) that dictates working shifts and equipment required, especially once the production phase begins. It is extremely important to avoid peak ratios as scheduling problems or capital requirements may become excessive.

In instances where the stripping ratio is such that equipment can be added or subtracted at a future date, it then becomes the average ratio of each situation that becomes the scheduling basis.

Timely Exposure of Ore Grade Material

Proper sequencing is achieved through incremental pit design. Each increment is directly related to mill requirements within specific time constraints. Mining sequences are tentatively established and then analyzed to evaluate the most logical total development program. The development program then assures a constant mill feed and also provides the basis for initially selecting higher grade areas with lower stripping ratios to maximize cash flow.

Reclamation Accountability

A point worth mentioning in these times of increased environmental awareness is that proper production planning and scheduling can minimize the costs associated with mine reclamation. The items most effectively handled by efficient planning are: (1) returning ground contours to approximate premine conditions, (2) minimizing surface depressions to the greatest extent possible, and (3) revegetating all disturbed areas.

Selected back filling can create new slope angles within the open pits, and in some instances, can completely fill portions of the pit to achieve original contour. Generally, back filling shortens waste haul cycles by reducing haul grades and distances. Waste dumps can be constructed on receding lifts allowing the creation of flatter than "angle of repose" slopes with only a minimum of dozer work required. Topsoil can be effectively stripped and stockpiled for reemplacement at a future date.

The principal point being that with the scheduling and design addressing the reclamation requirements, the ultimate cash flow can be substantially improved by increased operational efficiency. The equipment and operating personnel required to do the work can then be computed into the overall plan resulting in further definition of mine plant size.

Maximizing Effective Operating Schedule

The following constraints allow for more efficient scheduling and equipment utilization: (1) avoid excessive shovel moves, (2) minimize number of working areas, (3) work lowest number of benches possible at any one time, (4) reduce haul distances and ramp grades when practical, and (5) schedule round-the-clock seven-day-per-week operations whenever possible. Increased productivity results, and this has a direct relationship to equipment and manpower requirements.

Financial Considerations

The property life, based on production scheduling, directly influences the amount of capital expenditure required to produce a future income. While the evaluation of alternatives may single out an optimum size operation, it is wise to remember that mining plans are based on present-day parameters, and therefore, subject to change or modification as time progresses.

Influencing factors are: (1) capital and operating cost changes, (2) new mining and milling innovations, (3) increased knowledge of the deposit, and (4) changes in the commodity value.

Ore and low-grade stockpiles provide the flexibility for mill feed grade changes and spot sales. An improved discounted cash flow (DCF) may also be achieved through adequate stockpiling. The lower grade ore that was included in the marginal analysis can be set aside for processing at the end of the operation. Another method of optimization involves a form of limited high-grading as shown in the following example.

As applied to the example in the text, the following considerations are offered. During mine life, a spot sale, at an average selling price of \$50 per pound, has been offered over and above the commitments under contract. The extra pounds are only available as above average grade ore, as the mill is operating at the designed capacity of 3000 tpd at 0.058%. The spot sale contract would be for an additional 500,000 lb during the third year of a 20-year production life.

$$\begin{aligned} \text{ore reserves} &= 22 \times 10^6 \text{ tons @ } 0.058\% \\ \text{minus two-year production} &= \frac{-2.20 \times 10^6 \text{ tons @ } 0.058\%}{19.80 \times 10^6 \text{ tons @ } 0.058\%}. \end{aligned}$$

The requirements for added production are:

$$\text{mill capacity (tons) at } X\% = \text{original contract} + \frac{\text{spot sale}}{\%\text{ recovery}}.$$

3000 tpd @ X % = 1,143,180 (3000 tpd @ 0.058% × 90% recovery) + $\frac{500{,}000 \text{ lb}}{0.9}$ = 1,698,736 lb.

$$\text{ore grade required} = \frac{1{,}698{,}736 \text{ lb}}{1{,}095{,}000 \text{ tpy} \times 0.9} = 1.72 \text{ lb} = 0.086\%.$$

The increased production would reduce the overall remaining grade of the ore body:

19.80×10^6 tons @ 0.058 − 1.095×10^6 tons @ 0.086 = 18.705×10^6 tons @ 0.056.

Using the precalculated production cost of $21.60 per ton (Table 4), a comparison of cash flows is calculated from the beginning of production through the life of the property.

Value of ore per ton [grade × 20 (conversion to lb) × % recovery × selling price = value − production cost = net value per ton]. Refer to Table 8.

Case I. Uniform Production
Discounted @ 12% on a 20-year life:
$22,075,200 per year × 7.469 (cumulative PW factor) = $164,879,669 NPV.

Case II. Inclusion of Spot Sale

Years 1 and 2	$22,075,200 × 1.69 (cum. PW factor) =	$ 37,307,088
Year 3	47,019,300 × 0.71 (end of period) =	33,383,703
Years 4 through 20	20,498,400 × 5.067 (cum. PW factor 20th yr—3rd yr) =	103,865,393
	NPV =	$174,556,184

The time value of money is as important when optimizing production scheduling and resulting income as preproduction capital and resulting income as previously illustrated (Table 7).

The production scheduling guidelines provide the constraints for the alternative comparison by which the highest NPV is determined. The basic criterion for establishing alternatives is to vary the project duration (size) and the mill feed cutoff grades.

The following discussion on production scheduling will concentrate on the development of generalized operating costs for various sized operations which will be applied to the example included in the text.

PRODUCTION SCHEDULING

Scheduling Constraints

The production scheduling is limited by constraints derived from historical data, regional information, and estimates based on industry-generated records.

Electric Shovel Application

Diversity Factor: This refers to time lost due to moving, cleanup, and queuing of trucks. It is expressed as a factor of effective loading time. A commonly used factor of 0.83 (based on industry time studies) will be used in this chapter.

Effective Loading Time: This means full shift minus time lost to lunch and waiting on trucks.

Minutes per shift

	480	full shift
Variable	−30	lunch period
Variable	−10	beginning of shift truck travel time
Variable	−10	lost on each side of lunch period
Variable	−10	end of shift truck travel time
	420	per shift.

Truck Application

Job Efficiency: This refers to the minutes of actual production during a 60-min hr. Manufacturers' data which apply to all equipment except electric shovels are listed in Table 9. The average is to be used for all calculations until such a point that actual times are verified for a particular mine.

Production Time: This refers to operating time (full shift-lunch period) times job efficiency minus travel time (to and from parking lot); 8 hr − 0.5 hr = 7.5 hr × 50 min per hr = 375 min − 20 min = 355 min per shift.

Travel time (to and from parking lot) is optional. Many operations utilize the 50 min per hr as accounting for the extra travel time.

Table 8. Annual Cash Flow

Grade	Net value per ton	Annual value @ 3000 tpd, $
0.058	20.16	22,075,200
0.086	42.94	47,019,300
0.056	18.72	20,498,400

Table 9. Working Efficiencies

Classification	Working min per hr
Favorable	55
Average*	50
Unfavorable	45

*Table 11.

Table 10. Shovel Shifts (Based on a Seven-Day Working Week)

No. shovels	Shifts per week	Working shifts per week	% Availability	Designation
1	21	10	48	unacceptable
	21	15	71	acceptable
	21	20	95	unacceptable
2	42	30	71	acceptable
	42	35	83	marginal
3	63	45	71	acceptable
	63	50	79	acceptable
	63	55	87	unacceptable

Support Equipment

The operating time is also 7.5 hr but the production time will vary with property layout, scheduling procedures, etc.

Front-End Loaders

Production time is calculated in the same manner as trucks when the unit returns to the parking lot at the end of each shift and as a shovel when it remains on the work site between shifts.

Common Constraints

Annual Working Days: This section includes the total days per year (365) minus holidays and unscheduled shutdowns (weather, absenteeism, and major breakdowns). 365 days per yr − 10 holidays − 5 unscheduled days = 350 days per yr ÷ 7 days per wk = 50 weeks per year.

Available Shifts: The selection of the number and duration of working shifts per week depends on required production and the equipment utilized. Property life alternatives dictate annual production requirements which then become a function of equipment and scheduled shifts. The most efficient shift time allocation is 8 hr because it is then possible to operate 24 hr per day if required. Only 8-hr shifts will be used in the calculations in this chapter.

Mechanical Availability

Mechanical availability is expressed as the availability after mechanical repair, preventive maintenance, and servicing have been accounted for.

$$\text{shovels} \qquad \frac{\text{working shifts}}{\text{working shifts} + \text{maintenance shifts}} = \%\ \text{availability.}$$

A practical long-range availability for shovels would be in the range of 70 to 80% depending upon working conditions, maintenance procedures, etc. (Table 10).

Trucks: Trucks are calculated at 100% coverage of the loader capacity. Using an industry norm of 75% availability, the 100% coverage is ensured by adding the appropriate number of extra trucks to the fleet.

$$\frac{\text{number of trucks required for loader coverage}}{\%\ \text{availability}} = \text{fleet required.}$$

$$\frac{8\ \text{trucks}}{0.75} = 10.7\ \text{trucks or 11 trucks.}$$

Manpower: The basic number required divided by the availability factor gives the total men required. The availability factor is a variable using local averages of vacations and absenteeism. For example, use absenteeism at 5% and a vacation average of 2.5 weeks per year as 5%, making the total time loss 10%. The manpower availability factor, then, is

$$100\% + 10\%\ \text{loss} = 110\%\ \text{or}\ 1.10.$$

The total manpower required is

$$\frac{\text{No. working shifts} \times \text{hr per shift}}{\text{working hr per shift}} \times \text{availability factor.}$$

For example, assume there are 35 truck shifts per week:

$$\frac{35\ \text{shifts} \times 8\ \text{hr per shift}}{40\ \text{working hr per week}} = 7\ \text{men} \times 1.10 = 7.7\ \text{or 8 men required.}$$

In this case the extra man is assigned to the labor pool and is used as required (Table 11).

Table 11. Scheduling Constraints Recapitulation for Example Calculations

Shovels	Diversity factor	0.83
	Effective loading time	420 min per shift
Trucks	Job efficiency	50 min per hr
	Production time	355 min per shift
Working weeks per year		50
Working shift duration		8 hr
Coverage	Shovels	70-80%
	Trucks	100%
Manpower Available Factor		1.10

EQUIPMENT AND FACILITIES (FIG. 8)

Open pit mining is becoming increasingly capital intensive as pits are enlarging and the average grade of ore decreases. Contributing factors are: (1) increased mechanization to handle the larger tonnages, (2) lower grade ore resulting in higher plant investment per unit of refined product, (3) larger pits requiring greater preproduction cash outlays, and (4) environmental requirements.

The major capital requirements that are a part of the final investment analysis are: (1) plant, mobile units, and ancillary equipment; (2) preproduction stripping and mining; (3) parts inventory; (4) plant startup and working capital; (5) environmental base-line studies and monitoring; and (6) cost of capital.

The capital requirements and subsequent operational costs will be developed at length in the selection of loading and haulage equipment as discussed next. The main point to be emphasized is to minimize and/or delay capital expenditures as long as feasible. A reduction in capital will increase the discounted cash flow of any mining investment. A property can ill afford the capital expenses that do not lower production costs and increase earnings.

Equipment Selection

The selection of a fleet of equipment is accomplished by comparing the production requirements against the capital and operating costs of various selection alternatives. A pit haulage study is required to size and select types and number of haul trucks. Production requirements, rock type, and desired flexibility are the parameters guiding the selection of the primary loaders. Equipment size should be standardized whenever possible.

Loaders: Basic requirements for loader selection include rock characteristics, mine life, type of haulers utilized, tonnage factor, and swell factor. For example, from drill-hole log composities, the average composition, expressed as a percentage of the overburden, is determined and when applied to published tables, the weighted averages are made (Table 12).

$$\text{Swell factor} = \frac{1}{1+0.254} = 0.8.$$

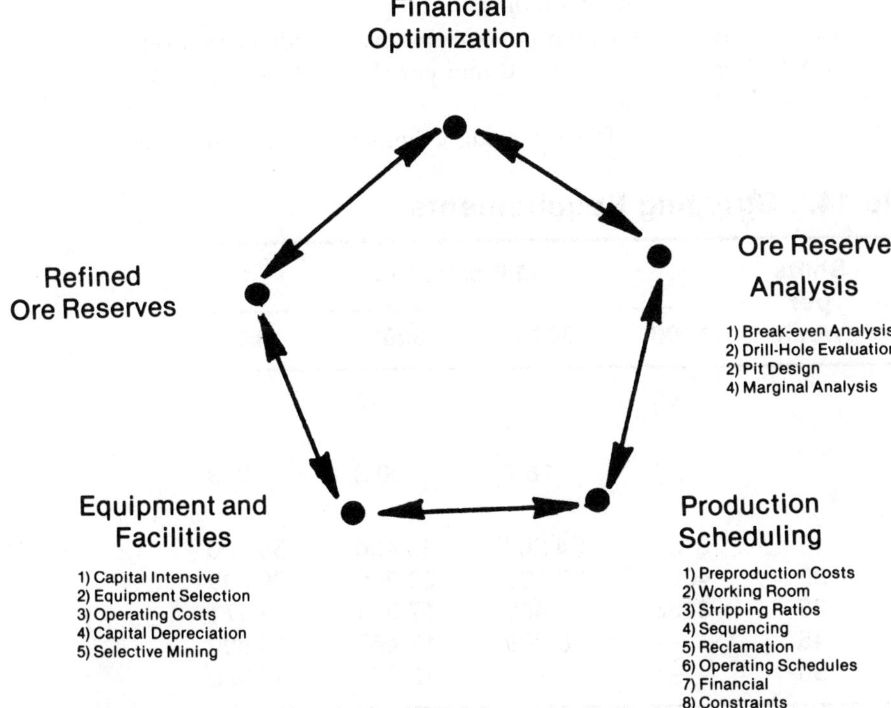

Fig. 8. Equipment and facilities.

Table 12. Overburden Swell

Lithology type	% of total	% swell (decimal equiv.)	Weighted average % swell
Gravel	20	0.12	2.4
Clay	20	0.25	5.0
Sand-loose	30	0.10	3.0
Sandstone-cemented	30	0.50	15.0
		Total	25.4

Table 13. Electric Shovels Manufacturer's Specifications

Shovel size	250 kw	375 kw	525 kw	625 kw	725 kw
Nominal dipper, cu yd	6	12	17	22	30
Nominal cycle time, sec	22	25	27	29	31
Tons/dipper	7.3	14.6	20.7	26.7	36.5
Nominal production per shift, tons	6940	12,198	16,035	19,278	24,626

The tonnage factor (TF) of broken muck determination is

$$\frac{\text{in-place TF}}{\text{swell factor}}.$$

The previously calculated TF is 16 cu ft per ton/ 0.8=20 cu ft per ton or ton per cubic yard=27 cu ft per cu yd/20 cu ft per ton=1.35 ton per cu yd. For the sake of simplification, only electric shovels will be evaluated in this exercise (refer to Table 13).

Consider the nominal dipper capacity when first sizing shovels. Once a shovel is sized, the dipper capacity is optimized to truck capacities. Using an average bucket fill factor of 0.9, the dipper capacity is:

bucket capacity (cu yd) × tons per cu yd × fill factor = tons per dipper.

shovel production = tons per dipper × diversity factor (0.83) × 420 min per shift × 60 sec per min cycle time (sec).

The following example computes the tons per shift

Table 14. Stripping Requirements

	Shifts per week	Mill size, tpd 1000	3000	5000	7000
Property life, years		60	20	12	8.6
Required stripping, million tpy		6.1	18.2	30.3	42.3
Required stripping, tons per shift	15	8133	24,267	40,400	56,400
	30	4067	12,133	20,200	28,200
	35	3486	10,400	17,314	24,171
	45	—	8,089	13,467	18,800
	50	—	7,280	12,120	16,920

Table 15. Dipper Selection, 250 Kw

Dipper size		6 cu yd	7 cu yd	8 cu yd	9 cu yd
Tons per dipper		7.3	8.5	9.7	10.9
Tons per no. passes,	4	29.2	34	38.8	43.6
	5	36.5	42.5	48.5	54.5
	6	43.8	51	58.2	65.4

required for various mill sizes and the number of shovel shifts utilized. The ratio between the stripping tonnage of 364×10^6 tons and 22×10^6 tons of ore (marginal analysis) will be utilized (Table 14).

Using the required stripping tons per shift, the shovel(s) is selected when nominal shovel production per shift is matched to the production requirements for each mine life.

60-Year Mine Life (8133 Tons Stripping per Shift) —Utilizing Tables 13 and 14, the stripping requirements are satisfied by utilizing 15 shovel shifts using one 250-kw shovel. The dipper size must now be optimized. Manufacturer's guidelines state that with the specified weight of this material, up to a 9-cu yd dipper may be utilized. In all dipper selections following, nothing below nominal capacity will be considered although in special instances, it may be advantageous to do so. Four dippers per truck is considered optimum, although up to six is acceptable under normal conditions (Table 15). By utilizing Table 15, it can be seen that trucks with capacities ranging from 30 to 65 tons would be applicable with the 250-kw shovel.

20-Year Mine Life (24,267 Tons Stripping per Shift)—Utilizing Tables 13 and 14, the stripping requirements are shown to be satisfied by utilizing 30 shovel shifts using two 375-kw shovels or 15 shovel shifts using one 725-kw shovel. Weight of the muck indicates a dipper range of 11 through 13 cu yd for the 375-kw shovels and from 28 cu yd through 34 cu yd for the 725-kw shovel (Table 16).

12-Year Mine Life (40,400 Tons Stripping per Shift)—Utilizing Tables 13 and 14, the stripping requirements are satisfied by utilizing 30 shovel shifts using two 625-kw shovels or 35 shovel shifts using two 525-kw shovels. In both instances, the nominal dipper rating is not adequate but proper size selection will give either one the required capacity. A third combination possibility would be 15 shovel shifts using one 525-kw and one 725-kw shovel. The final alternative will not be considered as equipment size should be standardized whenever other options are available. The weight of the muck indicates a dipper range of 15 through 19 cu yd for the 525-kw shovel and 20 through 27 cu yd for the 625-kw shovel (Table 17).

8.6-Year Mine Life (56,400 Tons Stripping per Shift)—Utilizing Tables 13 and 14, the stripping requirements are satisfied by utilizing 45 shovel shifts using three 625-kw shovels or 35 shovel shifts using two 725-kw shovels. Dipper selection can be made by referring to Tables 16 and 17. In normal practice, it is advisable to construct one reference table for each rock weight which should include all applicable shovel and dipper sizes.

Table 16. Dipper Selection, 375 Kw and 725 Kw

Dipper size		12 cu yd	13 cu yd	30 cu yd	31 cu yd	32 cu yd	33 cu yd	34 cu yd
Tons per dipper		14.6	15.8	36.5	37.7	38.9	40.1	41.3
Tons per no. passes,	4	58.4	63.2	146	150.8	155.6	160.4	165.2
	5	73	79	182.5	188.5	194.5	200.5	206.6
	6	87.6	94.8	219	226.2	233.4	240.6	247.8

Table 17. Dipper Selection, 525 Kw and 625 Kw

Dipper size		17 cu yd	18 cu yd	19 cu yd	22 cu yd	23 cu yd	24 cu yd	25 cu yd	26 cu yd	27 cu yd
Tons per dipper		20.7	21.9	23.1	26.7	28.1	29.2	30.4	31.6	32.8
Tons per no. passes,	4	82.8	87.6	92.4	106.8	112.4	116.8	121.6	126.4	131.2
	5	103.5	109.5	115.5	133.5	140.5	146	152	158	164
	6	124.2	131.4	138.6	160.2	168.6	175.2	182.4	189.6	196.8

Table 18. Truck-Shovel Combinations

Property life, years	Shovel size, kw	Shifts	Dipper size, cu yd	Tons per dipper	Cycle time per sec*	No. passes	Tons per load	Truck capacity, tons†	Tons per shift‡	Reg tons per shift§	% Capacity
60	1-250	15	8	9.7	23	5	48.5	50	8,821	8,133	108
20	2-375	30	13	15.8	25	5	79	85	13,219	12,133	109
	1-725	15	31	37.7	31	4	150.8	150	25,426	24,267	105
12	2-525	35	19	23.1	28	5	115.5	120	17,256	17,314	100
	2-625	30	23	28.1	29	6	168.6	170	20,267	20,200	100
8.6	3-625	45	24	29.2	30	5	146	150	20,358	18,800	108
	2-725	35	34	41.3	33	4	165.2	170	26,177	24,171	108

*Cycle time has been increased by 1 sec per 2 cu yd above nominal dipper capacity.
†Nominal truck capacity may be exceeded by up to 5% but increased repair and tire cost result.
‡Shovel production calculations.
§Table 14.

Trucks: The number of trucks necessary to move the required stripping tons per shift as dictated by mine life is calculated by analyzing incremental mine tonnage depletion and the corresponding incremental dump growth. Dipper sizing is a function of matching the tons per number of passes to the rated tonnage capacities of various size haulers. The maximum truck size per rated shovel production capacity is an effective objective although adverse grades, equipment availability, and cumulative bucket capacities also contribute to the final selection.

Although not limited to the truck sizes in Table 18, the following choices have been made to simplify the comparison process.

Truck Haulage Study—For determination of the travel and fixed times of various trucks, the following data have to be assembled: (1) pit layout (indicates the centroids of the pit sequences and dumps), (2) haul roads (lists road widths, passing lanes, intersections, and curves), (3) pit profile (details level hauls and plus or minus grades and the length of each), (4) dump geometry (lays out height, width, and traffic patterns), and (5) rolling resistance [RoRi] (a measure of force required to overcome the retarding effect between the tires and the ground).

Level compacted hauls with 40 lb RoRi per ton of vehicle weight:

$$\frac{40 \text{ lb}}{2000 \text{ lb per ton}} = 2\%.$$

Adverse grades with 20 lb RoRi per ton of vehicle weight:

$$\frac{20 \text{ lb per } \%}{2000 \text{ lb per ton}} = 1\% \text{ per } \% \text{ grade.}$$

Tire penetration or road flex with 30 lb RoRi per ton of vehicle weight:

Table 19. Flywheel Horsepower (FWHP) Ratios

Hauler, tons	Net vehicle wt, lb*	Payload, lb†	Gross vehicle wt, lb	Flywheel hp*	Net vehicle wt/ FWHP	Gross vehicle wt/ FWHP
50	75,300	97,500	172,800	608	124	284
85	112,800	158,000	270,800	755	149	359
120	175,200	231,000	406,200	895	196	454
150	211,000	301,600	512,600	1440	147	356
150	211,000	292,000	503,000	1440	147	349
170	212,500	337,200	549,700	1440	148	382
170	212,500	315,400	527,900	1440	148	367

*Manufacturer's specifications (variable).
†Table 18.

Table 20. Pit Haulage Study for 50-Ton Haul Truck, GVW/FWHP = 284, GVW = 172,800 Lb*

Segment length, ft	Grade, %	RoRi, %	Resistance, %	Max. vel. mph	Speed factor†	Act. vel. mph	Travel time, min	Remarks
200	0	5	5	11.5	0.4	4.6	0.49	@ shovel
200	0	4	4	22	0.65	14.3	0.16	
1875	8	3	11	9	1	9	2.37	ramp
3000	2	2	4	22	0.94	20.7	1.65	main road
200	0	5	5	11.5	0.33	3.8	0.60	dump
200	0	5	5	12	0.4	4.8	0.47	return
3000	−2	2	0	34	0.96	25†	1.36	cycle
1875	−8	3	−5	15	—	15‡	2.27	
200	0	4	4	34	0.64	21.8	0.1	
200	0	5	5	12	0.33	4.0	0.57	
					total travel time		10.04	

*GVW is gross vehicle weight and FWHP is flywheel horsepower.
†Factors that limit maximum attainable speed which include segment length, corners and other traffic restrictions, acceleration and deceleration.
‡Speed limit.

$$\frac{30 \text{ lb per 1 in.}}{2000 \text{ lb per ton}} = 1.5\% \text{ per 1 in. of penetration.}$$

weight + grade + penetration = rolling resistance.

Gross and net vehicle weights to flywheel horsepower ratio are given in Table 19.

Truck Cycle Time—Only one example will be generated using average centroids for the life of the pit and waste dump (Tables 20-23).

The number of trucks required to move the stripping tonnages for different fleet combinations can now be calculated (Table 24).

Table 21. Truck Cycle Times

Hauler, tons	Cycle time, min
50	10.04
85	9.81
120	10.57
150	8.83
170	9.23

Capital Costs

The costs in Table 25 are assigned to each piece of equipment.

The production fleet costs for the mine life and equipment combinations are given in Table 26.

The capital tied up in production equipment will be the basis for prorating the remaining capital costs. The breakdowns in Table 27 are averages from equipment lists sufficient to handle the total material generated from 10,000 to 200,000 tpd operations. Individual situations may dictate different percentages but can only be handled on that type of basis.

For the example included herein, the $\$4 \times 10^6$ will be added as an additional capital expense to cover offside ancillary power, water, access road, and environmental studies.

Drilling and blasting include rotary drills and explosive handling vehicles. Average rock conditions suggest the following relationships: One large diameter (9 to 12-in.) rotary drill per 30 to 40,000 tpd of stripping. Smaller drills (6 to 9 in.) will suffice for the lower tonnage operations at the rate of one per 10,000 to 15,000 tpd. Air tracks are required for secondary breakage in most operations. Under most situations, one air-track drill will suffice.

Table 22. Fixed Truck Times

Loader, kw	Hauler, tons	Spotting @ shovel, min*	Load @ shovel, min†	Turn and dump, min	Total fixed time, min
250	50	0.42	1.92	0.85	3.19
375	85	0.55	2.08	0.90	3.53
525	120	0.60	2.33	1.00	3.93
625	150	0.63	2.50	1.05	4.18
725	150	0.63	2.07	1.05	3.75
625	170	0.65	2.90	1.10	4.65
725	170	0.65	2.20	1.10	3.95

*Industry standards.
†Table 18.

Field and miscellaneous support equipment includes dozers, patrols, light duty and maintenance trucks, and cranes. The support equipment is very dependent upon local conditions making detailed relationships an impractical matter although the percent of the total capital remains fairly constant.

Power and utilities include power distribution, transformers, portable power, gas, water, and telephone. Here again, individual situations dictate the proportioning of the capital.

Shops and ancillary include the maintenance, tire, lube, and welding shops; office; changeroom; tools; compressors; tanks; etc.

The remaining capital costs will be lumped from this point to represent 30% of the total capital (Table 28).

$$\frac{70\%}{\text{production fleet}} = \frac{30\%}{\text{remaining capital}}$$

Table 23. Total Truck Cycle Times

Hauler, tons	Cycle time, min	Fixed time, min	Total cycle time, min
50	10.04	3.19	13.23
85	9.81	3.53	13.34
120	10.57	3.93	14.50
150	8.83	4.18	13.01
150	8.83	3.75	12.58
170	9.23	4.65	13.88
170	9.23	3.95	13.18

Operating Costs

Capital requirements alone cannot serve as the basis for comparing production and maintenance equipment. Operational costs, when combined with the capital requirements, provide a sound basis to arrive at a single fleet choice for each property life alternative. The resulting four fleets, as used in this example, will then provide the framework for the financial analysis and ultimate pit optimization.

The first step is to determine the operating hours per year for the loaders and the haulers involved in each option (Table 29).

The hourly costs are now generated based on actual field experience and manufacturer's guidelines.

Shovels: A 525-kw shovel will be costed out in detail. The operating costs of the other size shovels are calculated in the same manner, but will only be listed in Table 30. The primary costs include power, labor, and repair and maintenance.

Power Costs—Multiple kilowatt hours by rate to get the cost per hour. Consumption is rated at 525 kw per hr, and the rate is at $0.04 per kw-hr.

525 kw-hr × $0.04 per kw-hr = $21.00 per hr.

Labor Costs—Multiply operator rate and oiler rate by payroll burden and multiply that figure by 8 hr per shift divided by 7 productive hr per shift. Hourly rates are $8 and $7 per hr and the payroll burden is 40%. Payroll burden includes vacation, holidays, overtime, unexcused absenteeism, payroll benefits, FICA, health

Table 24. Truck Requirements

Hauler, tons	No. of shovels & size, kw	Truck cycles per shift (355 min cycle time)	Payload, tons	Shift prod per truck	Trucks to cover* Reg tons per shift ÷ truck production	Total coverage x No. shovels	Trucks† required (No. trucks @ 0.75 avail.)
50	1-250	27	48.5	1,310	$\frac{8,133}{1,310} = 6.21$	6.21	8
85	2-375	27	79	2,133	$\frac{12,133}{2,133} = 5.69$	11.38	15
120	2-525	25	115.5	2,888	$\frac{17,314}{2,888} = 6.00$	12.00	16
150	3-625	27	146	3,942	$\frac{18,800}{3,942} = 4.77$	14.31	19
150	1-725	28	150.8	4,222	$\frac{24,267}{4,222} = 5.75$	7.67	8
170	2-625	26	168.6	4,384	$\frac{20,200}{4,384} = 4.61$	9.22	12
170	2-725	27	165.2	4,460	$\frac{24,171}{4,460} = 5.42$	10.84	15

*See Table 18. The production limiting factor has been achieved by assigning the stripping requirement achieved through truck coverage as the target and not the shovel capacity. The latter is usually slightly higher.
†Round 0.0 to 0.29 to zero. Anything above 0.29 is rounded up.

Table 25. Equipment Capital Costs

Shovels		Trucks	
Size, kw	Cost, thousands $	Capacity, tons	Cost, thousands $
250	942	50	190
375	1471	85	330
525	1869	120	472
625	2296	150	610
725	3570	170	623

Table 26. Fleet Capital Costs

Mine life, years	Fleets	$ in thousands
60	1) 1 250-kw shovel and 8 50-ton trucks	2,462
20	1) 2 375-kw shovels and 15 85-ton trucks	7,892
	2) 1 725-kw shovel and 8 150-ton trucks	8,450
12	1) 2 525-kw shovels and 16 120-ton trucks	11,290
	2) 2 625-kw shovels and 12 170-ton trucks	12,068
8.6	1) 3 625-kw shovels and 19 150-ton trucks	18,478
	2) 2 725-kw shovels and 15 170-ton trucks	16,486

Table 27. Capital Cost Breakdown

Classification	% of total
Drilling and blasting	10
Excavation and hauling	70
Field & misc. support	10
Power and utilities	5
Shops and ancillary	5

insurance, unemployment insurance, and workmen's compensation. Labor costs, then, are:

$$(\$8+\$7)\times 1.40\times\frac{8}{7}=\$24 \text{ per hr.}$$

Repair and Maintenance—Calculate as the cost per ton multiplied by the tons per hour. Assuming that cost per ton is \$0.018 (manufacturer's recommendation) and production is 17,314 regular tons per shift divided by 7 hr per shift, or 2473 tph:

$$\$0.018 \text{ per ton}\times 2473 \text{ tph}=\$44.51 \text{ per hr.}$$

Table 28. Total Capital Requirements

Mine life, years	$ in thousands		
	Production fleet	Remaining capital*	Total capital†
60 (1)	2,462	1,055	7,517
20 (1)	7,892	3,421	15,313
(2)	8,450	3,621	16,071
12 (1)	11,290	4,839	20,129
(2)	12,068	5,172	21,240
8.6 (1)	18,478	7,919	30,397
(2)	16,485	7,065	27,550

*Preproduction mining is also a capital item, expensed for tax purposes, but will only be addressed as a function of production scheduling.
†Includes \$4,000,000 in additional ancillary capital.

Table 29. Equipment Operating Hours per Year

Property life, years	Equipment	Hr per shift	No. units	Shifts per week	Weeks per year	Hr per year
60 (1)	250 kw	7	1	15	50	5,250
	50 ton	7.5	8			45,000
20 (1)	375 kw	7	2	15	50	10,500
	85 ton	7.5	15			84,375
(2)	725 kw	7	1	15	50	5,250
	150 ton	7.5	8			45,000
12 (1)	525 kw	7	2	17.5	50	12,250
	120 ton	7.5	16			105,000
(2)	625 kw	7	2	15	50	10,500
	170 ton	7.5	12			67,500
8.6 (1)	625 kw	7	3	15	50	15,750
	150 ton	7.5	19			106,875
(2)	725 kw	7	2	17.5	50	12,250
	170 ton	7.5	15			98,438

Table 30. Shovel Operating Costs

Shovel size, kw	Operating Cost, $ per hr
250	57.24
375	70.20
525	89.51
625	97.57
725	107.86

The total cost (including power, labor, and repair and maintenance) is $89.51 per hr (Table 30).

Trucks: The operating costs of a 120-ton hauler will be detailed utilizing various experience factors and local wage rates.

Operator—Multiply the hourly rate by the payroll burden by the hours per shift divided by the productive hours per shift:

$$\$7.50 \text{ per hr} \times 1.40 \times \frac{8 \text{ hr per shift}}{7.5 \text{ prod. hr per shift}} = \$11.20 \text{ per hr.}$$

Tires—Multiply the cost per tire by the number of tires required and divide by the hours of expected life (tires are 30×51, 25-ply):

$$\frac{\$6275 \text{ per tire} \times 6 \text{ tires}}{3500 \text{ hr per tire}} = \$10.76 \text{ per hr.}$$

Fuel—Multiply gallons per hour times cost per gallon:

$$30 \text{ gal per hr} \times \$0.40 \text{ per gal} = \$12.00 \text{ per hr.}$$

Maintenance, Service, and Repair—The costing is based on the relationship between mechanics' wages, truck horsepower, and working conditions (Fig. 9). Utilizing mechanics' wages of $7.50 per hr multiplied by the payroll burden of 1.40, the rate is $10.50 per hr. Using Fig. 9, read vertically until the horsepower rating of the 120-ton truck is intercepted. At that point, read horizontally at the left. The factor in this example is 0.74. The historical factor (Fig. 9) is used by Wabco as standard procedure for estimating ownership and operating costs.

Fig. 10 shows the maintenance, service, and repair costs. Adverse conditions indicate long ramps, high RoRi, etc., and costs can be multiplied by a factor of up to 130%. Favorable conditions indicate light duty, level or downgrade hauls, low RoRi, etc., and costs can be multiplied by a factor of down to 80%.

Using the factor (0.74) obtained from Fig. 9, read vertically until the horsepower reading of the 120-ton truck is intercepted. At that point, move horizontally to the left and directly read the cost per hour assigned to maintenance, service, and repair. The cost of maintenance, service, and repair for the 120-ton truck becomes $21.80 when multiplied by 130% for adverse conditions.

Total Cost—The total operating costs for the 120-ton hauler is $55.76 per hr (Table 31).

The production fleet operating cost per hour for the various mine lifes and equipment combinations are given in Table 32.

The operating costs for the loading and hauling will be the basis for computing the remaining costs (Table 33).

The remaining operating costs will be lumped from

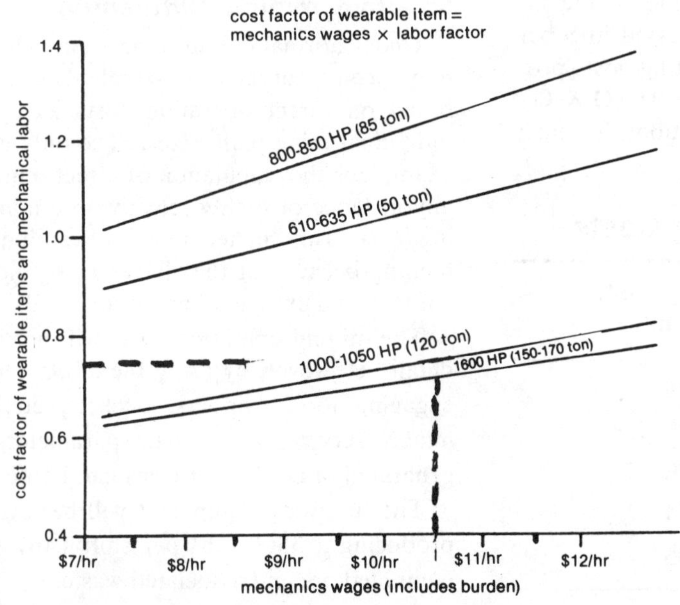

Fig. 9. Wearable items and mechanical labor factor. Metric equivalent: 1 st×0.907 184 7 =t.

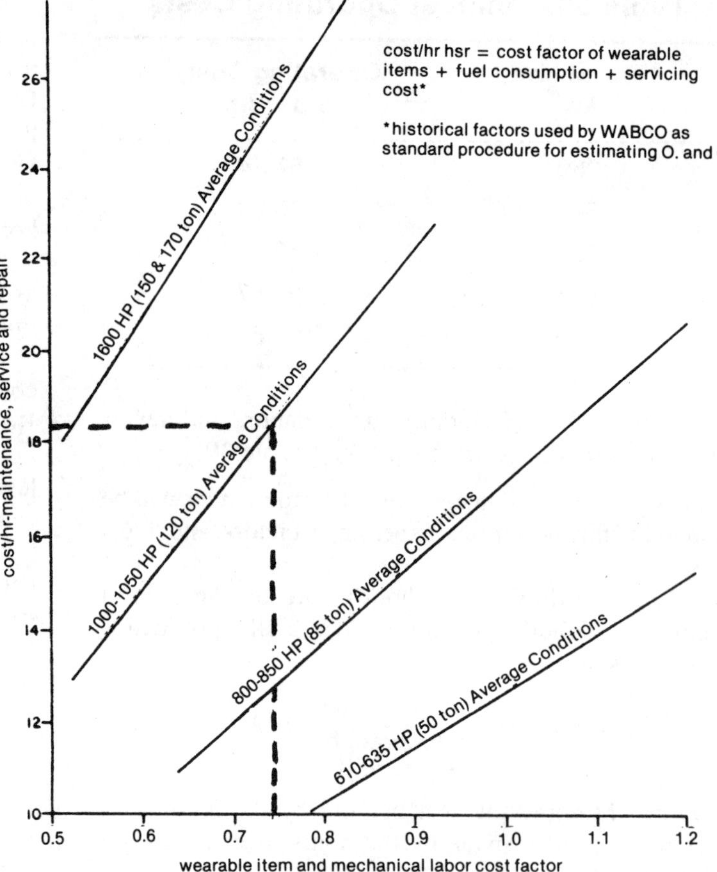

Fig. 10. Maintenance, service, and repair cost. Metric equivalent: 1 st×0.907 184 7=t.

this point at a value of 20% of the total (exclusive of selective mining):

$$\frac{80\%}{\text{loading \& hauling}} = \frac{20\%}{\text{remaining operating}}$$

Refer to Table 34.

The capital costs are now straight-line depreciated for simplicity over the expected useful life of the individual categories. The results are expressed in \$ per hour and when combined with the operating costs provide the total ownership and operating cost (O & O) from which a single fleet will be decided upon for each mine life. The 525-kw shovels and 120-ton trucks for the 12-year life (option 1) will be calculated for total O & O per hour (Table 35).

The total ownership and operating costs per hour for the various mine lifes and equipment combinations are given in Table 36.

Table 31. Truck Operating Costs

Haul truck, tons	Operating cost, \$ per hr
50	35.64
85	50.33
120	55.76
150	60.64
170	69.35

Selective Mining Differential

Under normal circumstances enough information has now been generated to recalculate the ore reserves based on direct operating costs as applicable to the individual mine plant sizes. The only exception in this example is the calculation of direct mining costs. As in any instance of highly selective ore mining, the operating costs run higher than conventional loading and hauling because of the slower rates, closer supervisory control, and extra equipment involved.

The mining operating costs will not be calculated in detail. However, by using the labor; fuel; tires; ground engaging tools (rippers, buckets, etc.); and maintenance, service, and repair parameters as previously generated, a new set of costs can be derived.

The following equipment will be rated as capable of producing a 3500-tons per shift (467 tons per hr) of combined ore and associated waste.

Table 32. Fleet Operating Costs

Mine life, years	Fleets	$ per hr
60	1) 1 250-kw shovel and 621* 50-ton trucks	278.56
20	1) 2 375-kw shovels and 11.38 85-ton trucks	713.16
	2) 1 725-kw shovel and 5.75 150-ton trucks	456.54
12	1) 2 525-kw shovels and 12 120-ton trucks	848.14
	2) 2 625-kw shovels and 9.22 170-ton trucks	834.55
8.6	1) 3 675-kw shovels and 14.31 150-ton trucks	1160.47
	2) 2 725-kw shovels and 10.84 170-ton trucks	967.47

*Table 24

Table 33. Operating Cost Breakdown

Classification	% of total*
Drilling & blasting	10
Loading & hauling (stripping & mining)	80
Roads & dumps	10

*The operating cost allocation is a highly variable item. Other items which might be considered would fall under engineering and development and a general category.

Because the mining usually has no bearing on the stripping fleet selection, the extra costs were not calculated into the total O & O costs. The mining costs ($ per ton) will contribute to the revised ore reserve and will be included in the cash flow analysis (Table 37).

$$\text{Direct operating cost per ton} = \frac{\$2708 \text{ per shift}}{3500 \text{ ton per shift}} = \$0.77 \text{ per ton}$$

or

$$\$2708 \text{ per shift} \div 7.5 \text{ hr per shift} = \$361 \text{ per hr.}$$

ORE RESERVE RECOMPUTATION (Fig. 11)

The direct operating costs will now be tabulated to derive new ore and marginal analysis reserves (Table 38).

The calculation of the new cutoff grade is essentially a multistage process. The first step utilizes the assumed cutoff (0.030) to establish the waste and low-grade to ore ratios to calculate a new cutoff grade. Secondly, use the new cutoff based on the first analysis to define a more precise cutoff grade and resultant ratios.

A sample calculation will outline the 3000 tpd option (Table 39):

$$\frac{17.79}{\$40 \times 0.90} = 0.49 \text{ lb} = 0.025\%.$$

Applying the new cutoff to Table 1 and assuming an even distribution of grade in this example, then 50% of the 4,000,000 tons from 0.020 to 0.030 would

Table 34. Total Operating Cost Breakdown

Mine life, years	Total operating costs, $ per hr	Ancillary operating, $ per hr	Total operations, $ per hr
60 (1)	278.56	69.64	348.20
20 (1)	713.16	178.29	891.45
(2)	456.54	114.14	570.68
12 (1)	848.14	212.04	1060.18
(2)	834.55	208.64	1043.19
8.6 (1)	1160.47	290.12	1450.59
(2)	967.47	241.87	1209.34

Table 35. Ownership and Operating Cost Breakdown

	Capital costs, thousands $	Life,* hr	Ownership, $ per hr	Operating, $ per hr	Total, $ per hr
Two 525-kw shovels	3738	100,000	37.38	179.02	216.40
16 120-ton trucks	7552	50,000	151.04	669.12	820.16
Ancillary, mobile 67%	3291	25,000	131.64	212.04	343.68
stationary 33%	5548†	96,000	57.79	—	57.79
			total		$1,438.03

*Variable.
†Mine life x 8000 hr per year (nonmobile). Also includes $4,000,000 in additional ancillary capital.

Table 36. Fleet Ownership and Operating Costs

Mine life, years	Fleets	$ per hr
60	*1) 1 250-kw shovels and 8 50-ton trucks	505.11
20	1) 2 375-kw shovels and 15 85-ton trucks	1142.49
	*2) 1 725-kw shovels and 8 150-ton trucks	832.84
12	*1) 2 525-kw shovels and 16 120-ton trucks	1438.03
	2) 2 625-kw shovels and 12 170-ton trucks	1436.68
8.6	1) 3 625-kw shovels and 19 150-ton trucks	2059.63
	*2) 2 725-kw shovels and 15 170-ton trucks	1749.02

*Fleet selection chosen for mine comparisons.

Table 37. Mining Operating Costs

Unit	Operating $ per hr	Multipliers		Operating $ per shift
		No. units	Hr per shift	
Front-end loader, 6-7 cu yd	44.00	1	7.5	330
Dozer with ripper, 350-400 hp	60.00	2	7.5	900
Trucks, 35-ton	31.00	4	7.5	930
Drill	25.00	1	7.5	188
Ore controllers & equipment	12.00	4	7.5	360
total				$2708

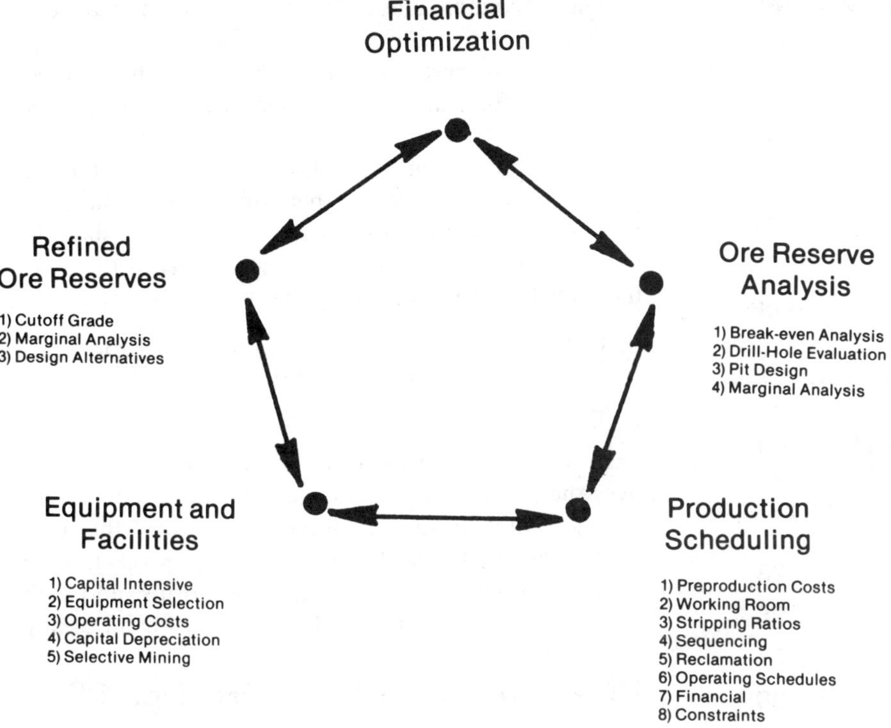

Fig. 11. Ore reserve re-computation.

change designation from low-grade to ore. Two million tons at 0.0275% would be added to the ore category resulting in a new stripping to ore ratio of 22.75:1 and a new associated waste and low-grade to ore ratio of 1.63:1.

The new ratios result in adjustments to the cutoff grade (Table 40):

$$\frac{16.76}{\$40 \times 0.90} = 0.47 \text{ lb} = 0.023\%.$$

Cutoff Grade Approximation

The following relationship exists where the grade distribution is uniform. The formula is assumed cutoff grade minus first cutoff calculation plus (the sum of the first n integers $n(n+1)/2$ of the difference between the first and second cutoff calculation) subtracted from or added to initial assumed cutoff grade equals final cutoff grade.

Utilizing the cutoff grade approximation, the following cutoff grade (3000 tpd option) results (this is only worth considering where the difference between the first and second cutoff calculation exceeds one significant integer, 0.002 in this instance):

$$0.030 - 0.025 = 0.005$$
$$0.025 - 0.023 = 0.002$$
$$0.030 - 0.005 + 0.002\left(\frac{0.002 + 0.001}{0.002}\right) = 0.022.$$

Table 38. Operating Costs

Operating cost categories	Mill capacity, tpd			
	1000	3000	5000	7000
Mill Operating $ per ton*	5.50	5.00	4.50	4.00
Directs & indirects*	4.90	4.50	4.10	3.70
Stripping (inc. ancillary) $ per ton†	0.39	0.23	0.29	0.24
Mining & assoc. waste, $ per ton	0.77	0.77	0.77	0.77

*Table 4.
†(Load & haul hr per yr × $ per year + ancillary cost factor)/tpy.

Table 39. Revised Production Costs

Category	Ratio	$ per ton	Total cost, $
Stripping	26	0.23	5.98
Assoc. waste	2	0.77	1.54
Ore mining	1	0.77	0.77
Milling	1	5.00	5.00
Directs & indirects	1	4.50	4.50
total			17.79

Table 40. Refined Production Costs

Category	Ratio	$ per ton	Total cost, $
Stripping	22.75	0.23	5.23
Assoc. waste	1.63	0.77	1.26
Ore mining	1	0.77	0.77
Milling	1	5.00	5.00
Directs & indirects	1	4.50	4.50
total			16.76

Table 41. Cutoff Grades

Plant size, tpd	Cutoff grade, %
1000	0.033
3000	0.022
5000	0.023
7000	0.020

The new cutoff grades for the four options are given in Table 41.

The ore reserves are different for each plant size option resulting in a possible modification in mine life (Table 42):

Proper project duration should also include the marginal analysis, although at this point, it is not necessary to determine whether or not the marginal ore should be averaged with the ore or milled separately at the end of the mine life. In the 1000-tpd option:

$$\frac{\$23.61 \text{ per ton} - \$13.21 \text{ per ton} + \$0.50 \text{ per ton}}{\$40 \text{ per lb} \times 0.9}$$
$$= 0.30 \text{ lb per ton} = 0.015\% \text{ (Table 43)}.$$

In Table 43, the mine life varies up to a plus 15% over the original calculations due to the lower marginal analysis cutoff grade. However, because of the uncertainties associated with any long-range projection, the 15% is considered within operating norms, and therefore, the 0.015% cutoff will be utilized for all options.

FINANCIAL OPTIMIZATION (Fig. 12)

Equipment is depreciated and replacements are scheduled by the expected usages. Operational costs used are the result of the detailed plant size calculations (Tables 44 and 45).

The next step is to set up a cash flow statement for the four mine life options (Table 46).

In this example, it can be seen that the net present value (NPV) continues to increase with an increasing size of operation. Using the NPV as the indicator, it would appear that further plant size optimization is probable. However, when the cash flow statement is prorated to 9,000 and 11,000 tpd operations, the net present value only increases by 2% and 0.3%, respectively. While the prorating does not take into consideration different equipment capital requirements and operating costs, the end result is usually within ±5 to

Table 42. Revised Ore Reserves

Pit size, tpd	Ore mill tons & grade	Low-grade & assoc. waste, millions of tons	Stripping, millions of tons	Ratios: Tons strip / Tons ore	Ratios: Tons low grade & assoc. waste / Tons ore
1000	12.8 @ 0.088	29.2	364	28.4	2.3
3000	17.6 @ 0.069	24.4	364	20.7	1.4
5000	16.8 @ 0.070	25.2	364	21.7	1.5
7000	18 @ 0.067	24	364	20.2	1.3

Table 43. Marginal Analysis

Plant size, tpd	Cutoff grade, %	Ore tons (mill) & grade	Low grade & assoc. waste, tons	Mine life, years	New mine life Orig mine life, % variation
1000	0.015	22.0 @ 0.058	20.0	60.0	100
3000	0.014	22.8 @ 0.058	19.2	21.1	106
5000	0.013	23.6 @ 0.057	18.4	13.2	110
7000	0.011	25.2 @ 0.054	16.8	9.9	115

10%, which is as good an accuracy as other price and operating projections.

With that in mind and also the fact that with an extremely short duration project, an inordinate amount of capital is required, most of which cannot be properly sized or fully depreciated. The equipment becomes salvage and must be transferred or sold. Also, the operating problems, such as traffic congestion, skilled manpower availability, etc., become magnified. The optimum plant size is, in reality, obtained by maximizing the NPV only to the point where the most workable plant operation exists.

Based on the preceding, the 7000-tpd operation is the optimum plant size for the example given although minor adjustments may be made by comparing 6000 and 8000-tpd operations against the 7000-tpd plant. Another circular analysis would be required.

The final point to be considered is how to optimize the cash flow in the 7000-tpd operation. The solution will be how to best utilize the marginal ore (Table 47).

The following calculation addresses the maximizing of the operational cash flow (Table 48):

Case I. Marginal analysis cutoff for uniform production discounted at 12% on an 8.6-year life

$60,962,000 per year × 5.184 (cum. PW factor—end of period) = $316,027,008.

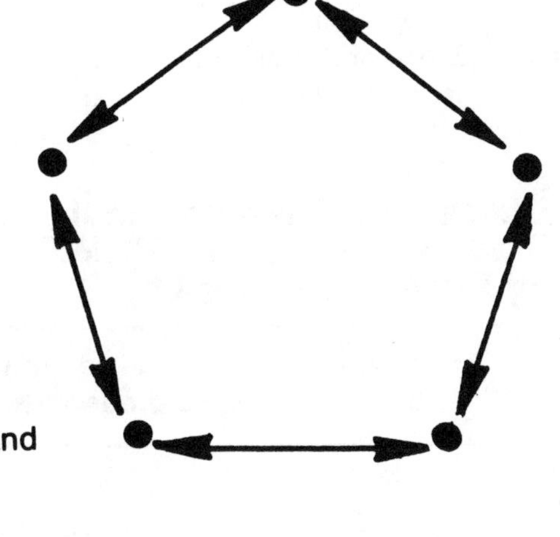

Fig. 12. Financial optimization.

Table 44. Summation of Capital and Operating Costs

Plant size, tpd	Life, years	Initial capital cost, $ thousands*	Replacement capital cost, $ thousands†	Annual Deprec., $ thousands	Total operating cost, $ per ton milled, $ Direct‡	20% of total ancillary	Total
1000	60	31,517	27,577	985	24.02	6.01	30.03
3000	20	46,071	22,927	3,450	16.23	4.06	20.29
5000	12	60,129	16,281	6,368	16.82	4.21	21.03
7000	8.6	77,550	11,230	10,323	14.32	3.58	17.90

*Tables 4 and 28.
†Based on the depreciated life of the equipment. The mill is depreciated over the life of the mine. A straight-line depreciation is then used against the total capital over the life of the mine for tax calculations.
‡Table 38.

Case II. Marginal ore at end of mine life. For years 1 through 7:

$71,561,000 per year × 4.564 (cum. PW factor—end of period) = $326,604,404.

For years 7 through 8.6:

$11,599,000 per year × 0.62 (cum. PW factor—end of period, 8.6 yr-7 yr) = $7,191,380.

The total net present value is $333,795,784.

As previously illustrated in Table 8 and related Cases I and II, the time value of money is critical when optimizing a property. Postponing the milling of the marginal ore results in a 5.6% increase in the total NPV.

Again, it must be emphasized that the time value of money is critical when optimizing the planning of a property. Postponing the milling of the marginal ore to the end of the property life results in an increase in the total NPV, if management can accept the small profit margin during the final phase of mine life.

The first circular analysis has now been completed with a resulting closely defined mine plant size and mill feed criteria. The degree of accuracy presented herein is adequate for most engineered feasibility studies. Further optimization will result by closely bracketing the target and doing detailed ancillary costing in addition to the loading and hauling costs. The cycle is repeated in exactly the same manner including the ore reserves which will be recalculated as detailed in the first analysis.

Sensitivity Analyses

It is appropriate, at this point, to mention the value of various sensitivity analyses to the optimization procedure just outlined. The most critical points in terms of effect on the evaluation are: (1) ore reserves, (2) mill recovery, and (3) commodity selling price.

Each can have a profound effect on the ultimate outcome of the property as each point has a direct influence on the total value of the deposit. Each carries at least as much weight as the sum of all operational considerations.

However, at any particular point in time, a decision must be reached on whether or not the planned project

Table 45. Annual Revenue: Utilizes the Marginal Analysis Cutoff for an Average Ore Grade of 0.058%, a 90% Mill Recovery, and a Selling Price of $40 per Lb

Plant size, tpd	Tons per year, thousands	Product, lb	Annual revenue, $ thousands
1000	365	381,060	15,242
3000	1095	1,143,180	45,727
5000	1825	1,905,300	76,212
7000	2555	2,667,420	106,697

Table 46. Cash Flow Statement ($ Expressed in Thousands)*

	Mill size			
	1000 tpd	3000 tpd	5000 tpd	7000 tpd
Revenue	$15,242	$45,727	$76,212	$106,697
Operational expenses	10,961	22,218	38,380	45,735
Operating cash flow	4,281	23,509	37,832	60,962
Annual replacement capital	460	1,146	1,357	1,306
Before tax cash flow	3,821	22,363	36,475	59,656
Tax Calculation				
Operating cash flow	4,281	23,509	37,832	60,962
Capital depreciation	985	3,450	6,368	10,323
Taxable income before depletion	3,296	20,059	31,464	50,639
Depletion @ 22%	1,648	10,030	15,732	23,473
Taxable income	1,648	10,029	15,732	27,166
Tax @ 50%	824	5,015	7,866	13,583
Net Present Value				
Before tax cash flow	3,821	22,363	36,475	59,656
Tax	824	5,015	7,866	13,583
Cash flow after tax	2,997	17,348	28,609	46,073
Cum. PW factor @ 12%	8.81	7.91	6.56	5.49
DCF after tax	26,404	137,223	187,675	252,756
Initial capital investment	31,517	46,071	60,129	77,550
Net present value	−5,113	91,152	127,546	175,206
Rate of Return				
Before tax cash flow	3,821	22,363	36,475	59,656
Tax	824	5,015	7,866	13,583
Cash flow after tax	2,997	17,348	28,609	46,073
NPV @ 12%	−5,113	91,152	127,546	175,206
NPV @ × %	10 −198	50 −3,568	60 −21	80 −792
ROR (interpolation)	10	48	60	79.8

*No preproduction stripping or mining costs or investment tax credits are used in order to simplify the calculations.

Table 47. Ore Reserve Comparisons

	% Cutoff	Reserves, tons in thousands
Marginal analysis reserves	0.015	22,000 @ 0.058
Break-even analysis ore	0.020	18,000 @ 0.067
Marginal ore	0.015	4,000 @ 0.017

Table 48. Ore Reserve Value Breakdowns

Grade, %	Net value per ton, $ per ton	Reserve life, years	Annual value, $ thousands
0.058	23.86	8.6	60,962
0.067	28.01	7.0	71,561
0.017	4.54	1.6	11,599

will become reality. Top management must rely on sound engineering judgment to decide on the basic merits of the proposition.

Financial analysis, at this point, can begin manipulating the sensitivities associated with the uncertainties inherent in any new project. It is the engineers' responsibility to provide as sound a basis as possible upon which a particular project can be weighed against other capital investment alternatives, each of which carries a rate of return sensitive to presently unknown factors.

ACKNOWLEDGMENT

The author would like to acknowledge the assistance given him by Dominic T. Arrieta, senior mining engineer, Minerals Exploration Co., in the Production Scheduling and Equipment and Facilities portions of this chapter.

REFERENCES

Buchella, F. H., Jr., 1973, "Open Pit Equipment Selection and Maintenance," *Proceedings,* UN Interregional Seminar on Application of Advanced Mining Technology, DP/UN/INT-72-064, pp. 268-291.

Douglas, E. J., 1971, "How to Make the Most of a Mining Investment," *Mining Engineering,* Oct., pp. 64-67.

Gessel, R. C., "The Basic Principles of Estimating," Wabco Technical Bulletin.

Lillico, T. M., 1974, "Economics and Open Pit Mine Design," *World Mining,* March, pp. 60-64.

Terex, 1970, "Production and Cost Estimating of Material Movement with Earthmoving Equipment," Gen. Motors Corp., OH, 77 pp.

Contents

The Committee, editors

22 Workshop Discussion

The Committee
Workshop Participants

The Role of the Drilling Contractor

BY FERRIS E. SAINSBURY

Ron Haxby, *Occidental Minerals:* What is the best method for drilling a 12-in. diam hole to a depth of 1200 ft and maintaining the deflection within 25 ft from vertical?

Ferris Sainsbury: In order to drill a 12-in. hole down to 1200 ft, in most places you would have to employ a rotary tricone bit. With a down-the-hole hammer, I expect by that depth significant inflows of water would be experienced. Even with foams, we would not be able to get the water out of a hole of that diameter. So the down-the-hole tricone method of drilling would be best. To keep it straight, we would probably employ the largest square drill stem compatible with the diameter of the hole.

Ron Haxby: Is it common in drilling contracts to put in a vertical hole clause?

Ferris Sainsbury: You can put in the vertical hole clause.

Ron Haxby: Does it cost more?

Ferris Sainsbury: Normally, if the hole deviates, the driller can correct it with deflection wedges and will charge it back to you at cost, plus 10% or somewhere along that range. Any drill hole has a tendency to go perpendicular to the dip, and there are times when stabilization systems are not successful in preventing the hole from turning into the dip of the formation. The only solution, in this situation, is to go in with an oriented wedge and kick the hole back out.

Ron Haxby: How long does it take to set the wedge?

Ferris Sainsbury: To date, we haven't set a wedge on a 12-in. hole. On a 4-in. hole or on an NC wire-line hole, we can usually set an oriented wedge in about 8 to 12 hr.

Pete Fowler, *Consultant:* How long does it take, as a rule, before information becomes public in the cases you mentioned concerning oil exploration drilling?

Ferris Sainsbury: The period of time allowed for the petroleum industry to maintain a tight hold on information varies from state to state. Quite frequently, most states in the petroleum-drilling industry will allow an oil company to hold the information on a confidential drill hole for a period which could vary by states. I think in Utah, if I remember correctly, it is one year before some information can be made public. Actually, Mr. Fowler, mining companies should get those drill logs because they provide good information on the formations to be drilled in that area. (This source of information is often overlooked).

Herbert Fine, *Mackay School of Mines, Reno:* I would like to know if it is possible to get some printed information I might be able to use to instruct students about the use of wedges, their placement, and similar kinds of information. Is there anything in print pertaining to what I am talking about?

Ferris Sainsbury: Very little. We are a backward industry in drilling, and I know of only two books that you may already be familiar with. Cummings has a book out in Canada called *The Diamond Drill Handbook,* which is available to the industry. Bill Acker of Acker Drilling recently put out a book on drilling methods and techniques. Right now these are the only two that come to my mind. Both are good books for students because they start out with a layman's type introduction and then proceed to more complex subjects.

?: What do you foresee as the most likely major breakthrough in exploration drilling costs?

Ferris Sainsbury: I think, as I said earlier, the major breakthrough is going to be a method of using the rotary air-type down-the-hole hammer or the hydraulic down-the-hole hammer if one can be developed. Such techniques would reduce the cost of extending a hole down to where core recovery is required. With the small rotary rigs we use, advances drop in the neighborhood of 50% when a switch is made from a down-the-hole hammer to a tricone bit. Even below the water table, it is possible to use the hydraulic down-the-hole hammer and obtain a 100% improvement in the rate of penetration now experienced with tricone bits.

In France, efforts are being directed toward developing an electric down-the-hole drill which is lowered into the hole on cables. The motor is equipped with pads which lock the mechanism to the sides of the drill hole, allowing it to turn a core barrel at the bottom of the hole. They were able to successfully complete one 2000-ft hole. In the second hole, the equipment was lost. I don't know whether or not development is continuing on the down-the-hole electric drill.

There were some studies made a few years ago by Christensen Diamond Products in trying to develop a hydraulic motor-driven core barrel. To date, success has not been achieved. It is possible to use a hydraulic motor to turn large diameter core barrels, but the fins on the hydraulic system are not capable of developing the necessary torque when scaled down to a 3-in. hole.

In Australia, they are doing some work with a down-the-hole bit that can be changed for core drilling without pulling the drill rods. The device locks into the inner tube on a wire-line system so that the bit can be extracted with the inner tube as the core is extracted. The bit comes to the surface on a cable and goes back in and locks into the bottom of the drill string. In certain places where low bit life is experienced, considerable savings could be generated by a system which would allow replacement of the bit without pulling the drill string. For example, the bit life was 20 ft for a particular property at a depth of 3000 to 4000 ft. Consequently, every 20 ft in depth we had to pull 4000 ft of rods, put on a new diamond bit, and go back. I think the cycle there was drilling 8 hr, pulling rods 8 hr, and putting the rods back 8 hr. Obviously, development of a wire-line bit system would reduce costs.

Kim McCarter: Can you suggest precautions for engineers or companies in regard to lubricants used on drill-rod connections? Specifically, are trace elements present such as copper, lead, or molybdenum?

Ferris Sainsbury: There certainly are. Some of the lubricants we use contain molybdenum. If you are engaged in a moly exploration program, you must be careful. One time when I was on a drill crew, we developed a beautiful scheelite project until we split the core and found out that the fluorescence was from our drill lubricant and not from scheelite. So there again, I say it is necessary to consult with your drilling contractor and ascertain from him what he intends to use as a lubricant. If the lubricant he intends to use contains trace elements which will interfere with your analysis, then a substitute lubricant must be used.

Pete Fowler: One more question about this matter of records. As you said, logs for oil exploration holes must be turned over to the state. Wouldn't it, in the long run, be to the advantage of the industry as a whole and to all mining companies if all hole logs were eventually turned over to the state? Such a practice would prevent information from being lost to everybody should a company obtain certain results and then abandon the project.

Ferris Sainsbury: I'm sure it would, Mr. Fowler. If a data bank were established for all drilling information, duplication of effort could be eliminated in drilling an area which has previously been drilled. I would think that such a data bank would be something the industry should develop. I can assure you that it will not be coming from the contractor.

B. H. Ecklund, *Minerals Exploration Co.:* Assuming similar conditions in the future, can the price increases over the last four or five years be used as a guide for the next four or five?

Ferris Sainsbury: Well, we're like everyone else. The inflation factor has been going up. Major increases have been experienced in the cost of equipment and labor. As you may be aware, in the last two years cost of tubular steel has increased 78%. In 1969, a rotary drill cost about $38,000. I ordered two of them this year, and they were $175,000 each. Our labor costs have been going up on the order of about 8% a year. If the trends continue, you are going to see about a 10% per year increase in drilling prices.

B. H. Ecklund: Is that what you have experienced on drilling contracts in the last four or five years?

Ferris Sainsbury: In 1975 we saw a jump of between 12% and 15%. At that time there was a seller's market for contract drilling, and we took advantage of it to catch up. In the last two years the price in our company has increased between 8% and 10% per year.

Development Drilling

BY RICHARD D. CALL

Pete Fowler: About a year ago I was laying out a preliminary program for a client (it didn't go ahead as it turned out), and I recommended an initial aerial survey. I found out that in the first year about $40,000 or $50,000 would be spent for the property. The estimate for the flight, an aerial photograph that would have contours on it, plus an independent contour map was about $4000, and I wanted this information to begin with. The company that was going to put up the money said, "We don't want to do that now." I just wonder what your evaluation would be of how soon aerial mapping should be initiated.

Richard Call: I agree with you—right in the very beginning. We just did a preliminary pit design for a property where the pit extended from one map area with good detail to another area where the scale was 1:10,000. We experienced a 50-m bust between the two maps. We had to do some real adjusting. This happened to be a high hillside, and so it really created some uncertainties in the tonnages involved in preproduction stripping. I have seen just too many hassles about location—where things are. In the long run you're well ahead to get good topographic control right in the beginning. You may spend a lot of time later on resurveying and redrafting because of inadequate original topographic control. So, I think that the payoff (for good topography) is really right in the initial stages of a property. In general, aerial mapping is the best way to go.

Bill Hustrulid: Rick, during exploration drilling, what other data should be recorded besides just looking for values? I know that you're talking about development drilling here, but perhaps you have an ore body somewhat defined. How early, or what should you be doing during exploration drilling regarding geotechnical information such as oriented core?

Richard Call: I think the RQD information is worthwhile. It is a fairly simple exercise. You are going to measure core recovery anyway, and the RQD can be measured at the same time. Photographing the core should also be a part of the record, particularly since the photographs can be a permanent source of information even after the core is split.

I think you are premature on oriented core unless it is critical to your ore search. There may be some cases where you are strata-bound or your geologic interpretation with regard to structure is critical in the ore search. In such cases you may want to go to oriented core, not so much from the geotechnical standpoint but from the ore-search standpoint.

I still think a few representative samples for rock strength testing are justified in the early stages just to give you some sort of a feel for the rock types and strengths with which you are dealing. They don't have to be tested; they can just be set aside to serve as a small library of samples. This library could be utilized should preliminary pit plans be desired to assess the viability of a particular deposit.

Principles of Ore-Body Modeling

BY BRUCE T. STANLEY

Pete Fowler: You mentioned having to get all the data in initially. Isn't it fairly easy with a computer to add to the data base if you generate additional information in the future?

Bruce Stanley: Easy? I don't know what to say about the ease of this task. If you want to include additional information at a later time, it must be compatible with existing program routines and format structure. It is best to include all the information which may be of value in the initial stages. At this point in time you may still have people available who were involved initially, and it is much easier to eliminate unwanted information than to expand existing programs or files. I do not wish to leave you with the impression that if data is omitted at the beginning it is totally lost. It is just more difficult to include.

Randy Weingart, *Baroid Div., NL Industries Inc.:* I have a question regarding what types of ore bodies are amenable to block model representation. In other words, would block modeling be best for a porphyry-type deposit but not as good for, say, a vein-type deposit? Is block modeling better for igneous or sedimentary deposits?

Bruce Stanley: It would be more difficult to use a block model for a vein-type deposit—more difficult because the boundary areas are more difficult to determine. With a regularized block concept, I would say that a block model would be easiest to use with a large porphyry-type ore body. A bedded deposit would also be quite applicable. You also want to include such things in your model as air-ground interface and try to estimate the amount of ore vs. waste in certain edge blocks. This is a concept that should be dealt with in following papers.

Randy Weingart: It appears obvious that this type of modeling would be best for large homogeneous deposits like porphyry. That's the reason why I had the question—to see if it is applicable or useful for other types of deposits.

Bruce Stanley: It is quite useful for other types of deposits. I'm not at all saying it should be limited to any particular type. I think each and every type of ore body can have its own block model. We are not only dealing with a type of mineralization, but we want to be able to deal with the handling of this information in making intelligent decisions on the property whatever type of mineralization it may be.

Bill Robinson, *Getty Oil Co.:* Does AMAX have a computer package that handles variable-sized blocks that they developed in-house for their simulation studies or did you go out and purchase a "canned" program?

Bruce Stanley: I'm familiar with two properties, Climax and Henderson. Climax is currently using the Mintec package for their ore reserves. Mintec is based now in Tucson. Henderson is now using an in-house package based on geostatistical techniques. I believe Climax is still using Mintec for pit generation outlines. Marc Lemieux developed a program for optimizing the pit limits at Climax. At Henderson, it's a different situation in that it is an underground operation. So it is necessary to use somewhat different parameterizing features. It's mostly self-developed. Now, there are a number of packages available. As yet, I believe the only package we have purchased is the Mintec.

Bill Robinson: Are you using variable-sized blocks in your studies now?

Bruce Stanley: Marc Lemieux's package uses variable-sized blocks.

Bill Robinson: Was that in-house?

Bruce Stanley: That was in-house. At Henderson, we do use the variable-sized blocks; however, we also have a basic size block in use, i.e., $80 \times 80 \times 50$ ft high. This size is used because of the mining plan and layout. The center of the block is directly over our drawpoints, and we can stack blocks up to create an ore column. Of course, this all pertains to an underground operation. The various blocks can be recombined into larger blocks for the purpose of modeling stopes and for such other things as financial analyses on the "ton of ore prepared." People in our home offices in Greenwich for instance like to deal with the prepared-ton concept. This concept considers a certain amount of money to prepare a certain ton of ore. It is possible then to obtain a good feel of how much of the mine is developed vs. how much of it is producing and have an insight into the cash flow.

Bill Robinson: When you develop your mineral inventories, what do you use for your cutoff? Do you have a break-even cutoff, or do you use multiple cutoffs?

Bruce Stanley: We deal with geologic reserves as being the whole mineralized area in the ore body. Once we have an estimate for an ore column or a series of blocks it is possible to determine what can be mined at a profit under our conditions. The acceptable columns become our ore inventory file, or the actual model that we are going to mine. Thus, geologic reserves are converted into minable reserves.

?: I believe that it was Michel David who raised the specter of the "vanishing tons problem" related to models of small blocks. Would you care to comment on that?

Bruce Stanley: The problem of vanishing tons? This is a problem dealing with geostatistics. I don't wish to infringe on papers yet to be presented, but possibly I could offer this explanation: Let's say we have a mineralized area, and it is drilled out on a specific grid. What appears to be a rich area may turn out to be an area of lower grade when drilled on a finer basis. The converse problem is also true. The problem here could stem from the fact that an adequate drilling program was not carried out initially in characterizing the spatial continuity of the ore. Therefore, we assume that a block of ore is high-grade just because of the neighboring values. If we had drilled the deposit out on a closer grid within the area of interest, within the range of the variogram, perhaps we would have come out with something different, perhaps not. The production phase also impacts this concept of vanishing tons. For example, if you predict 100,000 tons at 0.45% MoS_2 and you stop drawing ore at 80,000 tons because the grade drops below 0.2% cutoff (for instance 0.19%), you certainly will come out short on tons and short on grade. But if you go ahead and extract the other 20,000 tons, you are likely to come closer to your original estimate and actually come closer to obtaining the metal content originally predicted.

Kim McCarter: What practical limits should you place on subdividing a given volume into smaller and smaller blocks? At what point do you cease to generate meaningful information?

Bruce Stanley: The size of the block depends on the type of ore and the spatial continuity within the ore. If you try to determine continuity between 10-ft samples you may find that no interdependence exists. If you recombine the assays into larger blocks, you may find that the actual structure of the ore body is masked by the smaller sample. If you model on a 10-ft basis, then you should mine on a 10-ft block basis in order to obtain what you estimate. If all your estimates are based on 1-ft samples, then the only way you can expect to obtain what you predict is to mine 1-ft samples. Therefore, your mining scheme dictates, in part, the size of block that should be used.

Mineralization Interpolation Techniques

BY WILLIAM HUGHES AND RODERICK DAVEY

Bill Hustrulid: Would you explain again how the polygon is formed at the boundary when you run out of drill holes or sample points?

William Hughes: I kind of slid over that point because of what we do when we get to the periphery of our knowledge. We extrapolate, and I knew that further on in the paper I was going to warn everyone against extrapolating. That's really what is done with a polygon that is near the boundary; a predetermined polygonal form is generated. The one I like to use is an octagon at the chosen radius of influence. So as the boundary is approached, the radius of influence is measured out in the direction of the boundary. Then we would generate the polygonal form in that region. What we are really doing is extrapolating out to the radius of influence.

Kim McCarter: Another way of accomplishing the same thing is to draw a circle about the boundary point using the radius of influence. Then any polygon can be drawn tangent to that circle.

William Hughes: Right. The circle itself would also represent a polygon.

Kim McCarter: There is also another procedure. You can construct the polygonal mesh as previously described and then extend straight-line segments between peripheral sample points. The mesh with the extrapolated polygons would represent an optimistic tonnage while the smaller mesh, limited by the straight-line segments, would represent a more conservative tonnage.

Bill Allen, *Milchem, Inc.:* I work for an industrial minerals company which owns a predominantly bedded deposit. It is intensely folded, fractured, and really kicked around. We did some work with polygons and it got pretty messy. Management required the concept of "proven" and "indicated" ore, and we intuitively selected anything within 25 ft of an ore showing in a drill hole as proven ore regardless of geologic boundaries. We made one pass through the data by hand since we had to start from scratch with some technique of ore estimation. Previously, we had worked with cross sections, but I was intent on using the polygonal method which I have previously worked with and have gained some confidence in it. However, it is extremely difficult to handle a practical ore estimation problem by hand. The logistics of just moving the numbers around is tremendous. I would also like to know if you have ever encountered the problem with concentric polygons? Our deposit is quite "rivety" in certain areas. There may be as many as 25 intercepts of ore and waste in a 25-ft length of drill hole. In each case we were projecting these waste intercepts out to the polygon boundaries. A further complication is that we would consider the area within the first 25 ft of an ore intercept to be proven and the second 25 ft to be indicated. We also had a classification of indicated waste which further complicated the situation. I would prefer to employ geostatistics, and we are conducting some studies into this area. We will also eventually go to a block model method. Have you ever heard of the double polygon system?

William Hughes: I've only heard about it. I've not had any experience in trying to use it, but it would appear to me that in a complex situation such as you described, the best thing to do is to go to a more sophisticated technique, one that would show a gradient. Possibly a summarizing program could then be applied to the gradient to ascertain what is proven and unproven.

Ron Haxby: Do you maintain a constant exponent for the weighting function over the entire deposit or do you allow it to vary?

William Hughes: The only experience I have had is that with actual ore bodies we maintain a constant value and treat the difference in geology by using the equivalent distance method. However, I have done some experimenting with some theoretical ore bodies in which I have varied the value and have obtained results that are comparable with geostatistics because really we are doing the same thing. The number crunching process has changed somewhat, but the primary thing is that geology is considered as part of the interpolation scheme. To me, consideration of geology is the primary advantage of geostatistics.

?: How do you handle allocation of values to blocks when only one drill hole is available for information?

William Hughes: Once again preliminary sets of rules would have to be established between engineering and geology. Typically if there is just one drill hole available, and there are no other drill holes within two times the radius of influence, you are probably dealing with a waste area. If a hole encounters ore, the usual practice is to drill additional holes to ascertain the limits of mineralization. So, as you may suspect, the answer becomes academic. However, you could, and probably would in this kind of an academic environment, generate the entire polygon around that hole. Another way to handle this problem is to introduce "pseudo holes," holes that have never been drilled but holes that perhaps you know would be in waste areas. It would then be possible to use these holes to form a gradient which can be used to assign values to blocks.

Estimating Mineral Inventory

BY MARVIN BARNES

Pete Fowler: I would appreciate your thoughts on two subjects: First, would you discuss how changing conditions, such as prices, affect mineral inventory and ore reserves. Second, could you discuss how your strata-bound variogram would apply, say, to a lead-zinc replacement deposit in limestone.

Marvin Barnes: To answer your first question, a mineral inventory is readily convertible into an ore reserve when one devises a mining plan and selects only those blocks of the mineral inventory which can be extracted economically. This is the difference between a mineral inventory and an ore reserve. An ore reserve today may not constitute an ore reserve tomorrow. Unfortunately, many of us copper miners have just had this principle demonstrated to us. The object is to construct an accurate mineral inventory, then it is possible to apply any cutoff or any constraint to that data base to arrive at an ore reserve. Consequently, the mineral inventory will not change with economic conditions. An ore reserve changes with economic conditions. Some parts of the mineral inventory can be recovered only at higher prices. Now, back to answer your second question. Yes, I have generated variograms on strata-bound deposits, both in the new lead belt of Missouri and old lead belt. I have done this even in the Tennessee zincs, which, by the way, are extremely difficult to develop usable variograms for. The lead belt strata-bound deposits produce relatively readable and usable variograms.

Pete Fowler: . . . 1500 ft . . .?

Marvin Barnes: Well now, that is an example of a particular deposit. I've tried to make the point that each deposit is like a fingerprint; each is different. You can't assume that when you go from one deposit to another that you're going to have the same variogram. Now there is some hope that if they are genetically related and in close proximity to each other, their variograms will be similar. Consequently, if you have sampled one, you possibly can use the same intrinsic functions for drilling out the next one. This would give you a big leg up, but you won't really know the form of the new variogram until you compute it.

Jack Crawford: In planning further drilling, would it be appropriate to assume that your drill-hole spacing should not exceed the range derived from the variogram?

Marvin Barnes: Yes, that is an accurate statement. If you exceed the range, you have no correlation between samples. It is just like independent random sampling. It is desirable to establish enough correlation so that you can use the data to assign values to the volume of rock between the holes.

Jack Crawford: So in that regard, it would be preferable to have your distance between drill holes less than the range.

Marvin Barnes: Absolutely. The practicing geostatisticians usually agree that ⅔rds to ¾ths the range is usually a good assumption for drill-hole spacing. Of course the spacing must be tempered by economics. If the deposit is highly variable and is characterized by a large nugget effect, you simply may not be able to afford to sample it at distances of ⅔rds or ¾ths of the range. It may cost too much money. But that's a question someone else must answer. If you have that type of deposit, and you go ahead and develop it without knowing how that mineral is distributed, you can get yourself in deep water in a big hurry. I have seen this happen in industry more than once. So maybe the expenditure of more money on sampling earlier would be money well invested. Yes, your question is a good one. A good starting point is from ⅔rds to ¾ths the range to get your best estimates. Let me also make one other point. If the nugget effect in any deposit is over 50% of the total variance, kriging probably will not provide any better estimation than inverse distance squared or some other technique. If you have a well-defined variogram and small nugget effect, it will improve the estimation. Not only will geostatistics improve the estimation, it will also tell you how good your estimation is, which no other technique, of course, can do.

Bill Hustrulid: Practically, from a management standpoint, does the geostatistical approach shift the "hard" decisions from the engineer to management? For example, an engineer can quote the probability of failure of a slope. Management must then specify the risk they are willing to take. In the case of geostatistics, it appears that management must specify the risk they are willing to take that the values are not there. Does such a practice lead to more holes? What other problems does it present?

Marvin Barnes: Let me answer that question the best I can, having had much experience in this area. After running an estimation for the ore blocks, say, on a certain bench, then it is possible to contour the estimation of error associated with the variance. The resulting contour map is very useful. It is possible to use the trace of these estimation variance contours to determine whether a particular block should be classified as proven ore or probable ore or thrown into the possible category according to the confidence. So I would say that on a relative basis they are very meaningful. The absolute values may not be meaningful.

One of Bill's earlier questions concerned the extrapolation process from unknown areas. If the estimation variances are contoured outward from any of these holes on the periphery, they increase sharply. It is then possible to limit those blocks which will be accepted as part of the ore reserve to those whose variances do not exceed a specific relative value.

Ron Haxby: If you have a very erratic deposit, could you use a variogram to help determine the optimum pit bench height?

Marvin Barnes: Yes, it is possible. Of course erratic deposits are the most difficult kinds to deal with, but the variogram can become a powerful tool under such circumstances. If we look at individual samples, we get extreme variances, but if we look at samples in the deposit at the same distance apart, we may develop patterns that cannot be obtained otherwise. The smaller the block becomes, relative to your sample spacing, the higher the estimation variance. With high estimation variances, there is no certainty that the predicted grade will be realized. So, if you have a highly variable deposit, there's no need trying to estimate a small block. It is better to deal with the largest possible block relative to your geologic constraints and to your sampling interval. In other words, a block roughly equivalent to your sample interval would not be too large. In fact, if it is a very erratic deposit, you should have several samples in a block for which an estimate is desired. In the case of the gold deposit example in my paper, that's exactly what had to be done. The only way we could know the value of any particular block in the deposit was to have enough samples to develop a statistical variance under the terms of classical statistics, and that takes more than one sample. Geostatistics is a powerful tool, but it still takes good geologic common sense and judgment to apply it properly.

Mineral Block Evaluation Criteria

BY RODERICK K. DAVEY

Pete Fowler: You mentioned treating different parts of the pit as different costs. If you decide one part of the pit is too low-grade for your profit, then you're either stuck with mining it anyway and putting it out on the waste dump or else leaving it unmined. I've seen limestone quarries that were operated that way. They had a little dike here and there so that they had the quarry all chopped up, and it made just an impossible situation.

Roderick Davey: You have to average in some things, true? The size of increment that you are dealing with is important. You obviously don't leave small segments of interstitial waste in the pit. Its removal must be paid for by surrounding ore.

Jack Crawford: One thing one might do as part of that averaging technique would be to look at the cutoff grade, not only in terms with which Rod has dealt but also at a break-even standpoint, i.e., a zero profit. If material is put over on the waste dump, the loss may be greater than if milled. You may end up not actually having a break-even situation, but rather a lesser loss than if wasted.

Roderick Davey: Well, we get into an entirely different subject if we pursue the concept of minimizing losses. In a particular deposit, a vertical section (as sketched) depicts a small portion of the mineralization that is of subeconomic value in terms of the material that would have to be removed directly above it and laterally; it obviously gets included with the ore grade mineralization surrounding it if the economics warrant it. You don't leave hunks and bits of material for various operating and economic reasons. There is always the question of how small these elements are that you add together. You obviously at some point in time have to add pieces together into a reasonable operating plan. In some cases, the increment may represent shovel cuts. In other cases, the deposit may simply be subdivided into a long-range mining sequence. In a feasibility study, you may just divide the deposit up into very large concentric increments.

Bill Fall, *Getty Oil Co.:* In your work, when you are getting into your break-even cutoffs, do you ever get into estimating any reclamation costs?

Roderick Davey: It is a cost that has to be applied as any of your other processing costs. We have had one situation where this applied. For a pound of product, there was a cost applied for reclamation that increased the break-even point. That's all there is to it.

Cost Records of Open Pit Mining

BY ROBERT F. WINKLE

?: What are the principal differences between availability and utilization as you used them?

Robert Winkle: The use of availability and utilization has been bent rather badly out of shape. As far as I am concerned, mechanical availability is the one you hang your hat on. What this means in definition is that you take your total time available (let's say you are talking about a shift) and bring it down to a minimum area. You've got 8 hr in a shift and for 6 hr of that shift that machine has been available to operations. For 2 hr it has not been available to operations, and when I say not available, I am talking about service as well as lubrication time. As far as I am concerned, when a truck goes into the shop, it is no longer available to operations, even if it goes in for fueling and tire checks—maybe a 24-hr check. I think this is the one availability figure that means most to an operator.

I've talked with construction people. I was talking with some people from Peter Kiewit once, and I asked them what kind of availability they were getting on those dozers. They said, "Oh, we're getting pretty good. We require 94% availability." I said, "You're out of your minds; 94% availability is impossible." Then I asked, "Wait a minute. How much are you running them?" "Well, we only work five shifts a week, you know, day shifts." "Well, sure, with the rest of those shifts available (16 shifts) for repair, you ought to have something up to 90."

But for most mines that operate on a 7-day, 24-hr day schedule, the mechanical availability figure (which is strictly the time the operator can use it) is the one that I prefer. There are all sorts of other things. There is the physical factor, and the use of the availability is an extremely important thing. How much are you using it? If you don't use it more than, for example, 20%, then your availabilities ought to be way up in the 80's and 90's. If you are trying to match your utilization with your availability—just maybe a percentage point or two spread—then your availabilities are going to drop down. This must be taken into consideration.

?: In terms of utilization, what makes the difference, manpower, labor shortage, weather?

Robert Winkle: Well, there is always the possibility you might have too much equipment, too. And this should enter into any future purchases or original purchases. How much are you really going to be able to use this equipment? If you are buying it for a five-shift operation, then you should demand more out of it in the way of availability. If you are going to use it on a 21-shift-a-week operation, you are going to have to use it all the time when it is available; then you are going to have to accept a much lower availability.

Pete Fowler: You mentioned in the first place that you couldn't find records of this being done, which indicates that most companies don't do it. Perhaps the reason that many companies don't keep such detailed records is that it is an immense volume of paperwork, and then who's going to look at it and study it? You mentioned that it should be the first-line supervisor and I agree with that. But maybe what you have to do is have the computer analyze it to really make this thing effective, and then you can point out to people the more prevalent things they ought to look at.

Robert Winkle: When you refer to the tables included at the end of the paper, you will see that they start with a summary. The summary makes it possible for management to get a rather quick view of where they stand in the operation. Then there are the supporting documents. These make it possible for management to look back and find out where these original costs in the summary were generated. You say there's a lot of detail that will be developed in this, and you're right. On the other hand, if you don't have a rather decent detail, particularly on your mobile equipment, you are guessing when it comes to trade-in time.

Bill Hustrulid: Being familiar with many of the surface mining companies, I find that they keep fairly good operating maintenance repair records. The question for me is how they include major overhauls in those costs; these seem to be estimated on an as-occur basis. How do you spread those things over an operating cost and how do you examine major overhauls in terms of the performance of the equipment when they happen so sporadically?

Robert Winkle: The major overhaul cost is one that some companies will accept and some won't. Again, it's a philosophy of trade-in. There are those who think they should run a truck 30–35,000 hr and then give it a complete overhaul and maybe get another 15–20,000 hr on it. There are others who consider it a basket case at that time and say, let's trade it in and get a new one. At the present time, I have been looking at some of these philosophies, and I find people (in dozer trade-in, for example) who will go anywhere from approximately 10,000 hr, which Caterpillar advocates (they can sell lots of Cats that way) up to approximately 30,000, 35,000, 40,000 with some pretty major overhauls thrown in between. This is a philosophy which a company will develop. You can't say, "This guy's wrong, and that one's right," because it depends a lot on the condition. Let's say you're in an area where it is difficult to get good tradesmen, good mechanics, good electricians. Maybe your best bet is to buy components and change the components. If you're down where you have a ready labor market where you can get good mechanics, maybe you want to go a lot further. You may want to go further than in-frame overhaul of engines. Maybe you want to actually do a complete overhaul on an engine. Again, it's something a company has to develop for itself and usually it's by experience.

William Allen: I believe what you're referring to (and I'm not sure of this) is the cost accounting method of charging out major repairs over the life of the equipment. I've worked on large construction jobs with many pieces of equipment, and I am familiar with several companies that do this. Of course, the construction industry traditionally looks at engine hours or some time element as such and will charge what they call an equipment rate or rental rate. These rates supposedly reflect the upcoming costs of, say, a transmission or engine or something major, even though they are not incurred at that particular time. So, for instance, if you had a dozer or something like that, $50 an hr would be charged against operations, even though it might be brand new and might go several thousand hours. They really, in accountant's language, overstate cost. I understand this is quite a problem in that you might get into generating less profits, or you might get into a problem tax situation. But I know that this is what they do in the construction industry. You can find good reference books that will delineate these costs. I think one of them is *Construction Costs, the Index*. There is a certain company that puts this out. Incidentally, I talked to an accountant recently, and we went around and around on just this same thing. It sort of smooths out the peaks and valleys of huge layouts.

Robert Winkle: Yes, actually there is mention in this paper of utilizing a similar situation for an hourly cost charge-out, where a company will use charges over a year and keep them current by dropping out the 13th month and adding the next month, as they advance through the year, to give some kind of an hourly cost.

Jack Crawford: To further comment on this within my experience, the major overhaul costs within the cost recordkeeping approach that Bob Winkle mentions are immediately expensed in their entirety at the time they are incurred. However, with the hours and the work-order system, one could get a history on a piece of equipment where you can break out the overhaul costs from the normal running repairs that occur between overhauls. Also, you have the number of hours consequently occurring between overhauls. With that type of information, it is feasible to run it through certain types of financial analysis in order to determine the point at which you reach a trade-off between continuing to go through this overhaul procedure cycle or trading it in on a new piece of equipment. To that extent it falls into the realm of a specialized case of investment analysis.

Robert Winkle: One of the biggest problems most companies have is some sort of a formula to determine when they might trade in a piece of mobile equipment. Unfortunately, in most large organizations, time becomes an extremely important element. Most large companies are reluctant to allow a general allocation fund to be developed. They like a little better control of their money. They don't trust middle management to spend it without having some type of specific unit it's going to be spent for. And what happens too frequently is that you may have a fleet of dozers and maybe dozer No. 1 and dozer No. 3 look pretty bad to you and you say, "Well, we've examined the record and here we find out these fellows have been running quite a lot of time and they are accumulating a lot of cost. Let's get rid of those next year. We'll put in a request for expenditure to get rid of them." The time involved between the time the original papers are prepared and when they finally get to the board of directors to be passed upon is possibly anywhere from 8 to 12 months or more. They go to an engineering group that picks them apart and sends them back and then they come back to the operating group and then they may lie around in the main office in New York or San Francisco. Meanwhile, the operator out on the hill has to put up with this broken-down machine and finally gets to the point where he can't tolerate it any more. So he spends a lot of money, puts a new motor in the thing, a new engine, fixes the final drive, the tracks, the side frames, and he's got himself a nice piece of equipment. About that time the authorization comes from New York saying, "Yes, spend the money and trade that dozer No. 1 in." Well, he doesn't want to trade it in at this time. But very frequently there have been some atrocities committed in trade-ins because of this kind of system. This occurs because of the reluctance of top management to make a general allocation fund for trade-ins.

?: Generally speaking, experience has been varied as far as rubber-tired equipment in open pit as well as some underground LHD equipment is concerned. Can you address yourself to the subject of recapping tires for a minute or two? Have recapping techniques been developed to such an extent that they are even suitable for consideration, or are they not?

Robert Winkle: Recapping tires has been done by many people with varying degrees of success. Usually your best success occurs with the rubber tire rather than the haulage-truck tire. In order to recap successfully, you have to have a good carcass. If you have a cut in your carcass, most recappers won't even try to do anything for you. They won't give you any kind of a guarantee. If with haulage truck tires, for example, you try to get the ultimate out of the original rubber, you very frequently will damage your carcass to the extent that the recapper isn't too interested in giving you any kind of guarantee for the recap. If you, on the other hand, pull that tire with enough rubber on it to guarantee that you won't damage the carcass, you will probably find that you won't make any money doing it. Now, rubber-tired dozers, graders, and most of the slowly moving equipment are another thing. Usually recapping is a profitable thing.

Price Forecasting and Sensitivity Analysis for Economic Analysis of Final Pit

BY ALAN C. NOBLE

Herbert Spahr, *Johns Manville:* Will you attempt to forecast the price of uranium in the future?

Alan Noble: I cannot talk about uranium today. We have a very heavy lawsuit going, and I cannot answer any questions about uranium today.

Bill Hustrulid: Alan, I thought it was a very good paper. Something that I have seen in the past that people have done in trying to model those peaks and valleys is to do a regression analysis but to use terms like $ax + bx^2 + cx^3$, etc. I'm sure that you know this, but some of the people here perhaps don't know that this is a very good way for interpolation between values that you know but is a very bad way for extrapolation. Whatever you do, don't use those kinds of models to model the peaks and valleys and then predict what is going to happen in 20 years. These functions oscillate wildly and will give you terrible numbers, either terribly low or terribly high.

Alan Noble: To make a further point, one thing that is very common in statistics is trying to fit a curve which models these cyclic fluctuations. The reason I didn't do this (and I suppose I should recommend against it) is that when we look closely, the cycles aren't anywhere near the same length. If we try to follow the cycles and we are wrong—let's say, for instance, that we guess our price is going to rise and then we end up with the cycle putting us down—the amount of error in that case is going to be twice the error of the average trend.

Ron Haxby: An economic pit design involves both price and cost. How do you project your costs?

Alan Noble: The thing that is obvious is that we are going to have to play the same game with costs as we will with price. I think, basically, we have a little better handle on what cost is right now than what price is. The difficulty that arises is how do you compensate for differential rates of increase in price vs. cost. The answer to your question is that there really is no answer other than looking very closely at the situation and seeing if this kind of situation exists. Normally, what happens is that we assume that the price and cost increase at about the same rate. That might be true for the overall picture, but it certainly is something to consider.

Rich Hurt, *Mobil Oil Corp.:* I realize why you can't answer the question on uranium, but I would like to ask the audience if there is anybody who has had any experience in predicting uranium prices, because it is a pretty unique situation. Can anybody comment on that?

?: I think that if you could you wouldn't be in the mining business.

Alan Noble: I don't think there will be very many takers anywhere on that.

Pete Fowler: You mentioned your sensitivity analysis. Have you given any thought to applying risk analysis to this situation?

Alan Noble: In many respects that's implicit in what you are doing with the sensitivity analysis. Going back to the 80% confidence limit on those regression lines—if we make our first pit at the lower confidence level and then make our second pit at the average and then the high end—what we are essentially doing is, the low level is going to represent the pit that we would expect about 20% of the time, the middle pit about 60%, and the high pit another 20% of the time. There is a little risk partition built into that.

Marvin Barnes: Have you attempted to define a correlation, for instance, between historic copper prices and costs in the open pit industry? I have attempted to do this and have found in some cases a pretty good correlation if you can call a 0.64 or something like that a good correlation. But it is not. The interesting thing is that cost and prices are not totally unrelated; they are correlated. And since your open pit design is a function of cost-price, this does give you a bit of a leg on the problem. An increase in cost has the same function as a decrease in price. They are not totally independent variables.

Alan Noble: The thought I have on that is that if we could get a handle on what prices were running worldwide, not just domestic, then we should have a pretty good relationship between price and cost. The problem in using that and applying it for design of a particular pit would make it a rare circumstance that the cost and the cost increases for a single mine would parallel the industry costs at all, if for no other reason than that generally the stripping ratios on a particular mine are getting greater all the time, and then you have on the other side of the coin average grades getting lower for the world as a whole. So the cost trend there is being forced up in that manner.

Art Young, *Dravo Corp.:* It looks to me like the long-term trends represent pretty much the escalation. You have an escalation of costs and escalation in prices. Is the law of supply and demand ever taken into consideration? When I went to school, this was the law upon which we used to judge prices, whether your supply is low and your price goes up or vice versa. Do you take that into account at all? I think that would be very pertinent in uranium.

Alan Noble: The difficulty in defining the supply-demand relationship in the mining industry is that there is such a tremendous lag between having a nice high price out here and being able to get a mine in to do something about that high price. In fact, that's one of the things that tend to cause the cyclic nature of price trends in something like copper especially. Modeling of that is exactly what the complex econometric models attempt to do. They say, here's the situation that we have for all the properties that are in production or that could be fired up right now. Here's the worldwide demand, and they make a projection based upon demand, and they also make a projection based on new supplies from known properties that are in development and try and combine all these together to get the supply on one side and the demand on another side and get the price. Personally, I don't think that that is going to be terribly useful when you're looking out 20 years, because we're talking then about very hypothetical exploration relationships and what grades are going to be then. I just don't know how far we can take that for the purpose of designing the final pit. On the other hand, that's the way to go for finding out what kind of profit you are going to have out of the first pit increment that you open up.

Richard Call: In this type of statistical extrapolation of past trends, the assumption is that the conditions are going to be the same in the future as they have been in the past, and I kind of question this. Particularly, I am thinking of the world market where you have the nationalized copper production, Zambia, Zaire, Chile, where their response is sort of the reverse. As the price drops, instead of the normal response of reducing supply, they increase their production because they just need the dollars. I think that you have a change in the rules of the game here so that any extrapolation of past trends could be very misleading.

Alan Noble: That's an excellent point. That, in fact, is the main pitfall in doing a trend analysis. With respect to the current situation, I think it remains to be seen how much that will affect the long-term. It is a good point. When we do a trend, we are assuming that the market production framework that we are dealing with is going to follow the same trend that it has. If that changes, the trend is not valid. I might add that the backup for these price lines for copper and lead are in the paper, but these are intended to be instructive rather than something that we should charge out and use. I think that if you were really going to use a model for your long-range planning, it deserves a little more work than this has in it. This demonstrates the idea of confidence limits on your forecast, and it demonstrates the long-term projections, but it isn't really intended to be a solid model at all.

General Components, Data Collection, Remedial Stability Measures

BY BEN L. SEEGMILLER

Analytical Design

BY DERMOT M. ROSS-BROWN

Jack Crawford: You say you have been dealing with this as a probability from the standpoint of whether a failure will occur. But assuming you have 100% probability that a failure will occur, it takes a certain amount of time for a failure to develop. If one, for example, could mine to that slope so that he could get in and get out within a given amount of time, perhaps that failure wouldn't be a problem. Is there any way from a design standpoint to get the time factor into the evaluation of slope designs?

Dermot Ross-Brown: There are several ways in which the time factor can be incorporated into a slope design. The best one is past experience. For instance, it may be possible to study failures in a similar material at a nearby pit and observe how long it takes for the failures to develop. This time factor can then be incorporated into the slope design in cases where the life of the pit walls is expected to be fairly short, as might occur in an overburden stripping operation for coal or uranium.

With regard to analysis, time can be incorporated into a design using appropriate explicit finite difference or explicit finite element codes. Experience is still required, however, in setting up the constitutive models for the behavior of the rock mass.

From an operations point of view, I do not consider it practical to play with the time factor in the design of the majority of pits. While it is true that failures in relatively soft materials may take one year or more to develop, failures controlled by discontinuities in hard rock often give a warning of only a few minutes or less. By means of an effective slope monitoring program and a carefully controlled mining operation, it is possible to live with failures. However, I do not consider this to be a design feature. It is rather a means of taking a remedial action to mitigate the effects of an unplanned failure.

Bill Hustrulid: Dermot, I would like to strongly disagree with your suggestion that we can determine the rock strength, the cohesion, and angles of internal friction to within 5 or 10% and discussion of calculating stresses as to whether they are within 1 or 5%. It seems to me that this whole thing revolves around how we can pick strength factors to put into those models. I would like both you and perhaps Ben to comment on my comment as to whether you really think that you can get that kind of accuracy. We have a hard time reproducing results in the laboratory, much less saying how the laboratory results compare with what you've got with a massive slope.

Dermot Ross-Brown: I would agree with that. What I was trying to emphasize is that the accuracy of the available analytical techniques normally is far greater than the accuracy of the available input data. And that's true whether we are considering the properties of intact samples or the properties of the rock mass. For instance, it is usually necessary to scale down laboratory strengths to obtain rock mass strengths, and the potential errors are enormous. Simply, there have not been enough large-scale experiments performed. By a large-scale experiment I mean an open pit mine, for example, that has been completely monitored from the time it was started until the time it was finished, so as to obtain an understanding of how the rock mass deforms and fails and to enable the constitutive equations of the rock mass to be accurately written down.

Ben Seegmiller: The choice of input values, particularly those for the shear strength, can have a decided effect on the end results. For this reason I believe there is a lot to be said for a sensitivity analysis. Such analysis allows the effects of input data variability to be determined in terms of the safety factor and/or probability of failure. For example, the effect of a 10, 20, or even 50% change in the shear strength may be noted on the safety factor. If the safety factor is very sensitive to small changes in the shear strength values, perhaps we should be putting a greater emphasis on shear strength determinations. On the other hand, if the safety factor is relatively insensitive to changes in shear strength input values, it may mean that no further testing or strength analysis would be required. In other words, I am saying that the sensitivity analysis allows us to determine the importance of the variability of shear strength data, or for that matter, the importance of the variability of any of our input data. We may then concentrate our efforts on putting the emphasis on the parameters which will have the greatest effects on the end results.

Pete Fowler: In the discussion yesterday, Stanley Michaelson's paper referred to the Ruth pit. I understand from the discussion that it had a moderate slope which was eventually steepened up to 59°. Would you discuss this from the point of view of maybe starting with slopes that aren't too steep, and then finishing by steepening them, in order to get the last out and the relative factors of safety with time and so on? What could be done by starting with a flatter slope and then finishing with a steeper one?

Dermot Ross-Brown: I think it depends upon whether you are talking about small pits or large pits. If you are talking about a small pit, it is necessary to design the pit completely before starting to excavate it because you only get one chance to mine it. For instance, if you make a mistake in deriving the slope angles and a pit wall collapses, then either the ore is lost or else a lot more money is spent in stripping a new pushback. This can make the whole operation marginal or even unprofitable.

On the other hand, if you are talking about a large porphyry copper deposit, the situation is better. For one thing, the walls are rarely all waste, since an assay cutoff rather than a geologic cutoff is applicable. So what was considered waste at the design stage may be later put through the mill or leached to generate some additional income. The philosophy with a large pit is to start off with steep initial angles which you know are stable and which give an economic pit. Since several pushbacks are required to mine the ore body, the slope stability design is a progressive process. As the mine develops you can collect more data, you can observe the behavior of the rock mass, take measurements, take remedial actions, and refine the analytical techniques in order to fine tune the slope angles. By the time the end of the pit life approaches, you should have a good idea of how the slopes behave and what the optimum angles are of the final slopes.

Richard Call: I would like to comment a little on that, Dermot. One problem with starting with flatter slopes and going to steeper slopes, particularly if you do have a discrete ore cutoff, is that you can't put back what you shouldn't have taken away. But if you do start with a steeper slope and you do get evidence of instability, you can go back and strip and go to a flatter slope. There is an argument for actually starting steeper, and then if you do get evidence of instability, go to a flatter angle. So if you mine a slope at, say, 30° and you find you could go to 40°, if you don't have the ore at depth to go after, you've stripped material that you didn't need to. The economics are still there for steeper working slopes even in the porphyry copper case where you have a working slope and a final slope. We've been talking about final slopes, but if you look at the time factor of money and the discounting, often the angle of the working slopes has more economic impact than the final slope does. I don't think that you should restrict your stability analysis to the final slope design. This is so that when you look at the deferral of stripping during the life of the mine that you can achieve by working with steeper working slopes, you can show quite a difference in your cash flow.

One other comment and this is in regard to the choice of the probability of failure that you are willing to accept. I think this can best be approached by an economic analysis. There is a simulation method available, a program that Young Kim wrote up for the Canadian government for the CANMET manual, which is a simulation of the mine operations through the life of the mine. By taking these probability-of-failure curves and assigning a cost of failure to the failures, it goes through a simulation and comes up essentially with an economic evaluation. It samples the probability of failure schedule and generates slides during the life of the mine and assigns the cost of failure to those particular time periods when failures occur and then discounts this back to a present value. You ultimately come up with a net present value as a function of slope angle. There is an economic evaluation technique available to make this choice of what probability of failure you actually use. You don't have to do it just by looking at the curve and saying. "Well, I guess I can stand a 20% probability of failure."

Hand Methods

BY BEN C. KOSKINIEMI

Automated Methods

BY R. M. (MIKE) ROBB

Bill Hustrulid: You described this last one using the cone. Is that a volume method for looking at stripping ratios as opposed to the sort of boundary method that Ben talked about? What I am wondering about is, if you have waste zones within your ore body, in either of those two cases, are they taken into account? If you had a waste horst in the middle of your pit, how do you include that in terms of stripping? Is it included in your model? Is it included in Ben's model?

Mike Robb: It has to be included in your model. You make your basic model. It has to tell you what is waste. You've got a block in there with no grade or below your cutoff grade; that is your waste. As this pit is generated and you match it against your model, it is going to come back and tell you what the waste is.

Bill Hustrulid: Does it depend on the way . . .?

Ben Koskiniemi: Yes, it's also included. You're looking at both the waste and the ore.

Bill Hustrulid: I agree that you are looking at the waste at the pit boundary, but what happens if you have waste in the center part of the pit? How do you take care of that? For example, up at Climax you have that big crystal core in the middle of the pit. How do you take that into account?

Ben Koskiniemi: First of all, it is taken into account when you calculate your total ore reserves and total stripping ratio. You could come up with an overall stripping ratio that was greater than your break-even stripping ratio if you had enough interior waste. That is a very good observation and it is part of the designer's problem. In looking at each section, the presence of interior waste is usually apparent, and the designer must check the section to make sure that this waste isn't going to interfere with the economics, if it is going to be included in the mine.

Bill Hustrulid: What happens to your analysis after you've gotten the outline and then started looking at the sections with the stripping ratio?

Ben Koskiniemi: You calculate the overall stripping ratio for each section of your ore body. When looking at the break-even pit limits, the stripping ratios might at the boundary vary between 1:1 and 10:1 and even higher, depending on your ore grade at the pit limit. However, the overall stripping ratio for a section and/or pit might only average 2:1 or 3:1.

Jack Crawford: Ben, one of the things, too, in this type of thing (whether it be by hand or the automated approach) is that as you work from your initial surface toward your ultimate pit, you're taking increments called pushbacks or whatever. If those are sufficiently small, pick up the impact of these interior waste zones.

Ben Koskiniemi: That's correct, if you started as Mike showed us and you were working from an initial to a final pit limit. The way I explained it, you're really doing something similar by just visually estimating the location of the final pit limit. If you can hit the break-even limit on your first try, it speeds up your work. I've done by hand essentially what Mike has done with the computer, where you begin at an initial existing surface and take incremental pushbacks until an ultimate surface is reached. Pushback design is usually the second phase in hand designing a pit, and it is a necessary part of your detailed design for developing your annual mining plans. Usually the first step in designing a pushback is to pick as simple as possible a configuration for your bottom level in the pushback and design your levels above all the way to the surface. Each pushback is checked to determine if it meets the economic criteria to justify its inclusion in the pit limits. The pit limits will be expanded as long as a positive incremental cash flow is achieved. I think this points out that there are several methods by which the ultimate pit can be located.

Mike Robb: I think your stripping ratio often ends up being a function of your product price. A silver mine can carry a lot bigger stripping ratio than a sand and gravel operation. What you can sell it for is your whole criterion. I think something, too, to keep in mind (as Ben pointed out), is to get these increments and to start going back into the things. You've got to keep these increments in minable sizes. I can relate a horror story where a fellow designed an ultimate pit. He had very little operational experience and never left the corporate office. He came back with a 10-m pushback as the ultimate pit. That might be the ultimate economic pit, but you couldn't mine it with the size of equipment being used at the mine.

L. M. L. Klingmueller, *Lucky Mc Uranium Corp.:* I wonder if the last two authors could get together. On the one spectrum, we have a computer system; on the other one, we have a pencil or a little slide rule, or something like that. With the new instrumentation coming out (the hand calculator), maybe we can meet this somewhere. Would either one like to comment on that?

Mike Robb: I think that will be discussed by Tom Couzens who follows me. He has developed a very effective system that utilizes both approaches.

Kim McCarter: Just a comment on Bill Hustrulid's question. The conventional or traditional hand method for designing an ultimate pit is usually to determine the position in section of a straight-line slope where the ore grade will support the stripping. Once this position is found, you will have an indication of where you can place the three-dimensional limits. Once the three-dimensional limits are established, the volume defined between these limits and the surface is divided up into intermediate plans. If you run into a situation like Bill was talking about, where you have a waste area in the middle, then the stripping ratio of one or more of your intermediate plans will look very strange. Consequently, I don't think, that even with the traditional hand methods, that a large block of centrally located waste will evade detection.

Mike Robb: That's where you have to smooth your stripping ratios and allocate your equipment. You're right; otherwise you end up doubling your stripping in any given year, and we can't have that.

Pete Fowler: I get the impression that we are thinking almost 100% in terms of truck haulage. Would any of these things that have been talked about be influenced for a change in any way, if you were to consider, for example, going to a conveyor haulage system?

Mike Robb: I've used it in a pit where we did extensive conveyor work. Your slopes and your access ramps come into effect. Yes, it's certainly applicable. I know a case where it has been used in a rail haul, too.

Selected Aspects—Berkeley Pit

BY DON HENDRICKS AND ALAN DAHLSTRAND

Pete Fowler: Somebody pointed out a little while ago the importance of communications and being able to read blueprints. I've come to the conclusion that I don't understand the word pushback. Would you explain it for me?

Alan Dahlstrand: Let me flip back a bunch of slides, maybe back to my first one. Let's get back to this one, and we will work with the first two slides (see figures in the chapter). Look at this section or any one of these sections, for example. I can extend this line out, and let's say, for example, the first time we opened up the mine, we had an outer limit. By virtue of the outer limit and by virtue of the slopes we planned into it, we defined how deep we were going to get. Let's say that this is the remainder of that first pushback. If I extended these lines out in that area, that's what we mean by a pushback. It's a segment of the mine. In other words, we get into the mine with a final mine design, and the thing is so big that nobody in the world can afford to strip it all at one time. So we have to start with a small piece. We have to try to get into an economic position, an economic advantage, as fast as we can; that is, get into production as fast as we can. So we start with a small piece. We strip a small piece and get into ore. The size of the next push, the size of not only that pushback but every subsequent pushback, has to be such that it allows you to physically maintain your ore production also. So, theoretically, this one is still in ore and still has some ore left, so we start up at the top, and we start stripping another slice. It's just another slice of the pie. We take a little bit at a time so that we can minimize the giant stripping requirements that would have to be done if we mined the entire mine at one time. As Mike Robb mentioned (I can't remember all you said, Mike, because you rattled off about a half a dozen terms), basically, they all mean the same thing. Pushback or slice or phase plan are all related as far as their physical ties are concerned. They are all forecast out in time so that each time period of each pushback can actually be watched, mapped, and forecast throughout the mine life.

Tom Cherrier, *Climax Molybdenum:* Do you have a figure for average tons per shovel-shift for your 15-yd shovels, and also do you have a rule-of-thumb figure of number of broken tons that you would like to keep ahead and available for a given shovel?

Alan Dahlstrand: As far as the tons per shovel-shift for our 15-yard shovels, I think the overall average we're looking at is probably 11,500 to 12,000 tons per shovel shift. That's highly variable, depending on the age of the shovel and depending on the area of the pit that we have it in. Some of the areas such as high up on the north wall in the Berkeley pit are really hard, and our tons per shovel-shift drop like a rock there. As far as broken tons ahead, that's been quite variable over the years, too. There have been times when we've had a week's worth of broken muck ahead of us. In the last couple of years we've probably cut that down. In some places, we are only a day or two ahead of our shovels. That depends on the area also. It depends on shovel breakdowns; it depends on drill breakdowns. So the availability of the equipment has a lot to do with it.

Tom Cherrier: Maybe I should rephrase my question. What would you like to have?

Alan Dahlstrand: Right now, I don't know. You would like to say as little as possible from an economic point of view, but it's impossible to foresee when equipment is going to break down, too. So you run into trouble. No matter what you do, you're always surprising yourself. And it's usually the wrong way. The tons per shovel shift are based on the total mine. And, Tom, that's running an average of six shovel-shifts or six shovels to a shift.

Cliff Maddocks, *US Borax & Chemical Co.:* Could you restate your reasons why you went to 10- and 30-ft berms?

Alan Dahlstrand: Well, there are various reasons. We've had some differences of opinion on that subject even within our own operation and within our own planning staff too. But the alternating 10- and 30-ft berms were done partially for safety reasons. By going to 40-ft benches, some of the operators felt that this would keep the old berm configuration that we had before (of a berm between every other bench and wiping out the berms in between) and that we would be going to an 80-ft wall. This is kind of an ominous wall to be looking at if you are a shovel operator sitting below it. So, there were some thoughts about this. I think the main considerations were for safety. That's why we tried to break it up with a 10-ft berm, but we've been having some problems with blasting the berm off. Where we have left the berm in decent shape, 10 ft fills up awfully fast with raveling material. But even so, it's got its pros and cons. There are reasons why we like it and reasons why we don't. So we argue about these things among ourselves all the time, too.

Cliff Maddocks: We had the same situation at US Borax with its latest pushback, and we're pleased with our results.

Alan Dahlstrand: With alternating larger and smaller berms?

Cliff Maddocks: Yes, we had 60 without an intermediary bench, and then we went to a 10-ft intermediate bench, just like you did, and we're pleased with our results.

Alan Dahlstrand: That's interesting. I might make one more comment on that. I'm sure we are going to go back and reanalyze our berm configuration, how wide our berms are, and how often we leave them, after we get into our entire slope analysis a little bit more. As we find out that particular areas of the pit require certain kinds of slope angles to be able to mine the area and we find out a little bit more about what is happening structurally in the pit, I think that then we will have some variations not only in overall slope angles or interim slope angles, but we may run into some variations in our slope geometry, in sizes of berms, numbers of berms in a certain vertical interval, etc. So, we're trying to keep an open mind on this and just try to watch our situation.

Cliff Maddocks: Have you thought of taking it one step further and going to 20-ft berms consecutively?

Alan Dahlstrand: Twenty-foot berms. You mean every bench with a 20-ft berm? Yes, our operators are looking at that a little bit in certain areas in the alluvium right now. But as far as the rock is concerned, if we went to that over the entire pit, we'd probably be shutting ourselves out, causing ourselves a lot of problems as far as mine access is concerned (access for power, access for drainage facilities, etc.). So it's kind of nice to keep a combination if you can also please people as far as the safety factors are concerned in other areas.

Dominic Arrieta, *Minerals Exploration:* Alan, can you or Tom Couzens comment on what criteria you use for a catch-bench design? How do you size the catch bench?

Alan Dahlstrand: In other words, the width of our berms?

Well, most of it has been based on past experience and what we've seen work. In our old 45° overall angle, we used 33-ft benches. Often times, those filled up fairly fast. Sometimes it was a little tight on access to certain areas for power drops and for drainage, but we lived with it for a long, long time. We felt that under certain circumstances in certain areas, if we want access, we might leave especially wide areas. So it was felt that these 10 and 30-ft wouldn't really hurt us. The 30-ft would still leave us access; the 10-ft would give us a little extra margin of safety. Maybe, maybe not. If the 20-ft were clean, fine; if they are filled up, then they're probably just the same as an 80-ft wipeout. We have tried these on 38° slopes. We had 52-ft berms, and these worked very well as far as catch benches were concerned. They also gave us adequate access. But the 38° slopes in certain areas . . . it's nice where you need the room and where you can't hold the slope any steeper. But it's always nicer to go to a steeper slope even though you have to go to a smaller berm for economic reasons.

Jack Crawford: I would like to raise this question, not only to you Al but to anyone else in the audience. Has anyone had any experience with problems regarding this matter of catch-bench or safety-bench configurations in their dealings with state or federal mine inspectors, regulatory agencies, etc.?

Alan Dahlstrand: I never have personally, but our operating people have received a lot of questions and a lot of comments from MHSA officials coming through on their routine inspections, particularly in areas where we have had some failure and have had a lot of raveling. Usually it pretty much stays on a fairly friendly question and answer level. Basically, questions are on what we are going to do about it, and what we plan to do to insure the safety of the people working under it. But no real problems have come up yet, just lots of communication between the Anaconda operating people and the MHSA inspectors.

Jack Crawford: Have you made any actual changes in your configurations as a result of these discussions, and have you had to enter into any written commitments?

Alan Dahlstrand: No, not as yet.

Pete Fowler: You mentioned you were going to backfill part of the pit. Why do you backfill it and with what?

Alan Dahlstrand: We backfill it with mine-run waste. The reason we backfill it is because in that particular southeastern portion of the pit, we've reached the bottom of our ore body. Since we've opened that up, it's a nice short haul. I might add that most of our dumps and our tailings pond are up to the northeast. We don't want to build our tailings pond any higher; it's about a 600-ft lift. So this backfill is going to be a welcome sight because that is going to shorten our haul and reduce our life considerably.

Operating Layout and Phase Plans

BY THOMAS R. COUZENS

Tom Couzens: I won't try to pretend to be an oracle. Perhaps we could look at this as more of a sharing of experience. If someone does have a question I will be glad to try recounting my experience, but if others want to comment on the things I have said, that would be good, too.

Bill Hustrulid: I would like to know how you get started. You make a set of topography maps and plans and things. How do you get started designing the pit after you've got the outlines and information from people like Mike Robb who define the final pit limit? What do you do practically in sitting down and making a design?

Tom Couzens: I go back and I usually start drawing a lot of maps. You mention topography; you've got topography. You've got a set of bench maps; you should have bench maps. You may be working with some sections too, but in most open pit work, you're primarily working in a horizontal section of the bench-map configuration. You've got guidelines, some kind of guidelines as to production by this time. They're probably not the final answer because a lot of this is give and take; it's trial and maybe error. Maybe you'll hit it, but you will probably develop a pretty good feel about what ore you're going to be striking for first. I normally start to draw some hypothetical stages. If we have these phases developed (which we probably have) that release a certain amount of ore (it may be a one-year or a five-year approximate supply), and we know the stripping that is associated with that, we'll start drawing that out into a time sequence. We're laying it out on maps and measuring and doing adjustments until we get it into the parameters of the equipment capability. Am I addressing the question you are asking, Dr. Hustrulid? It's a paper simulation. You may go to the computer to get the integration of volume and the final statement of tonnage and grade. I hope you have some kind of equipment like this.

Jack Crawford: Tom, one thing we tried a couple of years ago at Nevada Mines Div. of Kennecott which seemed to help quite a bit in giving us some leads to targeting these intermediate range plan shells, working from the initial surface out to the final pit, was to generate in effect some ultimate pit plans inside of the final one that were based on higher cutoffs. This was done so that you got some feel as to the economic balancing between the geologic and economic parameters. If you are looking to maximize your present values, you'd like to hit the hot spot first and progressively work toward the lesser profitable materials as time goes on. And we found that this proved to be quite a valuable technique in at least giving you a preliminary look-see at how you might work out the more detailed operational planning sequences.

Tom Couzens: Yes, I think that's very good. You probably do have a pretty good feel for things, but you should certainly be open to alternatives.

Kim McCarter: On your detailed plans do you make a deliberate effort to make those end at a specific period of time or do you just make a geometric plan and then divide it up, for example, by year or by month? And if you do, how do you go about assigning an average to a particular time period?

Tom Couzens: There is more than one stage in this. I like, personally, to see time periods. And I probably have time periods in mind most of the way through this. But, of course, they are going to be adjusted quite a bit. The slope considerations come into the phase design very heavily. But pretty soon you override that when you're getting into this simulation of production. I often think you tend to override that. You're looking at, not the full extent of the slope, but at a series of steps. This is not always the case. In some pits, of course, you would go back to the final, the first shot. I think Dr. Ross-Brown referred to that when he said it would depend upon what kind of a pit it was.

Jess Martinez, *Great Canadian Oil Sands Ltd.:* I would just like to make a comment. From my experience, it sometimes pays to find out from the first-line supervisors whether they can understand the various plans. In some cases there are new supervisors who have come up from the ranks. They would never say that they don't understand and this becomes a big frustration to everyone when planning ends up in the waste basket for the simple reason that they don't understand it.

Tom Couzens: I agree with you. I know that that is a problem. Part of communication is to educate the supervisors and, of course, they are educating us too all the time because they are bringing their problems back to the planner. I would avoid as much as possible taking any particularly distorted computer output to operations. I don't like to see different scales. I think I take a very elementary approach to these things.

Richard Call: Tom, you skipped over the waste-ore ratio vs. time. I wonder if you could comment on that.

Tom Couzens: I would be glad to. I do have some transparencies of a couple of these if you are interested. This is an actual example of a large iron mine (see Fig. 1 in the chapter). Their management had committed the equipment and a large expenditure to an average waste ratio configuration. But when you actually planned out the time periods, you found out that this was the kind of thing that happened: They were able to get going at a 1:1 ratio, but then there were some years (and the best we could plan it) wherein it exceeded that ratio. In fact, over a period of time, it built up to a rather high peak. You can develop a graph of this sort as a useful tool to see what you're going to have to do in the way of equipment. That dashed line there is an average through (what I call) the heavy stripping period. After that it trails off, and you've got more than enough equipment probably anyway because most of the waste has been stripped after this large number of years. Actually that dashed line going up vertically probably isn't the way it will go, because you're not going to come out there magically with three more shovels some morning and jump the waste ratio like that. It's going to be stepped up. It's going to be more of an incline. This is one kind of tool the production planner can use, and it's the kind of an answer that he can give that makes the economic projections a lot more realistic. This is another kind of a graph. It is a communication tool primarily. This is after everything has been pretty well worked out. We're trying to explain to the people who are going to be making the investment the interrelationship between waste ratios, tonnage, throughput at the mill, and the amount of equipment.

Here we have a production schedule (see Fig. 2 in the chapter). The mill is this lower heavy dashed line. The mill was projected to start out at 23,000 tpd and then after the first year jump to 27,000. Then after the fourth year, it was to be expanded to 35,000; and finally after ten years, they wanted to go to 50,000. So the mining plan was based on that. You will notice that the upper line (the heavy solid line on top) is the total material movement. It goes up, but the waste ratio doesn't always because of the difference in the milling rate. The waste ratio kind of jumps around. Again we have a peak period there, and we've tried to gear it with equipment, how many shovel shifts per day we would have to work to meet these commitments, how many shovels we would have, and how many shifts per day we would be working to do this. If these people had just been looking at waste ratios alone, they would have a very different picture than if they looked at total material and amount of equipment.

Rich Hurt: When you're trying to establish an initial ore exposure to sustain your operation, are there certain considerations that you take into account to choose this ore, this part of the ore body? Do you take the highest grade? Do you take the shallowest part?

Tom Couzens: I think I understand what you're trying to say. Again, you're looking into the future and you're balancing a lot of things. Sure, you'd like to get high-grade ore as soon as possible in most cases. You're looking at what you can do.

Rich Hurt: So you would try to aim your initial pit at the highest grade ore and sort of work around that?

Tom Couzens: I think Jack was talking about an economic approach to this sort of thing. You certainly are going to try some alternatives.

Steve Winkelmann: I think it becomes an economic consideration. I don't think you can just say that you can go for the high grade. It all boils down to economics.

Honorio Narciso, *McIntyre Mines Ltd.:* I wonder if you can deal briefly with pit dewatering—whether there are effective and inexpensive methods of pit dewatering? Taking a hypothetical case like a mine that is about 200 to 300 m deep and located in a tropical country and 200 to 300 m below sea level. Are there some effective techniques already developed wherein you can effectively dewater this pit? I am talking of a pit which has about 150 to 200 in. of rain every year, and I am talking of about 20,000 gpm of rainwater and about 5000 gpm of seepages.

Tom Couzens: That sounds like a good consulting job for somebody! There are a lot of people in this room probably that have dealt with that kind of water more than I have. Bob Winkle has just come back from one where they have about 200 in.

Robert Winkle: In the Zambales area of the Philippines, the Dezoni property just about fits your description. Luckily, it's placed on the side of a hill, and we can tunnel in. So maybe you had better move that ore deposit up to the side of a hill.

Tom Couzens: You're going to have to move the water out if you are going to operate, obviously. There are a lot of ways to go about it, you know. I would say my answer is that I just can't, off the cuff, tell you just exactly what to do. You have to look at these problems a little bit.

Jack Crawford: I think one thing we might comment about regarding dewatering is the fact that not only does one have to be concerned about the quantity of the water and the depths, but I think one also needs to know something about the character of the water and its content, such as whether it is acidic, the amount of sediment you've got, whether or not you are going to be dealing with wood contamination, etc. Besides the scope of the project, you've got some very severe problems to possibly consider as you get into designing your components for actually moving the water.

Tom Couzens: You touched on something that you weren't touching on, Jack, but you brought it to my mind. That is also a consideration in dewatering, particularly in the states with environmental statutes. When you move water sometimes you have to be concerned about the quality of the water that you put someplace. I was talking to John Gould about this, and at Climax, they're on the dividing line between two major watersheds. They have some problems, quite interesting accounting problems, dealing with water because of the different quality requirements on different sides of the watersheds. I don't know, John, whether you want to expand on that. It's just an interesting point. You do have to keep drainage very seriously in mind, I think. Your roads are things you can use for drainage, but you have to also be careful how you achieve this sometimes. You can set yourself up to a point where you might use a road to drain the pit and create a large hydraulic miner, in a sense, because that water comes down the road in a period of, say, heavy rain in many inches per day. You may use the road to funnel that water to a place where it will wash out something that is very important to you. So it all has to be balanced as sort of a specific design problem.

Honorio Narciso: Would you be talking about drilling holes around the pit? I'm talking about trying to eliminate some water from getting into the pit, and I'm talking about seepage water. Would you be patching in some peripheral holes to pump out some of this water?

Tom Couzens: I have not done those nor have I really been associated with an operation that has done this. I have heard a lot about it, and I think Ben or Rick might know more about applications of these things than I do. That is a technique that I have heard discussed.

Ben Seegmiller: Dewatering in an open pit is an extremely important problem both from an equipment operating standpoint and from a slope stability standpoint. To comment on the first problem you have mentioned, that of surface runoff, I would say the best method of getting rid of it is to eliminate it from ever entering the pit. This can best be done by cutting trenches in the back of the pit crest to channel the water away from the pit. Such channeling will be necessary whenever a natural drainage runs directly into the pit or whenever the topography is such that a hill or a mountain runs parallel to and behind the pit crest.

The second problem, that of seepage into the pit, may be handled by at least four different techniques or combinations of techniques. The first one which may be considered is the one you have mentioned, the technique of peripheral wells behind the slope crest. I have observed that this has been a fairly common technique at many open pit operations. Depending on the type of deposit and the geological formations in and around the pit, the technique may be quite successful, or it may be only moderately helpful in dewatering. Usually the more complex the geology, the less total success is achieved with peripheral wells. In most cases vertical peripheral wells may be thought of as being moderately expensive.

Another technique that may be considered is horizontal drains. These drains have found wide applications in the civil industry, particularly in road cuts. I have been involved in a number of dewatering schemes and it has been my experience that horizontal drains offer a relatively inexpensive solution to dewatering unconsolidated materials which overlay bedrock. Many times, at least in the copper industry, we see alluvial gravel on top of bedrock. The bedrock often has had drainage channels cut into it prior to deposition of the gravels. Because the gravels have relatively high permeabilities, the channels tend to collect the water and transport it into the pit. By placing horizontal drains in these channels, we have in most cases been able to remove the ground water from the pit slope and dry up the bench face.

A third technique is to use underground drainage galleries with fanned drill holes. This is a very effective technique but is usually very expensive.

A fourth technique commonly used is to allow the water to flow into the pit, collect it in a sump at the bottom, and then pump it out. All of the techniques I have mentioned may or may not be effective depending on the geology and permeability of the open pit you are considering. I would say that each situation must be considered on its own and a dewatering method or combination of methods should be chosen to reflect the characteristics of each specific situation.

Richard Call: I didn't hear the question that was posed.

Tom Couzens: He was talking about peripheral holes for dewatering in a pit and the intercept to stop the water from coming into the pit. You may have designed or been associated with something like that.

Richard Call: I think that this is an applicable technique, particularly if you've got specific aquifers. In some of the Wyoming uranium properties, in particular where your ore sand is an aquifer, you can put dewatering holes around the perimeter and pump. And one of the things you'll find is that your pumping costs get pretty severe. So any time you can use gravity flow vs. pumping, you are better off.

In the case where you've got alluvium on the edge of the pit (and there are cases where alluvium is on the edge of a major basin), a fence of dewatering holes to cut off the drainage of this alluvial water supply is well worthwhile. Otherwise, you're going to get extensive quantities of water coming into the pit. This also has another ramification in that often that water belongs to the farmer out in the valley, and you can get into a nasty legal problem if his water ends up in your pit.

One aspect particular to this case where you've got high rainfall (although I think my recommendation in the case you posed would be to mine with a dredge) is that you need to go back to your planning. It would be better if in your layout of your mine sequence you actually plan in wider benches to pick up the stream flow and take it out of the pit at the upper levels. If you don't, you end up with the water running down into the bottom level of the pit. Then you have this pumping cost of taking water out from the bottom up to the top. Your economics look a lot more favorable if you can catch that water before it gets into the bottom of the pit and channel it out. We've done that in a couple of cases where we had a major stream or even a series of smaller drainage basins coming into the upper part of the pit. We actually laid out a specific drainage level that would capture these streams and bring them around, particularly on a side hill type mine where you can bring this around the edge of the pit and dump it outside the pit rather than just letting the water from this stream come on down into the bottom of the pit. Then you have the lift of bringing it on out. So I think this capturing of surface water is best designed into the pit in your pit planning. This is where this time sequencing is important (where you've got to have a time series) because you need to see what the design of this will be in any specific time period.

Tom Couzens: Yes, that's very frequently an important aspect of preproduction planning. In addition to just the amount of material to move is the problem of drainage relocation, and we run into this sort of thing frequently.

Production Schedules

MICHEAL B. KAHLE AND FRED J. SCHEAFFER

Curt Spahr: I asked Bob Winkle this question yesterday. I got a very plausible answer; but part of it was lower depreciation costs, which kind of bothered me. My basic question is centered around most operations, multishifting rather than single-shifting. If it's true that at night time production is lower, safety is lower, accidents are costlier, maintenance costs of equipment are higher on a multishift schedule, availability is lower, and some power has to be provided, is it true that depreciation is lower because you operate the equipment two or three times more than a single shift and you have to buy it in half the time or replace it in one-third of the time? In other words, does multishifting lower depreciation costs and lower overhead costs, which are common concerns? Does that offset these higher costs that I have mentioned? Is it economical to multishift rather than single-shift?

Fred Scheaffer: Well, I think the thing you have to remember on equipment is that in mining you're very capital intensive in most cases. In other words, if you go out and buy a 150-ton truck, it is costing you in excess of, for example, $600,000. And when you're looking at your operating statement you are seeing cost normally in cost per hour. It's either in an operating hour or scheduled hour. When you have such a large investment in a piece of equipment, one of the things that you've really got to shoot for is a high utilization. I'm not talking safety right now, and I'm not going to ignore it. I want to reply to that, too. But the point is in talking economics, you've got to have a high utilization on your equipment. One of the reasons we went through this calculation is to be able to demonstrate this. One of the considerations you have in a calculation like this is to maximize the utilization on your equipment as much as you possibly can. As far as safety is concerned, when you operate at night, I'm not so sure that is entirely true. Some of the best production records I have ever seen have come at night. Now, I don't know whether it was paper loading or actual loading. That's the only point I don't know.

I would have to say it is more dangerous in the sense that you don't have the light. However, with most modern mines these days, there is sufficient pit lighting to provide what I consider adequate lighting for safety. I think the biggest problem that you have on multiple shifts, especially graveyard (which is a shift that normally runs from 11:00 in the evening to maybe 7:00 in the morning) is the personnel themselves. One of the problems you will run into is people falling asleep. I've seen some rather tragic situations where guys have fallen asleep in trucks, large trucks, and have gone over banks and consequently died. Now, no matter how many times I've discussed this in safety meetings with people, people continue to fall asleep. We've got to the situation where we told drivers that if they felt drowsy, to stop their vehicle, get out, walk around, have a cup of coffee, and take it easy. Then, once they feel awake again, they can get back in their equipment and go. Obviously, I think that most mining operators have the employee well in mind, and I think these kinds of rules are generally accepted throughout the industry, whether it be coal, copper, or uranium. I am a personal believer in high utilization of equipment. Sometimes you can't do it. It's physically impossible for many coal mines to operate at night, but there are some contractual considerations that have been agreed upon in these situations.

Jack Crawford: To expand on this, I think that anybody who has been dealing with copper in the last four years has been faced with the question of multiple shifts and certainly has been looking at the five-day, six-day, and seven-day operating schedule, which is a variation, if you will, of multiple shifting. I think most properties probably will find that if they could market their copper, they would like to be running on a seven-day schedule because of the economic advantage of the cost per unit of production. This is largely because regardless of what you're operating at, whether it be multiple shift or not, you've got a fair amount of manpower tied up in certain routine and other types of work, which really are unrelated to the production rate. Consequently, if you slack off in your production schedules, you aren't going to be able to reduce all that much the manpower and the other costs associated with it. So, consequently, it is more economical throughout your organization to run the multiple shift, seven-day-a-week schedule.

Mike Kahle: I agree. The typical thing you will see in a coal mine is a dragline, maybe a 100-yd dragline, with an expense somewhere in excess of $20 to $30 million. Most coal operations schedule those things 365 days a year, three shifts a day. And it gets down to economics. You've got to put that machine to work; it costs you too much to let it sit.

Robert Winkle: Mike, I would like to talk a little more about this. We had a discussion earlier about it, but let's put it as an example. If you're going to mine X tpd (and this is what your sales department can sell, also), and if you're going to do this on one shift, it means you have to buy three times as much equipment essentially. So your capital investment is very high. Just the debt service on that capital alone should convince you that you had better stay on as many shifts as you can get out of the week.

Peter Forrest, *Great Canadian Oil Sands Ltd.:* I was wondering how you deal with the communication aspect? How do you get the commitment from the operations group (while you've done all the detailed work) that the plan will be carried out?

Fred Scheaffer: That's a problem. There's no need to deny it, and I really don't know what the answer is. I don't know if anyone has an answer. It depends on what the background of your operating people is. In the particular mine that I worked at, most of the people out in operations did not have a technical background. I'm not slighting that, but they do not understand the value of some of the planning or the techniques that go into planning. The easiest and the best way, I think, to handle that situation is a definite exchange (as somebody mentioned earlier), training these operating people so they can look at what you have done, understand what you have done, and appreciate what you have done. They should realize that it's a tool for them to meet their production requirements, that you as the planner and engineer are actually there to help them meet their production goals.

At M-K Construction, we're set up in a slightly different manner. We have an operating division that operates various mines, and then we also have our engineering group. These operating people have been very cooperative in providing the necessary information to the engineering group for use in the mine planning. Now, there's a problem on the reverse side of that, though. We take this information from them for a particular given situation. When you try to adapt it or project it for use in a different location or a different locale, and you make modifications to it, and then you go back to them and you say, "OK, this is the production rate that we have," there's a problem of communication. A lot of times they look at it and say, "Well, this isn't the same as what we used here. It's a different situation. It's a different application."

Mike Kahle: Peter, I think one way in which you might try controlling your operation (one of the ways we used to be reasonably successful in) is putting up lathe. In other words, probably the most important thing in mine planning which gets very little consideration is drill and blast. If you can control your drill-and-blast operation through surveying, lathe stakes, control lines, etc., usually you can control shovels. This is because they are not going to go into a hard bank. What I would suggest to every planning engineer is to spend some time taking a look at your drill and blast operations. I know that that is one of the things I didn't pay any attention to initially until I recognized that you've got to go back to where the whole thing starts—that's blasting muck.

Drilling Evaluation

BY R. H. HEINEN

Randy Weingart: I was wondering if all of these costs you are using for the bits and equipment are more or less 1977 costs.

R. H. Heinen: They are current costs. In fact, the cost information that I did use in this particular case was based on, as my footnote indicated, an article in *Engineering & Mining Journal* on blasthole drill economics. Most of the cost figures were derived from that.

Bill Hustrulid: I have two questions. First, do you believe those figures in terms of footage per operating shift? The reason I ask is that I recently made a study of copper and taconite mines. It seems to me that those figures are about 50% higher than what I got from the operating companies. So I would like, if possible, some feedback on that from you and maybe from the people here. Secondly, I wonder whether or not you include major overhauls in your maintenance and repair costs. Would you answer that please?

R. H. Heinen: In answer to your first question, you have to take a look at what they are basing their footage on. The information on the availability and utilization I put up there indicated that we had 5.1 operating hr per 8-hr shift. Now, the operations normally report those variables that I had listed as part of the utilization of that drill. Consequently, that in itself would lower the footage per operating hour. So you have to keep that in mind together with the operation that was outlined, the production rates, and the figures given from an existing operating plant.

Bill Hustrulid: In some operations, the maximum penetration rate is something like 70 ft per operating hr. So if it took 5 hr of operating time and 70 ft per operating hr, I get something like 350 ft per shift. For an iron mine, you're looking at a maximum 30 ft per operating hr, and I'm looking at about 150 ft per shift. So it makes a difference in your figures and in the end result. Your analysis is absolutely correct. I'm just questioning, for instance, the number of drills required for maximum footage.

R. H. Heinen: Well, like I said, you know every operation varies. They could have a larger percentage of harder material, like you indicated, in the taconite mines. As I said when I first started, taconites would be even harder than the hard material that we have up here; so consequently, your production rates would be that much lower. I think the main point is the fact of the evaluation itself, which is how I conceive the evaluation to take place. The production figures, as I said, were based on an actual operating mine that did achieve those types of production rates.

Shovel and Haulage Truck Evaluation

BY JOHN T. CRAWFORD

William Allen: Have you had any comparisons done with manufacturing people that have run these simulated computer runs?

Jack Crawford: We have, from time to time, not necessarily to prove our simulator as opposed to theirs. I think in any of these cases where you have an existing operation, if you can have a simulator and then verify it (or have an adjustments factor monitoring system), I don't think there is any great need to dwell on that aspect. Where you don't have it available in-house, I think it becomes very important to know what factors the manufacturers have got incorporated in their simulator, because you are going to have to make an engineering judgment as to how well that simulator represents the operating conditions that you are dealing with.

William Allen: That was what I was wondering.

Jack Crawford: I don't think there is an explicit approach to it.

William Allen: I was just curious if they were in the business for selling machines and if they were going to optimize cycle times. We have had several vendors approach us on truck fleets, and they look quite optimistic. Out of curiosity, I wondered if you had verified manufacturer's data with your actual operating statistics.

Jack Crawford: I think where we probably have more comparative data is in the area of shovels. And I know without a doubt that the manufacturers have been optimistic. This is why we've chosen to deal with this pretty much on our own.

William Allen: Concerning the productivity report that you had for the one month of December (I believe it was), was that truck going to any specific location? Do you have a dispatcher, or how do you account for every trip that he makes?

Jack Crawford: We don't use dispatching. What we do have is an equipment card which is filled out at the end of each shift by the haulage driver, and what he does is report the loads he hauls on each shovel-dump combination that he may have been assigned to in the course of that shift. He may have been working on one profile a shift; he may have had as many as three or four. This data is then keypunched into the computer and generates not only the report you were looking at there (which was principally a performance monitoring report) but also many of the other production reports that tie in with the cost statements.

Mike Kahle: Jack, you were stressing truck effectiveness, and I noticed on your charts that you had quite a wide fluctuation in truck effectiveness from the high 80's or 90's down to 60%. Is there any way to improve that effectiveness? I assume that if you can raise your effectiveness, you can decrease the number of units that you have or the number of trucks. Have you experimented with anything to try to increase that effectiveness rate?

Jack Crawford: That was out of one month. We started using that report at Nevada Mines Div. in December 1976, and the one illustrated was December of 1977. We've generated that report every month. We have the capacity to isolate it down to a day-by-day situation. Ordinarily you can, with that type of detail, track down and determine the association between effectiveness and factors such as rolling resistance, road maintenance, tight operating conditions. We started noticing some problems later in the year where there was an increase in the amount of one-way traffic that we had to handle because of some slide problems. We also had water problems which increased rolling resistance. They showed up immediately in the decrease in effectiveness. You need to look at whether or not your general operating practice is up to what you would consider a normal or appropriate standard to determine whether or not you can increase effectiveness. What this report really boils down to is a telltale or confirmation of some things which observant operators are probably already going to know.

William Allen: If you had a choice of overcovering or undercovering (for example, undercovering a shovel), would you do it to increase truck effectiveness?

Jack Crawford: I think I would not only do it for that but I would also look at it from the standpoint of overall cost effectiveness of the whole system. Mike Kahle is standing back there. He has pursued some rather interesting subjects that many of us at Kennecott have probed, discussed, and cajoled people for years. I have to agree with Mike in this regard that, if you are going to optimize your system of total loading and haulage costs (at the minimal amount of capital as far as your truck fleet is concerned), you're going to be better off scheduling an extra shovel shift if you find yourself extra trucks rather than overcovering shovels to get a resulting increase in shovel productivity. There is a secondary report in the series which deals with shovels. In that report we have shovel effectiveness, truck effectiveness, and truck coverage, and we can see the pattern of change in truck effectiveness vs. shovel effectiveness as truck coverage changes.

Dominic Arrieta: I have three quick questions. 1) What is the average effectiveness that you say you could use for a mine regulation feasibility-type study?

Jack Crawford: It's going to depend a lot on what variables you're going to put in your simulator. When I set this simulator up, we put in certain variables based on conditions where we were trying to start off initially with the standards file that would achieve approximately 100% effectiveness. It's going to depend a great deal on the factors you put in your simulator, because you can end up setting up a standards file with which it would be almost impossible to get better than 75% effectiveness.

Dominic Arrieta: (2) Do you use the 25 mph speed limit throughout your Nevada operation?

Jack Crawford: Our maximum speed limit both in operating standards in the simulator and in the field on flat grades is 25 mph. Now, as we get progressively steeper grades in the simulator, we start to impose more restrictive speed limits for trucks going downhill.

Dominic Arrieta: What is your downhill speed limit at, for instance, 8% grade?

Jack Crawford: At 8% we're down to an approximate speed limit of 15 mph instead of 25. If you're up dealing with 10 or 12% grades where you're beginning to get very close to the limits of your dynamic braking capacity, you may be looking at (and probably will be) 4 to 5, or maybe 6-mph maximums. You don't want to overrun your service brake capability.

Dominic Arrieta: Do you use ⅒th of 1% roundup? Will you comment on why you use that? I have a little experience with 0.3 and 0.5, and I would like to hear from someone else.

Jack Crawford: I think that it is just the gut feel that at times in any of these evaluations you've got certain elements of distribution or probability. You would like to have just enough conservatism in the size of your fleet that you in effect have some surplus capacity for those times when your availability goes down so that you can make up production. People may comment, "Well, isn't that excessive conservatism?" Remember that each year you're probably going to be updating your fleet requirement evaluations on the basis of changing operating conditions, replacement, etc. So the extra truck that you may incorporate in one year in your projections will balance out in future years. Remember we are dealing with approximately a ten-year life on a haulage truck.

Jim Nixon, *Guy F. Atkinson Co.:* I'm not sure I followed all your numbers. I think I did. But the one question I have is with respect to shovel capacity. As I read your numbers you've maintained capacity as a constant figure for a specific size shovel regardless of the size truck he is loading. I'm not sure that you did that, but I found that people in estimating work often times wind up with a mismatched truck-shovel fleet if they don't take this into consideration. Now, what I am speaking of specifically is the conclusion you came to: 20-cu yd shovel with a 170-ton truck. If your material, for instance, weighs two tons to the yard then you've got capacity to hoist 40 tons per dipper with your shovel. You've got to make 4.25 passes for your 170-ton truck where you would like to fill the truck with either four passes or five passes, not have a quarter of a dipper in the fifth pass. I'm not saying that this is missing here; I didn't follow it that well. But I've noticed that sometimes people neglect that factor.

Jack Crawford: I chose to maintain shovel productivity constant because ordinarily your shovel productivity would not vary much if you change sizes of truck provided there is sufficient coverage (and I stress the double spot condition in this situation). The other thing to remember (getting back to Mike Kahle's comment on truck coverage) is that in the overall optimization of our system, we are stressing our trucks as opposed to our shovels. We would rather keep a shovel slightly undercovered than overcovered. One would probably prefer to take up his variability in fluctuating shovel productivity rather than in fluctuating haulage truck productivity.

?: Do you have any history on hydraulic shovels?

Jack Crawford: No, this whole presentation has been based on electric (all-electric) conventional shovels.

?: How do you account for availability decreases over time on a truck fleet, and as part of that would that availability change vary within the different truck fleet sizes?

Jack Crawford: That availability, the way you would account for it, generally speaking, is when you are dealing with fleet productivity. If you want to have 35 million tons of capacity per year and your profiles and availability change, you're going to have to be looking at how many fleet units you need to add or replace in your fleet on a regular basis. Decreasing availability will reduce the annual productive capability of the individual unit.

?: I was wondering whether or not the availability with time would change between, perhaps, a 100-ton truck and a 150-ton truck?

Jack Crawford: There is no reason to think that the availabilities of each truck size are going to be identical. They will have different availabilities, most likely, and, most likely, no two operations are going to be the same.

Pete Fowler: I thought I understood you differently on two occasions. First, I thought you said it was desirable to have a shovel slightly overcovered for trucks and then later I thought you said it was better to have them undercovered. Would you clarify?

Jack Crawford: Truck coverage is obviously going to have a very heavy impact on shovel productivity, and going back to our target rule of thumb of 1000 tons per shift per cubic yard, we definitely need to have very good truck coverage. It's conceivable that under certain conditions that might not be the optimum cost-productivity target. That's why I came back and mentioned that for optimum cost of the combined loading and haulage system you may very well want to have each of your shovels slightly undercovered.

Charles Frush, *Colorado School of Mines:* I would like to ask how your costs and decisions vary if your trucks were a little bit over or underloaded?

Jack Crawford: Generally, you have to look at your record keeping. I don't know whether we really have variability analyses on how those factors will directly affect costs because, again, you have to have quite a history on how your payload pattern will affect costs. Weigh scales or other methods are needed to gather truck payload data. The same situation can be raised as to what is the cost function as you increase your lift. Some operations have said that the cost per hour is going to increase 10% per 100 ft of increased lift from some base reference. But I have yet to see in print anybody discuss how he actually determined that kind of thing. Your cost accounting requirements to gain that type of fine definition could be considerably in excess of the requirements that Mr. Winkle addressed.

Charles Frush: Has there been any way of putting strain gages on trucks, those that weigh the trucks automatically?

Jack Crawford: Yes, there have been. There's a company called Martin Decker that has put strain gage load cells in the anchor pins for the bed hinges and the area where the hoist jacks connect to the bed. While the truck is being loaded, the bed should be slightly elevated above the frame. Now, I'm sure someone is going to ask, "Well, as far as hoist jack maintenance is concerned, what kind of excess loading are you going to put on your hoist jacks?" I really don't know the situation; however, there is one option on the market that has a lighting system that faces the shovel. As the load gets progressively toward capacity, the lights go from green to yellow to red. When the red light comes on, that's all the muck you'd better put on that truck!

?: I've heard that some Canadian operators undertruck their shovels. I mean, they always make it a point to calculate the total number of trucks required and then pull out one truck. They say they get better performance from their drivers.

Jack Crawford: If one is dealing, perhaps, on a shift or per-shovel basis, this gets back to the matter of undercovering an individual operating shovel. I don't believe that this should be taken into account in determining how many trucks to have in the fleet to move the total tonnage, because that way you're going to be biasing yourself on the low side of your fleet capacity requirements.

?: What they mean is there is a psychological effect involved in this when the truck driver sees. . . .

Jack Crawford: On the individual shift scheduling basis, I would agree with you. However, I think that's something over and apart from how we determine the number of trucks we actually want to purchase and have on the property at any one time.

?: I was just wondering if there were some effect.

Jack Crawford: There may be, and that would show up in your haulage truck effectiveness factor.

Maintenance and Ancillary Equipment

BY DONALD C. MYNTTI

Robert Winkle: Don, at the copper company to which you were referring, you said you needed one bay for four 150-ton trucks, but previously you said you needed one bay for six smaller trucks. This indicates a 75% availability for the 150-ton truck vs. 83% for the smaller unit. Does this indicate you had mechanical difficulties with that 150-ton truck?

Don Myntti: Let's say that the primary reason for the change in this ratio is that similar jobs for small and large trucks require more time. For instance, changing the outside rear dual tire on a 65-ton truck takes an hour. A similar job on a 150-ton truck will require two men for almost a shift. For a small truck, it will take two men a couple of shifts (four man-shifts) to remove a bad-order engine and replace it with a rebuilt unit. The similar type of engine or power plant replacement in a 150-ton truck may take possibly two or three days. That's basically it. Now, these ratios, as I mentioned, are really based on the gut feel of the people running the shop. These ratios would make the maintenance management people just a little more comfortable.

Robert Winkle: But, on the other hand, your 150-ton truck is designed for rapid overhaul. You have modular components in it. You can take out and put in a new one very rapidly, which you weren't able to do with your older trucks.

Don Myntti: The quick changeout is being strived for. Eventually, at least in the changeout of the power plant modules, the time requirements should be reduced to about a day or a 24-hr period.

Jack Crawford: I don't recall that the shop people were particularly happy with that ratio at that time when Don said they had one bay for six trucks. That was something with which they had to deal and in designing the addition they said, "We can't live with this situation; we really need to work toward this one in four ratio." So, the inference on 83% availability against 75%, I think, might be misconstruing the facts of the situation at that time.

R. H. Heinen: Don, could you give us a rough idea of the cost per bay for the support facilities based on different truck sizes?

Don Myntti: In the shop facility that is in use at the large copper mine, the cost of the four-bay addition came to about $80 per sq ft. Now this is a 45-ft-wide bay, 130 ft deep, and with about a 60-ft eave line with two 35-ton overhead cranes. A lot of concrete is required to support such a structure.

Economic Evaluation of Open Pit Mines

BY FRANKLIN J. STERMOLE

Circular Analysis—Open Pit Optimization

BY GERALD C. DOHM, JR.

Jack Crawford: Would you speak about treating marginal grade material? Do you mill it so that it is current with mining or stockpile it and rehandle it? As a general rule, to what extent is the sensitivity of marginal analysis influenced by this matter of maintenance of stockpiles and rehandling charges? What is the advisability of stockpiling such material to the end of the mine life?

Gerald Dohm: This depends on your preliminary design layout. Generally speaking, if your stockpiles can be located in an area advantageous to the crusher and mill, the extra handling charges are minimized. You can include the extra handling charges, at least what you expect the double handling charges will be, in the marginal analysis. This determines the cutoff grade for this type of material.

Jack Crawford: So in fact you could really end up with two kinds of a marginal cutoff, one on a current basis and one on this deferred basis with the additional rehandling charges.

Gerald Dohm: Yes, that's right. I am looking at it, however, in terms of property evaluation. If 15 or 20 years down the road the economics have changed, you haven't lost anything by setting this material aside rather than throwing it away.

Jack Crawford: The reason for my interest is that on some of the projects upon which I have worked, you are thinking in terms of perhaps 10's of millions of tons of this type of material.

Gerald Dohm: The only thing to be taken into consideration here is that I am talking about a uranium property for which the tonnage is low in comparison to a porphyry copper.

?: Couldn't some consideration be given to leaching this low-grade material concurrently with the production of the standard grade going to the mill?

Gerald Dohm: Most definitely. Heap leaching of uranium ore has proven to be fairly successful. You are looking at an overall reduced recovery factor, but it is certainly worth looking at. You could also leach the low-grade materials that were beyond your pit cutoff limits. There are a lot of things that can be considered. It is just deciding the degree of detail that you really want to include.

Panel Discussion

W. A. HUSTRULID, JOHN T. CRAWFORD, Moderators

Bill Hustrulid: We are going to sort of shift gears now to where the workshop responsibility is not so much upon the authors but more upon you in the audience to come up with your ideas and your thoughts and your contributions. I would like to turn this part of the program over to Jack Crawford who will be the principal moderator.

Jack Crawford: I would not like to discourage any questions in the course of this discussion. At the same time I would like to rotate the field of questions so that we don't necessarily dwell on one subject for an excessive length of time. We'll go ahead now, and I'll ask for the first question. Anything we've covered in the last 2½ days or anything else you may have on your mind regarding open pit mining is fair game.

Curt Spahr, *Johns-Manville:* This question is addressed to Don and Jack about maintenance costs. Don, not so much facilitities, and Jack, production, is it common practice to overload trucks and is it economical to do so? And, if they are overloaded, to what extent, 5%, 10%?

Jack Crawford: Maybe Don might field the question, particularly in the area of tires, because I know that is a critical element.

Don Myntti: At a large copper mine, as the result of a lot of weighing of the haulage trucks, we zeroed in on a struck bed capacity which we included in our specifications. As I remember it, it was a 106-cu yd struck for the 150-ton unit. We learned as a result of the extensive weighing that the payload did average in the 150 to 155-ton range. We did acquire the haulage capability that we specified and purchased. The average payload was extremely close to the rated capacity of the truck. Now, however, to get an average of 150 or 155 tons, you are going to have some payloads that are much larger. In fact, I think we weighed some in the 200-ton range. I think that overloading is a common practice. I don't think it can really be avoided, especially if there are many different types of material in a mine. And we mustn't overlook the fact that trucks are also underloaded as well as overloaded much of the time.

Curt Spahr: If you have material of significant different densities, do you fully load the lighter density material and then light load the heavier?

Don Myntti: Generally what happens if a truck cannot be dumped, then the operating foreman instructs the truck dispatcher to talk to the shovel runner to cut the load, say, by one dipper.

Jack Crawford: I think one thing one might say, too, regarding this, is you are going to get a distribution of loading around the average. If one looks at the various and sundry cost aspects, manufacturer's warranties, etc., I think it is fair to say that as a general rule, it would not be advisable to make it a generalized practice to overload the truck relative to what you designed it and expected it to be in the first place. I think, Don, in the early days of the 65-ton trucks before tires were successfully developed to handle the weight, we were looking at as much as 33% and 40% or more overloading, even at manufacturer's specifications. There has always been a tire problem even in loading trucks to rated capacity.

The other thing to consider in this whole matter of overloading is relative to the wheel motors. At Nevada Mines Div., we have actually experienced getting to the threshold whereby, in order to stay within the electrical rating of the wheel motors, we have had to actually light load because of the time and load relationships on the wheel motors and because of the depths of our pit. So, I suppose if one wanted to come down to a broad generalization, it is that if you design your truck and decide you want a 100-ton truck for your operation, I certainly wouldn't advise planning on going ahead and loading it to 120 tons as a general rule, even if the bed size will accommodate it. I think your maintenance problems (operating problems) are just going to eat you alive on a cost basis. And your availability-utilization are going to suffer accordingly.

Curt Spahr: The subject of presplitting was mentioned several times. Could one or more people describe a little more fully how it's done, when it's done, where it's done?

Dermot Ross-Brown: There are several ways of modifying blasting when approaching the final pit limits. They all involve a neat line of parallel holes, preferably of a small diameter and spacing, along the required excavation line. Each hole is normally lightly charged, although some holes may be left uncharged as relief holes. Although simultaneous firing is essential, long lines may be blasted in sections in order to reduce the concussion effort.

Presplitting is one extreme involving a large burden and closely spaced holes which are fired before the main round. The object is to produce a split in the rock between the holes, which will minimize damage beyond this line when the main round is fired.

In cushion blasting, the final holes are fired after the main round using small burdens. The effectiveness of cushion blasting may be increased by reducing the burden and charge of the previous one or two rows, so that there is a buffer zone between the main blast and the final excavation line with each successive row blasting a little more gently than the previous one.

The results obtained from these different techniques depend very much on the rock type and the type of jointing that is present. Presplitting is very effective in hard rocks in certain situations. In less brittle rocks, such as those usually encountered in porphyry copper mines, good results can be achieved using cushion blasting techniques at much less cost.

Richard Call: I've been involved in a number of presplit exercises, and I can't say I am happy about them. They tend to be a high cost operation. I've heard numbers as much as $0.38 per sq ft for a presplit face. When you multiply that by the square foot of your final wall, you're talking about an excessive amount of money. And Dermot has a good point on both the rock type and the structure. If you've got fracturing running right angles to the pit face, fairly closely spaced fracturing, presplit isn't going to do anything for you because the split line from the hole that is shot runs out to the nearest fractures and stops. If you are using 6-in. holes and 6-ft spacing, and you've got fracture spacing at about a foot, you're going to presplit one foot and all that rock between the holes is not going to be split, so that you are not going to get any presplit at all that amounts to anything.

In the other case where you've got good fracturing parallel with the face, you're probably wasting your time doing presplit because it is going to break back to that fracture surface, anyway. So you have some real limitations, and broken rock in porphyry copper is a good example where you have close fracture spacing; presplit has very little advantage.

The other thing I have found with presplit is that it is not enough to design the presplit line; that's supposed to generate a crack but all too often you can get a nice presplit. You get your loading right and your spacing right, and you generate a nice crack along what will be the final face, and then with your buffer row of holes next to it, you proceed to blow it up. I find the design of the buffer row of holes next to the presplit line more important than the presplit line itself. From what I have seen, until you get up into slope angle ranges above about 57°, there's no real need to do presplit. Cushion blasting, just reducing the charge and spacing on your final row of holes, is satisfactory. You don't need the nice carved face that you see in power plants or highway construction. All you're trying to do is reduce the amount of backbreak, so you minimize the amount of raveling. And, if you're 5 ft off here and there along the face, it's not going to make any difference. It's only considered when you've used your main production blast right up to the final wall and then you've got a

30 to 40-ft backbreak of ravelly ground that is just going to give you a problem, and you can't maintain a decent catch bench.

Jack Crawford: I have a couple of questions that I would like to address to either Dr. Stermole or Gerry Dohm. In the economic area we have discussed the matter of discount factors, which boil down to an interest rate. My first question is, how might one approach what, for a given evaluation, is an appropriate interest rate? Secondly, in any of the work-ups of cash flow from an economic analysis, let's say as distinct from the financial analysis discussed by Dr. Stermole, how do we handle the matter of debt service charges, i.e., the interest we pay and, say, the repayment of the principal?

Frank Stermole: Well, the minimum rate of return reflects the other opportunities that supposedly exist for the use of money. In theory, at least, the minimum rate of return that is appropriate for a particular individual or company to use is the discounted cash flow rate of return which the company feels other existing opportunities would be earning money. And if we feel that 12 or 15% represents the other opportunities that exist for the use of money, from an economic viewpoint we are not interested in making investments that won't do at least that well. In practice, if you are looking at maybe a half a billion dollar project or a billion dollar project (the really large projects where maybe it isn't a question of having other opportunities to invest that amount of money at a different minimum rate of return), when you get into the matter of whether it's doing a big project with a lot of funded debt, a lot of borrowed money, or not doing the project at all, then it becomes a matter of looking at the profitability of the project from the viewpoint of what margin of profit will satisfy the investors combined with what margin of profit will satisfy the people who are lending the money which gets into your debt service ratios of cash flow to debt service, etc. But, in general, if it's internal cash flow, and other opportunities are considered to exist for the use of money, the minimum rate of return is represented by those other opportunities.

The second was on borrowed money. I presented my analysis. I guess virtually all the analyses we saw in the 2½ days have been from a cash equity investment viewpoint. Most major mining investments around the world do involve some level of borrowed money, often times a large level of borrowed money. Often at the time we are doing the economic analysis work, that is, what is the optimum way to plan the mine, we do not at that point often know how a project will specifically be financed, so the approach that I advocate and the approach that I find the majority of companies using in practice, is to evaluate projects from the viewpoint of what the economics would look like if I had the cash to equity finance the project entirely internally. If the economics looks like it's the kind of a project I would like to do if I had the money to finance the entire project. Then I know if I can go out and borrow money for an after-tax cost of money that is less than the cash investment DCF rate of return, then leverage will work for me and my leveraged DCF rate of return calculations will show me a larger DCF rate of return on my equity investment with leverage than without.

Pete Fowler: A couple of questions of Mr. Myntti. Basically, I would be interested in a couple of ratios if you could come up with them. What is the ratio that you keep of inventory of parts as related to the equipment inventory in the first place? Another thing is what would be the ratio of, say, the cost of maintenance work at a mine as compared to the other operating costs?

Don Myntti: In answer to your first question, the basic maintenance philosophy at many mines is that the maximum amount of component rebuild be farmed out to vendors so that they are able to keep the dollar volume of the inventory down and operate with smaller maintenance and warehouse crews and less warehouse and shop space. Other ratios may be of interest to you. We maintained many spare components primarily for haulage trucks. For instance, for the 150-ton haulage truck fleet, one spare power plant module was provided for every five trucks. In addition to the complete power plant module, one bare 1600-hp engine was provided. Another major component that comes to mind were the truck beds for the 65-ton trucks. One bed was provided for every 12 that were running. So as a result of experience for the particular conditions that prevail at the mine, these ratios did evolve and for that operation they were quite realistic and practical.

Pete Fowler: What about maintenance costs vs. operating costs?.

Don Myntti: Again, I will refer to the haulage truck fleet. That seems to be the most spectacular fleet that miners run. The maintenance and repair percentage of the total cost of running the truck fleet is in the 35 to 50% range, and this percentage varies considerably by truck fleets. Tire costs consume about 20% of the total cost. Fuel since 1974 has run about 10%; lubricants, 2%; depreciation in the 10 to 15% range. The rest would be operating labor and operating supplies. As far as the shovel fleet is concerned, the maintenance and repair expenditures amounted to about 60% of the total dollars spent to operate the fleet. In the case of BE 60R drills, maintenance and repairs were about 30%.

Dominic Arrieta, *Minerals Exploration:* I have a question for Mr. Myntti, too. This refers to sizing your shop. You spoke in one case of, a 150-ton truck. Now, what happens when you have your fleet sized up and you know exactly how many bays you will need? Do you have any comments or feelings for ratios as to the working space you need per unit?

Don Myntti: In the shop expansion that was undertaken, which was based on the use of the 150-ton truck, the bay size was determined to be 45 ft wide, as compared to 30 ft for the 65-ton truck. The increased space was due to a number of things. Space is needed for maneuvering by the tire handlers and in the removal of the wheel motors. The tire handler comes into the repair stall, removes the dual tires, after which the wheel motor can be removed. Then in the case of, say, the changeout of suspensions or hoist cylinders, you need access for maneuvering the overhead cranes, slings, and specialized lifting fixtures. The typical congestion in a truck garage can be eliminated by this approach.

Dominic Arrieta: What we do is evaluation-type work, and we need strict budgetary numbers. I have been using a ratio of for every unit you have you need about the same amount of space for the working area; this is for the total shop. Could you comment on that ratio and tell me if I am way off?

Don Myntti: Yes, a ratio of 1:1 would be very realistic. The ratio of a 150-ton truck and a 1-ft-wide bay is 1:1. I might add here that the size of the overhead door is 36 ft wide by 25 ft high.

Dominic Arrieta: OK, now we have a size of fleet. Can you comment on one more thing: size of storage or warehousing strictly for that size fleet? Would you say 5% of the fleet size, 10% of the shop area?

Don Myntti: I categorize warehouse space by three types: heated building, unheated building, and outside storage area.

Dominic Arrieta: I meant the type that you have to capitalize by building something over it, the heated area.

Don Myntti: Based on the experience of a large copper mine, the heated area is about 25% of the shop space. The unheated warehouse building is about 15 to 20%, and the outside storage area is about 30 to 40%.

Mike Robb: I would like to make one other comment about warehousing. I've been involved in a couple of overseas operations, and our inventories compared to the United States were just horrendous, as was the size of the warehouse we had to keep. I think the warehouse inventory and the physical size of buildings are very much dependent on just exactly where you are.

Dominic Arrieta: May I ask one more question, Jack? Again, I am still persistent on truck efficiencies. I would like somebody to comment again for budgetary purposes. If we use the simulation that an equipment manufacturer gives and leave all the criteria as they stand, there has to be some sort of an efficiency worked in there, whether it's 75% or 85% of the fleet. I know, Jack, you didn't like that much. I need a number.

Jack Crawford: Well, again, unless you can get into the depths . . .

Dominic Arrieta: Well, can we say shovel pit, deep pit, and no pit at all? I'm talking about something, say, 500 ft and over, 500 ft and under, and surface. Is there anyone who could make a comment on that?

Jack Crawford: The thing is, just like any kind of simulator, you've got certain standards or you've got certain data bases, and unless you know your data base thoroughly, there's really not much of a way of checking it. Now, the one thing about the equipment manufacturers is that they may have data where they have correlated their simulator with operations, maybe not quite as rigorously as what was discussed this morning, but they still may have some data to give you some feel as to that. But, if you go to WABCO for a simulation run and turn around and maybe take the same data and go to Unit Rig, there may be enough differences in their simulators that you cannot use the same correction factor to bring those results that you might get from each manufacturer into a comparable range.

Bill Hustrulid: I would like to ask a question of the panel on behalf of Stan Michaelson. I don't know whether the little presentation I gave at the beginning of this meeting on behalf of Stan was understandable, but the whole idea that Stan was trying to get across was that over, say, the last five years, the gap between cost and productivity increases has been widening at about 5% a year. He presented some suggestions as to areas in which we need to improve in the future to change this trend. His suggestions had to do with haulage and steepening pit slopes and things like that which require modifications in our design and planning procedures. We've got some bright and creative people sitting on the panel and I'd like them, on behalf of Stan Michaelson, to comment on how you as designers and planners are going to reverse the trend between productivity and costs. What did you think about some of those ideas he presented through me in that first paper? Mike Robb, why don't you make some comments?

Mike Robb: Talk about judgment day! I think that at least in US mining we have been kind of backward, particularly in innovative-type work, in developing and using new machinery, and this sort of thing. We as an industry do need to look at and come up with new mining techniques, ways to cut down dilution, better ways of extracting ore, etc. I think it is a problem that needs to be addressed industry-wide with some very basic research.

Tom Couzens: Of course we would like to develop methods to design which would compensate for declining productivity. Such things as steeper slopes are a step in that direction; but I think there is quite a limit in how far you can go. Jack, you came up with a figure in your example, I think, of something like 5.1 operating hr per 8-hr shift—sort of actual operating time for some piece of equipment. Do you remember the figure?

Jack Crawford: That particular figure was on drills, but pursue your avenue of thought.

Tom Couzens: Just look at it as a general thing, and maybe this is a social problem. I read a report about a large open pit mine (not in the United States) where over the last five years that figure had been a steady decline where it (I don't remember the exact figure) was something under 5 effective hr per 8-hr shift. This was a case where I think you have to look at management and union negotiations to pinpoint the problem. One thing that had happened was that they had had a number of raises in pay through negotiated settlements and they had also eliminated a production bonus so they rewarded everybody but took away all incentives to do any better. This kind of thing is really a serious problem; as I say, it's a social problem in addition to being a technical one.

Jack Crawford: There have been a number of things looked at and examined in the last several years in regard to the cost vs. productivity decline. One Kennecott Div., in order to reduce the effect of the shift changes, instituted for a time what they called "changing on the fly." They went out and relieved the shovel runners at the shovel, relieved the drivers on the road someplace, etc. Another thing one could look at is trying to coordinate your servicing with the lunch hour and having, perhaps, some maintenance crews in the field at that point so that you aren't taking double downtime. Your point on operating hours per shift is a very good one. For example, similar record keeping systems would seem to suggest that the shovels at the mine at which I happen to work were getting perhaps a ½ hr less operating time per shift than those at the mine to which Don Myntti was referring. Gaining an operating time increment and maintaining good productivity (in terms of tons per hour) throughout that period are very important.

Tom Couzens: I would make just one more remark, and this is not to get planners off the hook, because I still do think that planning is a very challenging thing, but we have to admit that operations are critical. Many times when companies think they have technical problems, they really have management problems or supervisory problems, and these are often much more difficult to resolve than technical ones.

Jack Crawford: An addendum to the question we could pose to Ben Koskiniemi is, from a productivity standpoint (manpower, energy, etc.), why hasn't the conveyor seen a greater growth as a materials-handling method in general open pit mining, whether it be for a new pit or retrofitting existing pits?

Ben Koskiniemi: I think one of the things about a conveyor system that you've got to appreciate is that you've got all your eggs in one basket is one way of putting it. When your crusher or conveyor is down, your productivity is reduced to zero if you don't have a backup system. Conveyor belt design has been quite critical, especially for steep grades coming out of a pit. Steel cable belts are used, and these have improved tremendously. In the past, belt failures were quite a serious problem. Other critical areas include drive motor, gearbox, and pulley design. At Twin Buttes, a cable belt conveyor is being installed to haul ore over relatively flat terrain for a distance of 6½ miles. This type of belt is quite a departure from the conventional type of conveyor, and in terms of energy I understand this conveyor will only require about 60% of the power of a conventional conveyor. I think there are a lot of opportunities for in-pit crushing and conveying, especially as pits get deeper. Trucks keep getting bigger, but realistically, there's got to be some practical limitations on the size of trucks, especially those required for the long grades out of deep pits. Conveyor systems, as is true of most mine equipment, require good maintenance, and that means keeping right on top of it. You should have people (belt runners) who keep a close check on the belts as they are operating, and you must allow adequate downtime for preventive maintenance and repairs. I think there's a future for conveyors, and we're going to see more conveyors and in-pit crushers used in the future.

Cliff Maddocks, *US Borax & Chemical Co.:* We're one company that's looking at the conveyor belt system, and I have a couple of questions concerning the belt conveyor system. What type of future trends do you expect in the belt conveyor system as compared to trucks in the fuel, labor, and maintenance departments, in terms of the present and the future?

Ben Koskiniemi: I would say that you have a better opportunity to develop savings with conveyors. Capital costs may be higher, but generally operating costs for manpower, energy, and maintenance are less. Conventional shovel-truck operations may be best for some operations, but I think each operation must be evaluated separately. Again, I think future trends will be towards wider use of in-pit crushing and conveyor systems.

Cliff Maddocks: The only problem is the fact that you have to do your planning around it. As you stated, if the system goes down, your whole system goes down, and if you can get around that, the system is better.

Ben Koskiniemi: One of the things you've got to watch is your maintenance program. A dual system would be ideal, but how many operations can afford a dual system? Our experience shows that if we are operating on a 15-shift per week basis, we can see the improvement in availability compared to operating 20 shifts. I think others have addressed this in commenting on comparing availability for operations working five shifts a week (one shift per day) vs. someone running three shifts a day, seven days a week. We see that conveyor system availability does increase if you have adequate time available for preventive maintenance.

Cliff Maddocks: For the Anamax project, there is some point where the haulage system has a break-even point for hauling by trucks and hauling by conveyor. Is that correct? From some distance beyond that, it is more profitable to go to belt conveyor.

Ben Koskiniemi: At Anamax's Twin Buttes mine, all the upper (top 400 ft) alluvium goes out by conveyor, and there is no truck haulage involved in this area. The crushers are located in the pit about 400 to 600 ft below the surface. Most of the truck haul consists of hauling to the crushers from the area of the pit below the crusher elevations. At Twin Buttes, the question is one of hauling vs. conveying to that last 400 to 600 ft, and I would say that in our case it is cheaper to convey. You are correct; there is generally some break-even point where it is cheaper to run one system rather than the other.

Cliff Maddocks: Could you comment on the alternative of building an incline belt (a stationary incline belt) compared to a movable belt for each pushback going down the ramp and then off the incline belt having truck lines?

Ben Koskiniemi: We've got a system that operates in a

similar way to what you're talking about. Previously we had scrapers which would haul alluvium to bins located over an inclined conveyor belt. Now we have a movable belt which is extended as we move along a level. This belt is loaded by smaller feeder belts, and the alluvium is discharged directly into the bins from the movable belt. Dozers are used to push alluvium from the bank into a loading pocket which feeds onto the feeder belts. The feeder or shuttle conveyors are each 106 ft in length, and each is an independent unit. These are used in series to move material from the loading pocket to the larger movable belt. Except for the permanent inclined belt, the movable and feeder conveyors are moved from level to level. We aren't using any type of portable crushers in conjunction with our movable conveyors.

Cliff Maddocks: I've visited both your property and Duval's, and I know your failings and Duval's. I'm curious if there is anybody else who has any comment regarding the future. I'm not saying you're biased, but I am interested in seeing what other people have to say.

Pete Fowler: There's one conveyor operation that I think of that uses a dozer to load it, as you mentioned. I went out to look at it a couple of years ago—Kaiser Sand and Gravel in Oakland or near Oakland. They were moving on a regular basis about 3 or 4000 tph, and it looked like a very smooth operation at a very low cost. Of course, they don't have any blasting, but it was a good operation. It's worth looking at.

Peter Forrest, *Great Canadian Oil Sands:* We don't move much dirt. We move about 100,000 tpd by conveyors. Well, I don't want to discourage anybody against using conveyors. I would just like to bring to your attention that the conveyor is a very useful item, but you have to find a way of getting your material on the conveyor. Don't get into a situation where you go from a periodic loading type to a continuous operation to a periodic loading—things like that. You will go broke. So, just exercise a bit of caution.

William Allen, *Milchem Inc.:* Incidentally, I did a little preliminary cost study myself on a conveyor system and found that the capital costs are quite tremendous, of course, compared to operating costs. There's a very interesting example down out of Tucson, Loredo Cement Co. It is very interesting if someone really wants to get into a study of an elaborate system.

I would like to comment and ask a question, and to do this we'll delve into possibly the realm of accounting. I think I touched on it briefly yesterday, and I would appreciate any comment from anybody. To start off, we're always talking about costs. It's pretty simple when you buy a new piece of equipment. You sit down, and you look at your monthly summary or your quarterly summary of costs. This new piece of equipment is reflecting mainly a small amount of maintenance, lube, or cutting edges or something like that, and you get a pretty uniform rate.

But pretty soon you go on down the road a few years, and you get into some major components going out, and you get these serious fluctuations in equipment costs. We see numbers in the literature saying, for instance, this piece of equipment costs $50 per hr, and we tend to assume that this is a quite uniform rate over the life, say, of this piece of equipment when, in reality, in the later stages, as I say, it's getting quite serious and may influence a decision to scrap that piece of equipment. But, my main concern is this smoothing out of these horrendous fluctuations by hourly costs, and we're all very well familiar with the concept of hourly costs. It's quite easy to use in construction work or mining work for budgeting or evaluating a little project. So we use this, for instance, this $50 per hr. But in reality it may not be that, and what I've used in the past (and I think other companies have done this) is a maintenance accrual for each piece of equipment whereby you would take your best shot at what you think this rate will be. Then over a long period of time you're putting in these major costs for repair. I know that there are companies that actually form subsidiaries, placing all their equipment in a subsidiary company, and it actually looks on paper like a leasing company to get these costs on paper. Then they will turn around and, for instance, on paper lease this equipment to the operating division.

The question is this: Accountants are horrified at this concept. They say, "Well, you're making a paper liability or accruing a situation that simply doesn't exist in actuality." That's accountants' terminology, you know. If the expense was incurred this month, that's when you expense it. It just horrifies them to think of accruing something like that, and if you did use it, it has a tendency, for instance, to overstate costs as far as tax purposes are concerned, I would assume. In actuality, if you, for instance, buy a new drill and you say, "OK, this thing is going to cost us $60 per hr," and you start right out from scratch accumulating costs against this piece of equipment at $60 per hr and actually nothing happens, say in the first quarter, then in reality your cash flow is quite higher than what is showing on paper. I just wondered whether anybody had any comment on this.

Jim Nixon, *Guy F. Atkinson:* I've thought about this problem quite a bit and have gotten involved in it to some extent. But my feeling is that you can approximate what you feel is going to happen to you on a time relationship with respect to your maintenance expenses. In other words when your equipment is new, it's not going to cost you anything for a while, and after a while, you will see a gradual increase in your maintenance costs. If you could make some kind of a model that would approximate what you anticipate in that regard and then reverse that model for your in-house depreciation schedule or ownership costs for the equipment, you might be able to, in many cases, arrive at an hourly cost that is fairly uniform over the life of the machine by leveling out the high early depreciation with the low early maintenance costs and then the reverse towards the later years of the machine. A lot of times what we'll do is we'll give the maintenance department a budget for operating that machine on a per hour operated basis, I should say, for maintaining the machine. We will charge the maintenance department an in-house ownership charge, and they'll be charged that as part of their maintenance costs. It will be a curve, which hopefully is a reverse of the actual maintenance expenses to be incurred. Then the operation itself, for example, the actual mining of the overburden or the ore, is charged a flat hourly rate.

Jack Crawford: From a mining operator's standpoint, it seems to be preferable to charge off your expenses directly as the dollars are expensed and are being spent. And then you've got your hour history. If you're going to spend the dollars you've got a tax effect, and so bookkeeping is not generally going to affect your cash cost position. Whether you have an internal bookkeeping system or not (that's different from what you do for taxes) isn't going to affect the financial condition of your operation. Secondly, your overhead required to maintain your books can go up quite sizably. We mentioned this morning about accounting costs against true economic costs. I think there is something to be said that the less we do with the data other than keeping a record of what we accrued, what we had in actual operating hours, and how much money we spent, the better off we are. If we can keep the data fairly simple in that regard, we've certainly got a cleaner record. I'm inclined to think toward evaluating the need for, say, overhauls and replacement of equipment. This smoothing of costs from an accrual standpoint would tend to create a picture where you will be inclined to defer replacement of equipment beyond the point of a true economic cycle.

Delfor Cadario, *Argentina:* I am the acting superintendent of a 14,000 tpd operation, and I have four questions which I think are related to the paper that John Crawford has presented. One refers to the shovel-haulage truck evaluation. There is the graph referring to availability. It was said that more or less after ten years the forecast for availability is reduced to 60% from an initial 75%. The question is: When your equipment availability is less than 50%, you can assume that you have to change your equipment or that you have to revise your maintenance policies or usage of spare parts, etc. What I wondered is that our availability is not as high as was stated; it's not 75, nor 60%, so from your experience can you say it is a question of maintenance or is the equipment not adequate?

Jack Crawford: I think there have been comments about this in some maintenance meetings. Don Myntti, I'm sure, thinks along much the same line. The Southwest Master Mechanics meetings often get into the subject of availability. You can find availabilities on very similar equipment which appear as though somebody fired a shotgun at a piece of cardboard because they will be all over the lot. I think two things cause this. One may be, as you point out, a problem of maintenance, a problem of inappropriate application of the equipment. Secondly, it can be a bookkeeping matter. But to say explicitly that it is a problem if you've got availabilities down below 50%, I don't know that one can say for sure that there is a problem. One might be inclined to lean in that direction. I think one would have to have a far greater understanding of the specific character of the operation with which you're dealing.

Delfor Cadario: The second question is: Does it make sense to consider from the economic point of view paving some main roads of the yard where the trucks are operating, or is it nonsense to consider paving part of the main roads in an open pit or underground mine in order to reduce tire consumption and in order to reduce truck maintenance?

Jack Crawford: There are several pros and cons on that. Some tire people have said that if you have a black-top road with some of the off-road tires, with the heat in the road being absorbed in the tire and the fact that the road doesn't have a certain amount of cushioning effect, that you're going to get more internal tire problems. The other thing is that if you have a paved surface, you're going to have to make sure you keep it clean of spill, because it is an excellent foundation on which to put that little pyramid of rock, which just loves to get in there and chew up a tire. At Nevada Mines Div., we have one section of road that is asphalted, and it has proven very effective. But, again, we've had to look rather carefully at the matter of keeping it clean of spill, and it is also being used in an area where the haul isn't particularly high speed. It's a downhill haul loaded, so we do have some speed limit restrictions. It has been quite helpful when you get a great deal of alternate freezing and thawing.

Delfor Cadario: OK, I did understand that you have some cases where you have paved the main roads.

Jack Crawford: This is just one section of about a mile and a half of road, and we haven't expanded it, and I'm not aware at this point (at least within Kennecott) of any other artificially paved haul roads. I don't think you can say it's totally inappropriate to use paved haul roads but it is not a common practice.

Delfor Cadario: The third question is: When you are doing the forecasting for the availability of front-end loaders, do you better compare with the truck fleet or with the shovel fleet?

Jack Crawford: Generally speaking, front-end loaders aren't going to do as well as either. And I can only suggest that one look at past history of front-end loaders specifically. I don't think there is any way to infer from haulage truck or shovel performance as to how well a front-end loader will do.

Delfor Cadario: Well, the last is just a general question. In cases where you have a surface operation, an open pit, when can you say it is preferable to consider a bucket wheel, or draglines, or shovels, if some blasting is required? For the mine in Argentina, some blasting is required.

Mike Kahle: Normally with a bucket wheel, there isn't a great deal of application in the United States. Some of it is in the midwest. A bucket wheel is a high production unit. You're talking in the order of anywhere from a 1000 yd to 20,000 yd per hr. So the first thing to look at is the production rate that you're up against. Now, as far as the dragline is concerned, normally a dragline in most operations is in shallow deposits. What I mean is, it's not what you copper miners normally think about. A dragline operation on the average will go down, let's say for a 75-yd machine, maybe 150 ft, 200 ft at the maximum. It all depends on your boom size and boom size can go up to 300 ft. So, generally when you look at the three classifications of equipment, you have to first look at the general description of your ore body. In other words, is it reasonably shallow, is it 100 ft or 200 ft from the surface or less? Then the second thing is your production rate. Those are the two elements you normally look at when you are choosing between truck and shovel and dragline and bucket wheel. Also, bucket wheels and draglines are extremely expensive units. In Jack's paper this morning, he mentioned that a 25-yd shovel was in the order of magnitude of $3 million. I think a 75-yd dragline costs about $30 million. So we get back to comments like Ben was making and Frank was making and Gerry was making about the capital you have employed in your initial operation. So it is something to consider. Draglines when they are operating will operate more cheaply than a truck and shovel operation, and a bucket wheel will operate cheaper than either dragline or truck and shovel. But you have that deposit to look at, you have your investment to look at, and you have your production rate to look at. Does that help you out?

Jack Crawford: I would like to call on the gentleman from Canadian Oil Sands to comment on the compatibility of bucket wheels with blasted material.

Peter Forrest: Well, just to set the scene. We are mining some of our overburden with it, and it's a glacial deposit with large boulders and the lower part is silt from an old ocean bed. Below that there is the tar sand, and that is mined by other systems. We recently bought a new bucket wheel for the overburden, and we found that we have had numerous problems trying to get it operating up to its efficiency. Some problems are design, and others are related to the nature of the boulders in the clay. These boulders could be as big as, say, 6 ft in diameter, and the size of the bucket would be, say, 3 ft. So when you hit that boulder in an overconsolidated mass, you can imagine the forces that are going through the machine. The time it takes for a bucket from the time it hits the bank to the time it comes around the back end is such that you can't possibly stop it. So, if you get a boulder in that bucket, it comes right down around the back, hits the main structure of your bucket wheel, and you've got a real problem. You've got bent beams and everything. This has really been a problem to us. We got it cheap; we got it for $11 million, and today it would probably cost you $25. So we have a real task of trying to make that thing work at a reasonable cost. We are going to have to go into a blasting program whereby when we do hit these boulders, they will move a little bit, and won't get caught up in the wheel, and do subsequent damage. You're saying innovations—that's one of the things we're going to have to look at and see if we can make it work. Otherwise, maybe we'll get approval for a big expansion and be able to take the wheel down into the pit, the main pit, and buy something a little more compatible with the type of operation. Selection of equipment depends on the type of deposit that you're going to work with and the materials you're going to handle. There aren't any really clear-cut answers or guidelines. You may have to study a number of cases.

I would like to ask a question of Don on maintenance and operating schedules. We are operating on seven 24-hr shifts. Our maintenance crew works five 20-hr shifts and on Saturday and Sunday the shop operates the same 20-hr shifts but with half the manpower. I was wondering whether you think that is bad practice or do other people operate the same way, or do they match maintenance with the operating schedule?

Don Myntti: The manning of the copper mine shop was based on a 21-shift per week work schedule and as a result the shop was never shut down except for several holidays. One basic premise for the operation of the maintenance crew was that the employees work a 40-hr week with minimum overtime. In order to go from a five-day to a seven-day work schedule, you multiply the number of people by 40% to get the additional coverage needed for the other two days.

Peter Forrest: You mean you would cover it by straight time, but your men would work on a shift basis?

Don Myntti: Yes, that is correct.

Peter Forrest: The reason we operate this way is that it's contractual and hopefully we'll get out of that situation.

My second question alludes to the point you were trying to make at the end of your paper that you should design for the future. We thought we had designed for the future. We're now getting into the 150-ton range of trucks, but we built a number of bays for the 75-ton. We are looking at phasing out the latter, and I was wondering what you thought about going to the concept of a boxed-down facility—modify the doors and use the building. What benefit would you get, maybe 30% of an ordinary bay? Or, would you even consider this to be an economic situation?

Don Myntti: Again let me refer to a situation at the copper mine. I mentioned that the steel skeleton in the original structure was on 30-ft column spacings and in our modification program we enlarged every other door. We went from the roll-up slat-type door to the vertical lift door and in the 30-ft column spacing we were able to get a clear opening of 29 ft. This permitted the use of the structure for the 150-ton trucks. We put as many of the large 29-ft doors into the original structure as we could. Then as we received more 150-ton trucks, we went to the high bay additions. At this mine, we were able to utilize the original structure to a large degree. That meant that for every bay with a large door we had every other bay with a small door. We repaired the smaller trucks and support equipment in the bays that had the smaller doors, or we turned those areas into storage space. Storage space is always lacking in a truck repair facility.

Peter Forrest: Could you, just off the cuff, say what percentage of maintenance you could do with the small bay vs. the high one? Is it normal maintenance?

Don Myntti: In the original section of the shop, the overhead crane could not pass over the elevated bed of a 150-ton truck, so that was a serious restriction in the flexibility of the operation.

Peter Forrest: You mean the bay would permit you to put the box up? Would your bay permit you to put the box up at all?

Don Myntti: It would, yes, but the elevation of the crane rails for the overhead crane was such that with the 150-ton truck bed in the air, the overhead cranes could not move from one area to another. In the truck repair section of the shop, there were three 15-ton cranes in the 510-ft length of the original shop.

Peter Forrest: I see. The problem we have is that our smaller bays wouldn't even permit us to put the box up, so from that point would you guesstimate how much maintenance we could do?

Don Myntti: With the truck in the shop with the bed down, I would say that the amount of maintenance that can be done effectively is restricted considerably. You are able to reline truck beds, etc., and change radiators. In smaller trucks you could change engines and replace transmissions. In a repair facility designed for the rear-end dump trucks, the overhead cranes must pass over the elevated beds. This arrangement gives you a large degree of additional flexibility.

?: I want to address this to anybody who has had experience in reclamation planning. In mine planning we have to live with reclamation. I wonder what the waste dump lift regulations are in the United States? What is the slope to be obtained in reclaiming these waste dumps? What happens to existing dumps having lift increments of 200 ft, 400 ft, and over? Could anyone give us an idea about reclamation costs per acre or what a certain person paid for operating costs?

Mike Kahle: I would like to take a stab at that one. Reclamation costs vary somewhat from state to state, but let's take some of the more stringent ones. Wyoming, for instance, right now has a regulation on reclaiming slopes of approximately 3:1. There is some talk of taking that slope to 5:1, which doesn't make anyone too happy. In most mining permits that you see now, whether you are talking about coal or uranium or any of the surface mines, with the new US surface mining law you're going to be faced with a maximum of about 3:1 that you will be able to go to on reclaiming slopes. This includes your pit slopes as well. Reclamation costs will vary considerably, depending upon the kind of vegetation and terrain with which you are involved or where your mining operation is located. But generally it will run anywhere from about, let's say in present-day dollars, $2000 an acre up. It's not uncommon to see $10,000 an acre.

?: The second question is what happens when you have an existing dump, say, of elevation of 200 ft or, let's go further, 400 ft?

Mike Kahle: Sometimes they won't allow you to do that. In other words, is it the question of whether you can build higher and higher dumps?

?: Yes; probably I should give more information. Of course, anybody will suggest that you build your waste dump from the base up. But, in a lot of cases this has not happened, so an operator, if he could get away with it, or if he was able to get away with it in the past, built his waste dump 200 ft higher. Now he is faced with the dilemma of reclaiming such a waste dump. For instance, in Canada you are required to reclaim your waste dump. The latest requirement I heard is a 150-ft slope distance at a 27° angle or 2:1 for every lift or for every terrace. I would assume that this would be a very expensive process, and it would make the whole venture not viable.

Mike Kahle: The best rule of thumb you can follow is this: The US surface mining law requires you to reclaim to the approximate elevation of preexisting mine topography. As far as reclaiming dumps is concerned, you have to be able to develop a slope at an elevation that permits you to replant the dump. In other words, you've got to put grass or trees or shrubs or something back on it. If it's 200 ft high, you may have to cut an intermediate bench in the dump elevation and do a lot of reclamation for a period, of let's say, five to ten years after your mining operation is completed. Essentially, then, you have to cut down the slope of the dump, revegetate it, irrigate it, fertilize it, and give it the full treatment, that is, bring it back to its preexisting environment. Does that help with your question?

?: Yes. Another thing, is there any attempt by the industry here to soften the impact of stiffer environmental regulations?

Mike Kahle: Yes. We have a tendency to go to a lot of subcommittee meetings. Unfortunately, not as many of us go as we probably should. We don't get there enough and talk with our local officials or state officials or federal officials. We seem to hold back a little bit and wait for that law to come out. Then we sit around crying. We should have been acting a lot more before that legislation was put through and voted on. This is a pet peeve of mine.

?: With regard to reclamation costs, I recall that at the AIME meeting in Atlanta in 1977, kaolin producers were estimating $1000 an acre, but their costs were lower because they were dewatering the pits and forming recreation lakes. And their costs were lower than the coal industry which in 1977 was saying about $5000 an acre average.

Randy Weingart, *Baroid Div., NL Industries Inc.:* I have some comments to make on reclamation activities, being from Arkansas and being involved in some mining activity there. We had several leases on BLM-Forest Service land and I would like to point out that the arrangement that was agreed to on reclamation was that we had to break the slopes to a 32° overall slope. In order to do this and to reclaim surface areas, we spent in the neighborhood of between $4000 and $5000 per acre for reclamation because of the actual amount of labor involved in breaking the pit slopes. Now, I don't know where MHSA or BLM or anybody got that magic number (32°), but that's what it came out to be.

Of course, I certainly believe that the impact in reclamation has not been felt yet. Generally, we are talking about a lot of operators who are here in the west. They have a lot of land available for possible use. In the east, of course, it is a little more difficult to operate. Lands are held by private individuals. You don't have the dump grounds you may need. In addition, water requirements are certainly more important.

I would like to ask the question regarding the requirements encountered by other operators here in terms of their waste piles or dump areas. I'm certain a lot of you are trying to leach your piles and recover some copper contents. However, we're doing just the opposite. We're trying to figure out ways whereby we can build the dump areas in such a manner that there is minimal leaching going on, because the ground waters generated discharge into the streams. Then we as operators are responsible for these streams, particularly since there are massive runoffs of 45 in. of rainfall per year. We're talking about several thousand acres on which we'd have to treat water. I can tell you that water treatment for us is quite an expense, and we try to minimize it in all possible cases.

Does anyone here have that problem or know anything about the interior activities of dumps or any research that is going on to minimize leaching?

What are the operators presently trying to do to contain water around any of their dumps for treatment purposes before it is discharged into the rivers or streams?

Al Dahlstrand: At the Berkeley pit in Butte, we do have a leaching operation. It's temporarily shut down at this time, but it's been going on for probably about 15 years. The earlier leach dumps were set on asphalt pads and prepared bases, and our later leach dumps, which are upslope from the original leach dumps, were just laid on the surface of the ground with no preparation. We do catch the water; how much of it, I really can't estimate. We probably catch most of it. Unfortunately, the rest of it probably comes through the wall of the pit, because we see all kinds of copper and iron stains in our pit walls. So we know that some of this is coming from our leach dump and precip-plant activity, particularly the iron stain material. But, we do this on purpose, and unfortunately we have the problem of not being able to keep enough of our interior dump areas wet. You're not trying to leach and you probably get it all wet. I don't know how that works out. But we do have a fair amount of success in controlling it, and we keep probably about 75 to 80% of our water in a continual closed circuit. As runoff increases from snow melt or some other source, or if we have a sudden inflow of water over our normal that we pump out of our underground workings, we pour that through the tailings pond and use our tailings pond as the major step in our final treatment system. We do have one final small pond before we turn it loose into the stream valley below. Most of our water is kept as much as possible in a closed circuit. The water that we do turn loose is, of course, of concern to us because it's getting very costly.

Jack Crawford: Alan talked about a prepared dump base to prevent the water from going from the dump on down into the ground. One thought might be to seal the surface of the dump and collect the water up there so that you can control it so that it never has an opportunity to get into the dump itself. It isn't going to do any leaching if it doesn't get down into the dump in the first place. I suppose there is a combination of things one might do in terms of physically sealing the dump or, in fact, doing a certain amount of chemical treatment of the water on the dump to the extent that maybe it causes chemical reactions near the surface of the dump to form impermeable layers. If you run haulage trucks enough over some of these dumps, you get some pretty thick impermeable layers across the top that will not let that much water go on through. As you say, leaching facilities generally have led to some rather extensive design and evaluation work regarding water collection itself and routing it and how to handle it at a reasonably inexpensive cost.

Alan Dahlstrand: One other small factor we're looking at in the not too distant future is a requirement that's been laid out in hard rock mining regulations in Montana whereby we catch all runoff or as much runoff as possible upslope from our operations to keep it from coming onto the dumps and causing the same problems you're looking at. We're looking at the possibility of having to construct a drainage ditch that catches as much runoff as possible above our operations, which would be above our leach dumps, our waste dumps, and tailings pond. This is all well and good for cutting down the amount of water that we allow to enter the system and raising our reclamation costs or our treatment costs on that water. However, we have a problem of the place that we have to put this ditch. It's probably going to be a five- or six-mile ditch. It has to be in a position where you can get the water away from the operation or around it, and we're probably looking at lining the ditch so the construction and maintenance of this ditch is going to be really expensive. Now it's down to a trade-off of, you know, do you do the construction of some fancy facilities to divert this water around the operation or do you take that amount of water into your operation and swallow the treatment costs. We haven't completed this evaluation as yet, but we're looking at it right now.

Randy Weingart: Union Carbide has a vanadium mining operation in Arkansas, which, of course, is the only vanadium mining operation in North America. Their expenditure recently for water containment has been formidable between (I don't know exact costs) $1 and $2 million. They had to mine a pit very rapidly in order to have the containment area to treat a 24-hr rain. They were estimating around 600,000 gpm. That's what they were going to have to be treating in terms of water. A natural watershed goes right through their pit area, and they have to collect every bit of that water and treat it before they can discharge it. If you live in the south, you have a lot of rain to contend with, and you're going to go out there to start spoiling waste in some areas. I would like to know a little bit more about which direction we are going government-wise. Are we going to be responsible for everything that falls on the location?

Mike Kahle: Essentially, what it boils down to is that any mining permit that's being issued here in the west (and I guess you would have to say throughout the United States) says, essentially, that you will contain and catch all water. To give you an example, in Texas (which is controlled by the Railroad Commission), all your water has to be caught and contained for a 25-year, 24-hr event. That amounts to a lot of water. Essentially the way most mine designs are going is that you're catching all the water, all the runoff that you possibly can before it can get to a pit. You're routing that water around your operation. As the gentleman from Anaconda was saying, that could be several miles—five or six miles worth of dike work. With the permits you're having to do that now. That's just a requirement. Your ditches can amount in magnitude up to four or five million yards of material moved before you're allowed to mine, just to build ditches. These ditches have to be sloped. They have to be able to contain the water; in other words, you have to compact them in most cases. And these ditches get to be quite substantial in size. It's not unusual to see in many mines, or in many mine designs now, 8 to 10-ft culverts, and this is for an event that may occur once during a 25-year period. As far as any water that you catch on a spoil pile or in a pit, what you have to do is pump that water out and put it in a retention pond, and depending upon the state and the regulation, you may have to go to a situation of no discharge of that water or to a very elaborate treating system. I think, Gerry, in the uranium mines, that's a common fact. You mine designers and mine planners out there, not only are you going to spend time worrying about pit slopes and dumps and benches and so on, you're going to be spending a lot of time looking at ditches.

Kim McCarter: Maybe I would make one more comment concerning dumps, particularly dumps associated with copper mineralization. I think that in order to keep the dump from leaching, you could do everything you don't do when you're trying to get it to leach. For instance, the dump could be designed in the form of a monolithic structure which is not open on the sides. If you want to leach a dump, the ideal geometry would be a finger dump which is a long, low dump that's open on all sides to promote oxidation. So, if you're designing a dump that you don't want to leach, I think you would want to tend toward the monolithic type that is closed.

E. H. Eakland, *Minerals Exploration Co.:* We are in the process of putting in a uranium mine in Wyoming. We had designed the tailings pond to withstand a storm that might happen once in 2000 years. The officials came back to us and said that wasn't good enough and that they wanted us to withstand it for the possible maximum flood. Well, I don't know what the possible maximum flood is, but they gave us a figure—16 in. of rain in 24 hr. Well, we said in agreement that we were going to deduct 30% for seepage into the ground. They said no, that this follows a storm that has already saturated the soil! We can't really complain about the escalation of costs, particularly in the reclamation area (if it happens to us) unless we've written a letter once a month to our congressman and gotten other people to do the same. If you look at the new Clean Air Act regulations that are going to come into effect sometime this summer (you don't have to reduce them to an absurdity, but just say what some energetic regulator could impose), nobody will be allowed to be born in this country until somebody dies, particularly in nonattainment areas, because you can't add the additional pollution of a person breathing. I'm not saying it's that absurd, but we are getting close. Uranium miners are now facing the fact that they have to put all their tailings back in an abandoned part of the pit. I don't know what you do the first four or five years when you don't have an abandoned part, but I asked the question and they said you are going to be required to dig a pit and bury them in the pit and then put them in your abandoned part of the pit. We've gotten into an area here.

Mike Robb: I'd like to comment on the water discharge and the environmental issues. By most standards, we don't really make much water out of our uranium mines, as compared to most base metal mines, but we're now in a situation that the water we discharge has to have an IX plant built to treat it. So we're now required to discharge cleaner water than the surrounding wells produce, and the intermittant streams run naturally. This has reached the point of absurdity. Basically, the Indians have been drinking this stuff for umpteen hundred years, and they don't glow in the dark. But, we can't stick it back in the creek, you know.

Steve Winkelmann: Randy, I don't know whether this would be applicable to your problem, but I have heard of some cases where they use evaporative techniques to get rid of excess water in the southwest, and I don't know how successful they are. But you might be able to get rid of large amounts of water through atomizing sprays and evaporation.

John Spencer, *Cities Service:* Before we started Pinto Valley, we had a very large amount of water we had to get rid of. And we installed quite an elaborate spray system and abandoned it after a few months. It didn't do a bit of good. We are in a zero discharge situation. We can't let any water off the property, but so far most of it is being circulated either directly back into the mill system or through the old cement launders to strip any copper out and then being stuffed into the tailings underflow, neutralized again in the tailings, and pumped back. It all gets into the mill system. But, the evaporation cost us a great deal of money, and we couldn't see any progress whatever.

Jack Crawford: I appreciate all the questions we have had in the areas of maintenance, operations, and reclamation. I would like now to give attention to anyone who wants to raise a question either in the geotechnical area or in some of the earlier stages of the ore reserves evaluation process that we discussed in the earlier parts of our workshop.

L. M. L. Klingmueller, *Lucky Mc Uranium Corp.*: I would like to address this question mainly to Gerry Dohm and Frank Stermole. When we look back through our evaluation process, emphasis during the last three days seems to have been on operational cost savings to the point where I feel they are much more insensitive to the actual reserves we are predicting. In this respect I would like Gerry to make a few comments as to where in the circular analysis he would put risk analysis. We know that the fluctuations in metal price, in mill recovery, and in the grade we are predicting are much higher than any fluctuation, for example, in resistance of truck tires to pavement or something like that. I just wanted to know where all this kind of thinking should be introduced.

Gerry Dohm: That's a good question. I think basically what you're looking at in mine planning is trying to optimize your scheduling and your design so that you can average out these fluctuations. Particularly in uranium you want to stockpile availability to cover changes of grade throughout the mine. As far as the unknown commodity values are concerned, you have to make some assumptions, at least from a preliminary point of view. You've got to allow yourself enough latitude in your thinking so that you can adjust as you go.

Are you thinking in terms of an operational problem or a mining planning problem?

L. M. L. Klingmueller: I am talking about mine planning because your uncertainties are much greater. For example, you have a 5% change in either commodity cost, or grade, or mill recovery. That might be equivalent to a 200% operational cost. What I'm trying to say is that on the one side we are looking at things in such fine detail and on the other side we have a lot of possible variations we can expect. For uranium in particular I see problems of equilibrium with grade distribution and mill recovery.

Gerry Dohm: These are a lot of problems you are addressing as one question. You shouldn't be opening a uranium mine if you have a problem with the equilibrium.

L. M. L. Klingmueller: I think you missed my point. The variations in these areas can be very small in comparison with all other corporate areas. This goes back to risk analysis. I feel that it is very important to these factors in the form of a risk analysis to really cover all aspects. You might say that if you have a 7000-ton operation, you do have a figure, but a 5% change in grade from what you predicted might change the outcome.

Gerry Dohm: I know what you are saying. The only thing I can say is that you have to do your homework ahead of time. You base your planning on the best available information. If the project looks good from the financial end, then you start working on your sensitivities. What will a reduction in the selling price do to the operation? What will a change in the equilibrium factor do? Again, I'm not really well versed in sensitivities, and I'm used to dealing with the today situation. But you can make a parallel with financial projections, cost of capital equipment, and changes in the commodity values. But I'm not so sure how much meaning that type of sensitivity analysis has in the preliminary mine planning.

Marvin Barnes: I can't resist getting into this fray a little bit at this point. Having done a great many sensitivity studies in my career and almost inevitably whether it be porphyry-copper or whether it be uranium, whether it be asbestos, or a large bulk of silver mine, gentlemen, in almost every case in looking at the sensitivity of these operations, the most sensitive parameter is the price. I can't think of one instance in which the greatest area of sensitivity to your ultimate profitability, to any profitability index you want to use, but particularly DCFROR, does not have to do with the price that's paid for that commodity. Now, price is absolutely a linear function, whatever your recovery is, both through milling and smelting or any other process; price is absolutely a linear function of your recovery or of the grade. Price and grade have the same sensitivity. I know at least two major multimillion dollar (over $50 million) investments that have been made based upon grade estimations which later on were not substantiated. These projects that should have given a high rate of return became instead a perpetual loss which to this day has not been recovered.

Now, what's the problem? The problem goes much further back than what we've been talking about here the last two days, and has to go back to a time of sampling and the determination of what we really have in the ground in terms of a mineral inventory. I think too often we overlook this problem, and we tend in those early stages because of the expense (the so-called expenses of sampling) and because drilling is expensive, to shortcut accurate mineral inventory preparation. But it's a lot less expensive than losing your shirt down the road a little ways, which some people have done. I hate to sound like I'm making a pitch, but if there's a better way of sampling and a better way of knowing the distribution in reality, the distribution of that mineral in place, then we ought to be using these techniques. If we're going out there to mine that high-grade area in the early years to maximize our ROR and to minimize our payback period and thus relieve the chairman of the board of the worry of risk, we can predict in the short-range probably a little better than we can in the long-range. We would like to be sure that we'll find the ore we expect to find. If we can predict grade distribution better and go in with our mine planning and find what's predicted, we're going to be a great deal more successful than if we go there and find that the ore that we planned to mine in that stage isn't there, at least to the extent you thought it was. Such a situation can raise hob with the DCFROR as nothing else can. What I am saying is that if there's a better way to do it, and if there are questions and doubts at some stage because of inadequate sampling, then perhaps we better reconsider. You, as future operators, present operators, and mine managers, ought to make sure that you've got the best handle possible on that mineral inventory or on that ore reserve, because a 5% variation or a 10% variation in the estimated ore grade is going to have a lot more effect than a 15% or 20% variation in your operating costs.

Gerry Dohm: One thing I would like to add here and you said it a lot better than I did, is do your homework before you start. Sensitivity analysis is really one of the factors about which you're not so sure. The more detailed engineering work, geological work that you do ahead of time, the less sensitive the whole project becomes.

Pete Fowler: I certainly would have to agree with all that's been said, but I can't help wonder if management doesn't have to come in with a decision here as to what precision they want in terms of what they want to pay for it. I remember reading an article on risk analysis a year or so ago in which the author pointed out that when you consider all these engineering factors and the extent to which various factors can influence the ultimate cost and then consider what the top manager does by saying, "Well, instead of 15% discounted cash flow return, we want 18%," the engineers' cost calculations look almost ludicrous. I think we have to weigh the cost of this information against what we're going to buy with it.

Richard Call: I would like to raise a point in this context with regard to bulk sampling and pilot plant tests for metallurgy, particularly now that we're going to deeper deposits and buried deposits. This is getting to be a rather expensive exercise to sink a shaft or drive an adit and collect a bulk sample and run off the bulk metallurgical testing. I would like to kick this out for discussion a bit as to the alternate of doing more of your metallurgical testing from drill-hole information. What I see is that you usually can't afford to sample the whole ore body with bulk sampling. What happens is you get a very precise handle, perhaps, on one part of the ore body with your bulk sampling but you don't know very much of anything about the other part. This is particularly true if you sink a shaft, sample some of the deeper parts of the ore body, run your pilot plant, develop a mill design based on that, and then you start mining and hit the oxide up on the top and your mill just doesn't work. So, I think there's a point of doing more bench testing and getting a better distribution of samples throughout the ore body rather than just one bulk sample of one part of the ore body.

John Spencer: Rick, at Pinto Valley, we shortcut the thing. We did our bulk sampling entirely by drill hole, using a 9-in. down-the-hole hammer to the ore zone and then took 6-in. core, because we wanted grindability tests. I don't remember how many holes we had, but it was quite a few in all parts of the mine, so we sampled what we thought were differences in the ore zones. This whole thing was run in all the various combinations through a pilot plant, and I think it gave us excellent information. Now we're running 89% in ore recovery, and we're quite happy with the results.

Jack Crawford: We'll take a few more questions, and then we would like to get some closing remarks from Bill Hustrulid.

Bill Robinson, *Getty Oil:* I would like to address this to Dr. Stermole. You commented on preproduction stripping, writing that off if you have other revenues coming in vs. carrying it on. Aren't there some other considerations like, perhaps, the time value of having the cost as it is in today's dollars and then applying it against your revenues after your two years of preproduction at current revenues? Is there any tradeoff there?

Frank Stermole: In evaluation work you want to account for the costs when they are incurred. You want to account for the tax savings from the deductions on those costs when they are incurred, whether you are expensing the costs in the year incurred against other income or whether you are getting the deductions by depreciation over the following years. I gather that what your thinking relates to is the question of whether to do the analysis in today's escalated or constant dollars. You must be consistent in your analysis. If you are going to evaluate this project and other projects for comparison, you want to do everything either in escalated dollars or everything in constant dollars. And, as I indicated in my talk, if you use today's dollars, you are considering that using today's dollars is a good estimate of your escalated dollar situation, that any escalation of your operating and capital costs will be offset by similar escalation of revenues. Rather than actually project those escalations, when you use today's dollars, you just assume that that overall effect will be a washout. If you stick consistently with escalated dollars all the way through your analysis, then you are accounting for the time value of money on your costs vs. when your revenues are realized in your present work calculations involving your discounting of cash flow.

Bill Robinson: What I am talking about is deferred accounts for preproduction stripping. Is there any advantage to other than writing it off in that first year to carrying it on into your production stages?

Frank Stermole: For tax purposes, your economics will always be enhanced by getting your deductions as quickly as possible. One of the things you want to keep in mind is, for evaluation purposes you want to use what your company is doing for tax purposes, not what your company is doing for book purposes. Generally, your development costs will be deferred and deducted ratably over the mine life or over ten years for book purposes. We don't care what the company is doing for book purposes, and by book purposes what I am talking about is annual report purposes to the shareholders and so forth. It's what's done for tax purposes that is going to determine when the company really receives the tax savings from the deductions, and that's what's relevant to our economic analysis considerations.

Bill Robinson: So expensing it would be better.

Frank Stermole: Expensing it is always better than capitalizing and carrying forward, unless you're afraid of the seven-year limit on carry-forwards. If you don't use loss carry-forwards in seven years, you lose them. So if you're concerned about that or if you're a company with a lot of other loss carry-forwards from previous years that may run you past the seven-year limitation, then, in that case, you may want to capitalize and defer the deductions because of fear of losing them under the statute of limitations.

Bob Johnson, *AMAX Exploration:* I would like to address a question relative to increasing productivity. I think, Jack, you're probably in the best position to answer it, but there are others here who could also contribute. The Swedes have been particularly forceful in automating a lot of their underground operations to the greatest possible degree to increase productivity. At their Kiruna mine, they've made tremendous advancements in terms of automated rail systems underground in which they virtually are in the process of taking everybody off the system other than in the control room. Not too many years ago this was almost unheard of or unthought of. At the same time, ASEA is making overtures to the effect that they have a similar control system suitable for truck haulage in open pits. I do not know the details of this ASEA system relative to surface operations. I have some knowledge of what they have done at Kiruna, but I would like to pose the question of what we are doing here in the US relative to maybe automating truck haulage either partially in the pit to the ramp or at least manually to the ramp and then getting on some kind of an automated system up the ramp via trolley or whatever.

Jack Crawford: Well, let me just comment on a couple of items. Probably in truck haulage, the immediate thing that comes to mind is the ASEA wire-line system on which they are working with Unit Rig for truck control and guidance.

Concurrently with the question of automated truck haulage, you get into the whole matter of the power source, and one can raise the subject of the trolley line power system for the trucks as opposed to the diesel electric systems presently in use. Haulage trucks might benefit from the same energy savings that canveyor belts do, namely, a cheaper source of power. To compare that, say, with a rail haulage system where it has been done, as you say, to a greater degree, I think one has to consider that a rail system is a lot more rigid with certain constraints. It's somewhat tied down, to, let's say, more rigid procedures of traffic control. Perhaps the reason why it hasn't really been worked on (why there haven't been any major breakthroughs that have turned a buck) is that the big element of truck haulage and why people have gone to it over the years has been its flexibility. Automation in a way can be construed as requiring a certain amount of reduction in flexibility to respond to changing conditions. I would like Alan Dahlstrand to comment on this.

Alan Dahlstrand: We talked to the Unit Rig people and the Saabskona people who were a couple of years ago (and I believe still are) looking at this automatic wire-line system that Jack refers to. I can't agree with you more in your comments about the flexibility of the truck haulage system and the reduced flexibility of automating some of these things. We've taken just a brief look at the automatic truck control, the Saabskona-unit rig system, and its possible applications in our large truck-shovel operation, and what we find is that our flexibility is all of a sudden gone. There are certain areas, geographical areas, say, from pit exit areas to certain dump areas that might have a good application, but these haulage profiles are usually so minor when we consider the whole overall scope of our haulage profiles and the changes in our haulage profiles from now through the end of the pit life that we did not consider it worth pursuing. I believe those companies are still pursuing this application for more simplified operations. Our problem was that we just had too many different dump areas. It wasn't like leaving from one pit exist and going to one dump area that just grew and grew and grew. We're building leach dumps; we're building waste dumps; we're building a tailings dam with our waste and leach material. We've got all kinds of pipelines laid all over the dumps for leaching operations, and it just gets to be a big rat's nest. The more complex the operation gets, the harder it is to apply. The same kind of thing can apply to a trolley system (the trolley assist system) that a couple of companies have looked at and that one company has actually operated up in Quebec. In Quebec, they had a fair amount of success, and the only reason it isn't running now is because the pit has been mined out, and they are mining in some other area where most of their haulage is downhill, so they are not looking at trolley assist. There are certain attractive things about it but anytime you get on a wire-line system like automatic truck control or a trolley system, you have to plan very carefully that you put it in that part of the operation that is going to have the least amount of change from now through the end of the mine life. This is what you really need to justify some of these systems, especially if they are fairly capital intensive like a trolley could be. They are interesting concepts, but you have to have the right kind of operation for the right kind of applications.

Jack Crawford: In railroads we have found two things that seem to have worked out quite well over the years in terms of reduction of manpower, and I think that this is one of the main things you are alluding to regarding automation. Several operations with railroads have gotten down to single-man crews on the locomotives by virtue of remote radio control of the locomotives. The other thing is centralized radio or cable traffic control. A thought that comes to mind that might be considered is the question of whether or not it would be feasible to control haulage trucks with radios such that a person through a network of transfers might conceivably be able to operate more than one haulage truck without really changing the basic configuration of the design of the haulage truck.

Bob Johnson: That was exactly what I was referring to in terms of some of the thoughts that have been expressed during the discussion. I was just unsure as to whether anything of this was being done. It's been thought of relative to underground operations where you might be able to control a load-haul-dump working in an open stope. Again, I'm not so sure that there are any practicing mines using it, but it is a means of increasing productivity.

Jack Crawford: I am not aware of any studies of radio controlled haulage trucks in the manner that I mentioned.

Bob Johnson: Would it be possible, or has there been any investigative work done, say, even to control drill rigs so that maybe a single operator could be monitoring drill rigs?

Jack Crawford: Not on the matter of multiple drill rigs, but with pigtail cables and a control box there are single-man crews on drills whereby the man can move the drill while he's off the unit to spot it over successive drill holes. There has been a lot of evolution over the years going from two-man crews on shovels and two-man crews on drills to various combinations of crew mix. In the case of drills, we may have three-man crews handling two drills or three-man crews handling two shovels to allow a certain amount of the benefits of operator relief and current servicing on the units. There are some areas that have been looked at and changes implemented at relatively minor costs with significant manpower reductions but without substantial capital investments in modifying the equipment but just by changing management practices.

Bill Hustrulid: I might say something, having worked for an equipment manufacturer for some time. Mechanical engineers can do wondrous things for you in terms of automated equipment, but often they don't understand the conditions under which our equipment is expected to operate, both underground and in the pit. So sometimes you're getting rid of one operator and you're adding one mechanic or one electrical engineer or someone else to try to keep this running for you. One example of this with which I am familiar in this country is an automated train which has caused problems because it has to run outside the mine for part of the run. The snow and other environmental conditions that we normally have to put up with have presented serious problems, which are now being corrected. I think we cannot forget the conditions under which we must work and the kind of flexibility for which we are asking. A mechanical engineer can do those things for us, but at a very high price.

Richard Call: Basically, the more complex a system, the greater the probability of failure. So when you start adding electronic components and automated systems, your availability can go down pretty rapidly. Even though you eliminate some of the labor cost, your truck is going to be in the shop more often, not because the truck won't run but because the electronic gear that's controlling it won't operate. So I think you reach a trade-off pretty quickly. And I think this manpower aspect is particularly true. You eliminate the truck driver and you hire three electronic technicians that you have to pay more to anyway. You see this in the whole computer accounting area, that is, you buy a big machine and you get rid of all these accountants. Sure, you get the machine and you get rid of some of the accountants, but then you've got all sorts of programmers around, more programmers than you had accountants.

William Allen: I'd like to add something to that, something that maybe Mr. Myntti could appreciate. When it comes to the maintenance of these trucks, the driver, believe it or not, comes in pretty handy sometimes because he can see and smell and hear and feel. When it's on a wire, or one thing or another, why the system could conceivably wind up running trucks to destruction on occasion when it's not at all necessary.

Peter Forrest: I will just mention the company that had the overhead wire, Quebec Cartier. They are implementing a computer-controlled dispatching system, and some of our people were up there a couple of weeks ago. They're paying something like a million dollars, and they expect payout in less than a year just by the increased productivities they expect for their trucks. I don't know whether it will all occur or not, but if anyone is interested, maybe he can follow it up.

Jack Crawford: One thing I would like to mention in that regard. I believe several companies have looked at computerized dispatching, including Pinto Valley. But the one thing I think is most illustrative in this area is, if you've got radios in your haulage trucks, I would strongly advise anyone as a first start in this matter of truck dispatching to look at Ed Mueller's paper * on the Pima system. It is based on an analog computer, and you can build it in your backyard. You probably can put the system in if you've got the haulage trucks with radios in them for less than $500 and the work that Pima's done with it probably got 90 to 95% of the benefits of haulage truck dispatching without any electronics to speak of, without any fancy hardware, technicians,

* *Mining Engineering,* August 1977.

etc. Some of these more sophisticated systems may be valuable. You may want to go into them ultimately. I think for first generation dispatching, I would strongly recommend looking at the Pima analog approach.

At this point, I would like to thank everyone for the questions and discussion and ask Bill Hustrulid to make some closing remarks.

Bill Hustrulid: I think it is a real tribute to you, the audience, that you've stayed now for three days until well after four o'clock on a Saturday afternoon to listen to the papers and to the discussion. I think our authors have done a wonderful job. They followed the instructions we presented to them about concentrating on the engineering principles involved in open pit planning and design. They have taken the time and the trouble to write down the thinking process that we as engineers go through. It's very easy to avoid that by saying, "Well, it's a difficult problem; we've gone through it in detail; and this is our best judgment." With this material as a basis we have something concrete upon which to build improvements in this field.

For the organizing committee, I would like to extend special thanks to Marianne Snedeker. She has been the real worker on this workshop, organizing and scheduling us all and even assembling the workbooks. We are very much indebted to her and the other conscientious people working for us in the headquarters of the Society of Mining Engineers.

INDEX

Stanley J. Lefond, compiler

Index

D

G

H

N

O

Q

R

S

T

U

V

W

Y

Z